THEORETICAL MECHANICS OF BIOLOGICAL NEURAL NETWORKS

This is a volume in

NEURAL NETWORKS: FOUNDATIONS TO APPLICATIONS

Edited by Steven F. Zornetzer, Joel Davis, Clifford Lau, and Thomas McKenna
Office of Naval Research
Arlington, Virgina

A list of titles in this series appears at the end of this volume.

Theoretical Mechanics of Biological Neural Networks

Ronald J. MacGregor

Aerospace Engineering Science
University of Colorado
Boulder, Colorado

ACADEMIC PRESS, INC.
Harcourt Brace Jovanovich, Publishers
Boston San Diego New York
London Sydney Tokyo Toronto

This book is printed on acid-free paper. ♾

ACADEMIC PRESS, INC.
1250 Sixth Avenue, San Diego, CA 92101-4311

United Kingdom Edition published by
ACADEMIC PRESS LIMITED
24–28 Oval Road, London NW1 7DX

Library of Congress Cataloging-in-Publication Data

MacGregor, Ronald J.
Theoretical mechanics of biological neural networks/Ronald J. MacGregor
p. cm.—(Neural networks, foundations to applications)
Includes bibliographical references (p.) and index.
ISBN 0-12-464255-1
1. Neural circuitry. 2. Neural networks (Computer science)
I. Title. II. Series.
QP363.3.M33 1993
591.1'88—dc20 92-43460
CIP

Printed in the United States of America

93 94 95 96 BB 9 8 7 6 5 4 3 2 1

This book is dedicated to
the memories of
Plato, Aristotle, Kant,
Newton,
Bach, Haydn, Schönberg,
and all others who have given of their life's energy to
the pursuit of enduring forms.

Contents

Preface

This book attempts to define a biologically rooted theoretical mechanics for describing the neuroelectric operations of naturally occurring neural networks. The goal is a universally applicable hierarchical theoretical structure wherein the essential neurobiological governing processes are centrally expressed and determinative, the significance of important biological parameters can be clearly seen, and the varied rich detail of neurobiology can be seen and placed in its proper perspective of naturally functioning networks. The work aspires to a physically based mechanics of neural networks akin to the Newtonian formulation of classical mechanics, the Navier–Stokes formulation of fluid mechanics, the engineering sciences of thermodynamics, elasticity, and electromagnetic theory.

The book attempts to define comprehensive theoretical formulations and the corresponding mechanics of operation at each of four basic levels of neuroelectrical signalling: membrane and ionic mechanisms of neuroelectric signal generation in single neurons; firing rate sensitivity and transfer in single neurons; coordinated firing patterns in local neural networks; and systemic operations of coordinated firing patterns in the confluent synaptic junctional matrices of composite neural networks.

The first chapters root the approach in the physical processes of neurobiology. Membrane mechanisms for controlling transmembrane ionic fluxes are utilized within the membrane equation to generate and govern synaptic and action neuroelectric signals. The second section develops a theory of dynamic similarity to provide universal comparative characterization of neuronal firing sensitivities for different types of neurons. The third section introduces and develops a theory of coordinated firing patterns in neural networks, called the *sequential configuration theory*. The fourth section develops theoretical interpretations of the operations of composite networks in terms of the actions of synaptic junctional matrices on these coordinated firing patterns. The composite networks of the hippocampus and the cerebral cortex are considered explicitly. The final chapter of the book presents a speculative top-down theoretical view of essential systemic organization of nervous systems, so as to provide contextual guidance for considerations of operations of composite networks.

The work presented here is almost all original and almost all new or recent work. The initial formulations regarding basic mechanisms of signal generation in the first section and the applications to the hippocampus and cortex in the latter sections rely, of course, on basic neurobiological science and experimental observations. Nonetheless, even here the basic formulations are conceived and constructed somewhat differently than is typical. The primary distinction of the present approach is its roots in engineering methodology and perspective. This implies in particular an emphasis on succinct delineation of central governing physical processes.

The work presented here is the culmination of 27 years of research in this area. That of Part I was performed mostly during the period of 1965 through 1975, with extensive computer simulation programs produced through 1986, but with theoretical refinements produced continuously through 1992. The dynamic similarity theory of Part II was produced from 1987 through 1992. The work on local neural networks dealing with internally sustained activity and sequential configurations described in Part III was produced from about 1976 through 1990. The work in Part IV on composite networks was produced from about 1990 to 1992 and continues.

The approach to membrane ionic fluxes developed in chapter 3 has a more immediate physical character than that followed in most neurobiology books. Similarly this book discusses the Maxwell displacement current in terms of structural distensions of the membrane lipoproteins, which provides a more direct and physically immediate interpretation of the membrane capacitive effect. The book emphasizes the membrane equation rather than basic biophysical rules such as the Nernst equation as the central governing equation of neuroelectric signals for reasons that are discussed thoroughly in the text. The book also emphasizes the compartmentalization approach to describing and simulating activity in dendritic trees although the continuous approach is also introduced and illustrated. The book emphasizes a state-variable approach to ongoing repetitive firing in neurons rather than detailed mathematical descriptions to wave forms of individual action potentials as described by the Hodgkin–Huxley or Traub models, but includes descriptions of the latter approaches as well. Chapter 7 is a summary of how to incorporate the mechanisms discussed in the previous chapters into comprehensive, universally applicable computer simulation programs for neuroelectric signalling in neurons, neural networks, and composite neural networks. General purpose programs for each of these levels are included.

Some of this work in Part I, but without the deeper theoretical foundations and rationale has appeared in my previous books and in papers in *Biological Cybernetics* and the *Biophysical Journal.*

The dynamic similarity material in Part II is almost all new. Some preliminary ideas in this direction were published in the *Journal of Neurophysiology* in 1988.

Chapter 14, which introduces the sequential configuration theory for coordinated firing patterns in local networks is an elaboration of material presented in abbreviated form in *Biological Cybernetics* in 1990. Chapter 16 is a greatly elaborated version of memory capacity for sequential configurations in local recurrent nets presented in greatly abbreviated form in *Biological Cybernetics* also in 1991. The theoretical characterization of local and composite neural networks of Chapter 13 and the theoretical model on disorganized activity in Chapter 15 are new. The theory in Chapter 15 originates in a fine old paper by Harth and colleagues published in the *Journal of Theoretical Biology* in 1970. I and my students have worked with and elaborated this model over the many years since then. Only now, against the more recent theory of the sequential configurations do I feel I can see this earlier work in the correct perspective. I see the earlier model as representative of disorganized activity, whereas the sequential configuration theory deals with structured patterns by which significant meanings are likely represented.

I should like to thank my esteemed colleagues George Gerstein, Moshe Abeles, and Valentino von Braitenburg for helpful commentary and feedback regarding the sequential configuration model, its relation to cortical operations, and to Abeles's "synfire" model for dynamic patterns.

All the material in Part IV is new. I believe the distinction of local versus composite neural networks is in itself a significant distinction, one that has not been generally appreciated in the neural network community. The theoretical view of hippocampal and cortical organization presented in Chapters 18, 19, and 20 are rooted in the communal knowledge (and communal ignorance) regarding these composite networks, but are nonetheless basically original both in their essential underpinnings and in the view of their operations.

In this context I should like to thank my esteemed colleagues Larry Adler, Greg Gerhardt, and Robert Freedman of the Department of Psychiatry in the School of Medicine at the University of Colorado in Denver for feedback on the hippocampal theory presented here and, in particular, for the information regarding the malfunction of the septal input to the hippocampus in schizophrenia.

The highly speculative material of the last chapter on systemic organization is greatly influenced by the theoretical views of MacLean, Penfield, and Luria. Nonetheless the synthesis and the views of systemic engineering discussed there are original.

I should like also to thank the reviewers of this book for very thorough and constructive commentary resulting in an improved final product. I should also like to thank my editor at Academic Press, Kathleen Tibbetts, for constructive supportive guidance through the labyrinthine mechanics of book publishing.

I thank *Biological Cybernetics* and Springer-Verlag, Inc., Francis Schmitt and Pergamon Press, Wilfred Rall and the American Physiological Society, the Biophysical Society, Behavioral Science, and Academic Press for permissions to abstract information or figures from previous publications.

I thank the University of Colorado for its support of this work over the last 22 years.

Finally, I should like to thank my wife for her patience as I have struggled with this task over the last years.

Chapter 1 Introduction to a Theoretical Mechanics of Neural Networks

The goal of this book is to define the foundations for the theoretical approach to the electrical activity of the biological neural networks of the nervous system and brain. The book deals progressively with the physical basis of neuroelectric signalling; comparative input–output firing levels in neurons; coordinated dynamic firing patterns and memory representation in local neural networks; and structural–functional principles of organization at neuronal junctions and composite networks, including tentative theoretical models of the composite networks of the hippocampus and cerebral cortex.

Primary Orientation of This Book

The book strives to provide the conceptual and mathematical foundations of a theoretical mechanics for neuroelectric signalling over these several essential levels of organization.

The primary qualities offered by this work are

1. its foundations in explicit physical processes,
2. its formulations in terms of operational engineering design, including the focus on the few central causal driving agents and hierarchical relationing of these agents,
3. broad theoretical syntheses, sometimes over wide areas and sometimes clearly speculative

Each of these qualities speaks to essential needs within the scope of the current neural and brain sciences.

Perhaps the single most characteristic quality of the structure and operation of the brain is its tremendous richness and complexity. Virtually any of its operations, ranging from molecular through psychological levels is endowed with overwhelming richness of highly ordered detail, variability, and tantalizing combinations of autonomy and dependence on other levels of functioning. Any theory or model for any neural process is immediately guilty of the omission of some subset of ingredients that influence the system in some contexts. Sorely needed are clearer perspectives for placing the various neurobiological agents and their rich worlds of detail within carefully structured comprehensive, yet succinct, theoretical formulations, so that the operational significance of the particular neurobiological ingredients and their parameters can be seen in their proper places within the scheme of things. This concept of engineering operational design, with its focus on the few central causal physical agents and their hierarchical relationships, carefully housing the important neurobiological parameters, permeates all sections of this book.

For example, a serious central weakness across much of current brain research is the absence of common foundations among different contributing disciplines. In the field of neural networks for example, there is a very wide chasm concerning notions of foundational starting grounds between experimental neuroscientists and computational neural network theorists and simulators. This book claims that this is precisely because the central operational physical mechanisms that drive the neuroelectric processes studied experimentally have not been adequately incorporated in the computational neural net field in explicit relationship to the physiological processes they represent. The necessary knowledge has been present for some years, but the significance of the physical foundations for providing a common meeting ground and linking computational and theoretical studies to the physiological experimental world has simply not been recognized.

The rich complexity and high diversity of the various individual fields and disciplines sometimes obscures our vision of the essential operative principles, and their implications. This book attempts to strike a critical essential balance between the world of overwhelmingly rich biological detail and diversity, on the one hand, and obscuring operational, computational, and mathematical complexity, on the other, by resting on the critical fulcrum of succinct and simple, but accurate description of the essential physical processes.

We live in a time when perhaps 90% or more of all the scientists who ever lived are currently active. This has produced a tremendous degree of specialization. Many bureaucratic and some scientific pressures reinforce this specialization. Broad theoretical syntheses across disciplines are rare, particularly in the neural and brain sciences, where any such attempt is certain to contain some errors of detail and likely to contain some substantive conceptual misjudgments. Further, the neural and brain sciences are virtually 100% experimental. Current research, textbooks, thought processes, conferences, entire fields are conceived within the milieu, momentum, and constraints of experimental facts, experimental equipment, experimental paradigms. In the conventional scientific wisdom of "theory guides, experiment decides," there is a minimum of broad speculative theory in the neural sciences. The situation seems somewhat analogous to the state of the science of motion and mechanics before Newton. The deepest ambition of this book is to provide the initial formulations of a Newtonian-like mechanics for neuroelectric signalling.

At even broader levels in the neural and brain sciences, we should look forward to the time when psychologically directed theories, in, say, the various streams of psychiatry and psychoanalysis, can be linked more directly to operations of neuroelectric signalling in the brain. For such broad theories to be truly well grounded, their descriptions of brain operations should be rooted fundamentally in operative physical agents within the physiology, cast in the appropriate hierarchical structure.

Even short of these admittedly grand objectives, good, broad speculative theory can excite the imagination to visualize more broadly, and penetrate more deeply. The view of this book is that scientific knowledge within a given area is best contained within and in relation to succinct theoretical formulations of the essential operative processes in the area. The view here is further in concordance with the Popperian view that in a developing field it is the responsibility of theories to push conjectures over places of uncertainty so as to encourage the subsequent sharpening or perhaps replacement of the original speculations with new ones more in accordance with observations prompted by the original speculative views.

Structure and Themes of this Book

The first part of the book develops the conceptual and mathematical structure of the mechanics of the electrical signals of single neurons in terms of their underlying physical causal agents. The goal here is to provide a Newtonian-like formulation of this most fundamental level of the dynamics of neural networks.

Chapter 2 discusses the conceptual hierarchical engineering design envisioned for this formulation and its parallel to the Newtonian formulation of mechanics. The chapter also provides a brief, qualitative overview of principles of neural signalling. Subsequent chapters of the section deal with the biophysical principles of ionic fluxes and balances and the membrane equation for neuroelectric signals. This equation is seen here as the central governing equation of neuroelectric signalling, and the analog in this domain to Newton's second law of motion. Chapter 5 shows how to produce systematic theoretical formulations of neuroelectric signalling in entire neurons with arbitrary dendritic morphologies. This effectively integrates the membrane equation over space. These early chapters show that spatially and temporally localized selective modulations in membrane permeabilities are the central ingredient used by the nervous system in controlling its neuroelectric signalling. Chapter 6 shows several ways to bring descriptions of these permeability modulations into the theoretical formulation. This includes a particular and special but useful representation of the mechanism of repetitive firing in neurons based on principles of accommodation and postfiring refractoriness (adaptation). Chapter 7 shows how the theoretical mechanics developed in this section can be formulated in succinct, computer-efficient simulation programs.

The second part develops a theory of dynamic similarity for the comparative input–output firing behavior of neurons, expressed in terms of a few universal nondimensional characteristic parameters. This theory is prompted in part by the recognition that, although the nervous system utilizes great elegance in specifying particular classes of cell types to various regions, differing in size, shape, physiology, and distributions of input, we generally have very little idea as to the comparative principles of design at work here. Why are neurons involved in a given function in one particular region of the brain different from those in another function in another region? The dynamic similarity theory approach allows one to characterize and compare neurons in terms of universal nondimensional numbers, which represent the effective mechanical combination of their constituent physiological and morphological parameters.

Part III presents a model for the coordinated firing patterns used to represent meaning in neural populations. The current general view of neuroelectric signalling remains unduly limited to single-neuron concepts. The model presented here is intended to explicitly represent the language by which neuronal populations speak to each other as organic wholes. It is called the sequential configuration theory. Part III applies this model to describe the anatomical representation and dynamic recall of memory traces in local recurrently connected networks. The theory is used to predict the memory storage capacity in such local networks.

Part IV outlines a general theoretical mechanics for characterizing the operations of neuronal junctions. Feedforward excitatory junctions are seen as switchboards; inhibitory junctions (feedforward and recurrent) are seen as providing the foundations of essential spatiotemporal sculpting of information; recurrent excitatory junctions are seen as providing the means whereby local networks can gain autonomy by transcending the influences of their input systems (these nets can exhibit self-determined, internally sustained activity, for example).

These ideas and the view that information in neural populations is represented by sequential configurations provides the theoretical basis for considering the combined interactive activity of composite neural networks. Theoretical models are presented that see (a) composite networks of the hippocampus as representing abstract qualities of the external world in reverberating coordinated firing patterns, and selecting those for embedding on the basis of significance measures mediated largely by input from the medial septum, and (b) the composite networks of regions of cerebral cortex as collections of interacting local recurrent circuits each signalling a particular state by a particular sequential configuration. The part includes a chapter discussing the general experimental and theoretical field of cortical organization. Finally, the part and book conclude with a chapter containing a speculative top-down view of system organization in nervous systems and an interpretation of the systemic network operations suggested by this view.

An afterword contains a look at the value, completeness, and degree of validity of the theoretical developments of the book and recommends further theoretical and combined theoretical–experimental cultivation of the ideas developed here. The afterword also contains passing comments on higher dimensions of brain and mind.

TABLE 1.I. Fuels of Functional Organization in the Brain.

Global (conscious awareness, behavior, truth, beauty, conscience, and so on)
Systemic (autonomic, sensorimotor, sensory, motor, instinctive, affective, representional, volitional, cognitive, and so on)
Neuroelectric networks
local, composite
neurons
ions, membranes
Chemical (neurotransmitters/synaptic transmission; monoamine transmitter systems; chemical neuroregulator systems, and so on)
Molecular (molecular neurobiology)

Natural Levels of Functioning in the Nervous System and Brain

A central quality of the operational organization of the brain is its hierarchical array of multiple levels of functioning. A representation of this is shown in Table 1.I. This feature of hierarchical levels of ordered functioning can be seen in cosmological structures, is most pronounced in biological systems, and is probably more developed in the brain than in any other known system. The essential ingredient is that each individual level (say, conscious awareness, appetitive behavior, neural network activity patterns, neural transmitter or neurohormonal systems, certain genetic predispositions) seems to have its own principles of operation, logic, and cohesion, but at the same time seems to stand in some significant influential way(s) (either dependent or supportive) with the levels below and above it. One of the most fundamental and difficult questions one can ask about the functional organization of the brain is about the degrees of relative upward and downward influences and autonomies between and within these levels.

This book focuses on the mechanics for neuroelectric signalling as seen in the context of Table 1.I. Coordinated patterns in local and composite neural networks are seen in terms of arrangements of primary neuroelectric signals in neurons, which, in turn, are mediated by ionic currents gated by active modulations of membrane permeabilities. Thus, this book focuses explicitly on the hierarchical driving of coordinated firing patterns in neurons and local and composite networks, as driven fundamentally by selective modulations in membrane conductance, localized selectively in space and time, and integrated according the structural features of individual neurons and networks.

The dependence of these membrane modulations on the chemical and molecular levels of organization and the neurobiological influence of these levels generally are not treated thoroughly in this book. They are given only a brief introductory overview for the benefit of beginning students in Chapter 2.

The broad field of systemic organization of the nervous system is symbiotically related to the field of network operations dealt with in this book. Consideration of the systemic context within which a neural network operates provides essential guidance for considering and understanding its operations and functions. This important field is not substantially considered in this book. However, the final chapter does give a broad speculative view of systemic organization and briefly discusses its implications for operations in constituent composite neural networks.

The higher functions of mind and conscious awareness are mentioned in passing in the afterword.

Throughout the text the symbol **NBM** will refer to *Neural and Brain Modeling*, Academic Press, 1987, by the present author.

I

Theoretical Mechanics of Neuroelectric Signalling

Chapter 2 Introduction to the Mechanics of Neuroelectric Signalling

The first part of this chapter defines the basic structural rationale of the theory developed in this section. It is essential for grasping the point of view of this book. The longer second part may be skipped by advanced students and professionals. It is an introductory overview of basic neurobiology for beginning students.

Toward a Newtonian Formulation of the Mechanics of Neuroelectric Signalling

The central purpose of this first part of the book is to provide a hierarchically structured and mathematically formulated description of the physical processes underlying the electrical signals used by neurons and neural networks in representing information—a theoretical mechanics of neuroelectric signalling. In brief, we will assert that the membrane equation, when properly formulated on the neurobiologically grounded equivalent circuit for local areas of neural membrane, constitutes the proper central governing equation for neuroelectric signalling; that the preponderant neurobiological activation factors are naturally represented in this central equation in terms of their selective modulations of membrane permeabilities; that the integrative behavior resulting from spatial morphologies of entire neurons and entire neural networks can be represented by careful piecing together of multiple neuronal compartments, each represented by a governing membrane equation; and that this approach as a whole provides a natural and comprehensive physicomathematical housing of the underlying neurobiology and the influence of its various parameters in the most direct operational perspective.

The approach, as developed here, is regarded as analagous to the Newtonian formulation of classical mechanics. The membrane equation is analogous to the second law of motion. Permeability modulations are thought of as a "final common path" through which a great range of neurobiological diversity and detail must funnel to effect their central driving influences on neuroelectric signalling. In this perspective, elaboration of the individual idiosyncrasies of these richly varied neurobiological factors (various different chemical synaptic transmitter–receptor couplings, molecular gating processes for sodium, potassium, calcium subserving action potentials, and so on) are seen as fundamentally important but conceptually subsumed within the permeability terms of the membrane equation in the same way that, for example, various force fields (gravitational, electromagnetic, springs, dashpots) are subsumed within the broader central governing roles of the second law of motion.

Equations such as the Nernst equation for equilibrium and the Goldman equation for the resting potential, for example, are seen in this perspective as essentially background relations that define parameters of the active operational dynamical functioning of the system, quite distinct from and subordinate to the central governing equation that guides its ongoing signalling. By analogy, descriptions of the equilibrium positions of springs are necessary for a complete description of the functioning of a mechanical system, but certainly do not provide the correct perspective for describing the overall governance of its dynamic behaviors. This larger purpose is served by the force-momentum principle embodied in Newton's second law.

The approach developed here is further thought to provide a direct description of the natural physiological laws of neuroelectric signalling, in the same way that Newton's laws in mechanics, the Navier-Stokes equations in fluid mechanics, the first and second laws of thermodynamics, Maxwell's laws in electromagnetic theory, and so on, are thought to represent the fundamental physical laws of nature in these physical sciences. That is, the approach is not thought of as merely a collection of analogies or "models" in any of the various weaker forms of that word.

The equivalent circuit, in particular, is to be seen as a convenient shorthand representation for the rich and actual neurobiology it describes; any feature of the neurobiology represented in this circuit can be elaborated more fully whenever desired. The component parameters of the circuits and their membrane equations, when properly formulated and properly understood, show clearly and succinctly where such elaboration is to take place within the overall structure of the system being studied.

Few or none of the individual components of the approach developed in Part I are new in themselves. However, both the essential interrelations of the various parts in terms of the whole and the fundamental significance of the whole are clearly not recognized in the current multidisciplinary milieu of the neural and brain sciences. This book asserts that precisely the physicomathematical structuring of the physiological processes of neuroelectric signalling advocated here is required to link the overall integrative signalling of neural networks and composite systems with their underlying neurobiology, and thereby provide the means for linking the experimental fields of neurobiology with the computer-based fields of computational neural networks, and to provide the necessary foundation for theoretical speculation regarding higher levels of biological neuroelectric signalling.

The remainder of this chapter presents an introductory overview of the essential ingredients of neuroelectric signalling and their context within the overall neurobiological operations of the brain and nervous system. Advanced students and researchers may proceed immediately to the more rigorous treatment initiated in Chapter 3.

Essential Ingredients of Neuroelectric Signalling

Surface Level View of Neuroelectric Signalling

Figure 2.1 illustrates the primary neuroelectric signals used by neurons and neural networks in representing information. One may succinctly characterize the main characteristics of this signalling at a surface level as follows. The typical neuron consists functionally of dendritic tree, a cell body or soma, and an axon with its terminal projections. The dendritic tree and soma serve as input regions of the neuron, where characteristic input signals are manifested; the soma serves further as a triggering section, where input

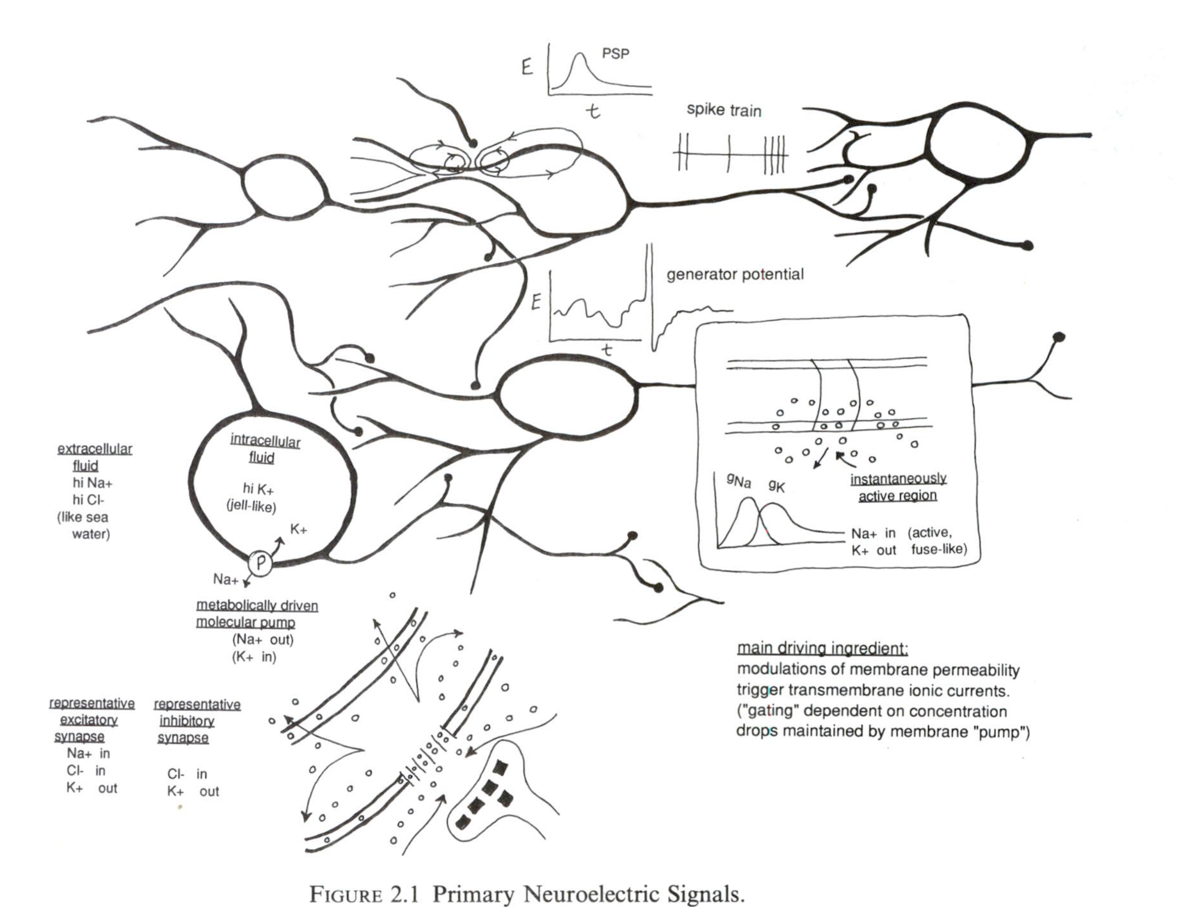

FIGURE 2.1 Primary Neuroelectric Signals.

signals are converted to output signals; and the axonal regions serve as output regions, where characteristic output signals are manifested.

The characteristic output signal of individual neurons consists of individual action or spike potentials and their collection into temporally variable sequences called *spike trains*. To a first approximation, one thinks of action potentials as discrete, unitary events that have essentially the same time courses and amplitudes whenever they occur. That is, the significant fact about action potentials is that they occur and when they occur. The view is analagous to the firing of a bullet from a gun: the event projected following the pulling of the trigger is sensibly identical from one event to another. Action potentials are sudden excursions wherein the transmembrane potential rises some 100 mV or so above its resting level, inside relative to outside.

The characteristic potentials at the input ends of neurons are PSPs (standing for postsynaptic potentials). PSPs are much smaller than action potentials and are continuous, graded, analog signals as illustrated in Figure 2.1. They have a short rise time (usually about 2 msec or less) and a longer gradual decay period (usually approximately exponential with time constant about 5 to 15 msec in vertebrate neurons). Individual PSPs are the unitary input responses in neurons; each PSP is the response originating under a synapse in a neuron to a single action potential in the presynaptic terminal. The rise time of PSPs corresponds to the period when the synapse is active; the decay period corresponds to the natural relaxation of the membrane potential to its resting level after the synapse is closed, and sometimes to a decay period of this synaptic conductance modulation. The rise time of the PSP directly under and close to a synapse usually corresponds closely in time to the duration of the input action potential.

PSPs spread out in time and decay in magnitude as they are conducted passively along the membrane away from the synapse. PSPs initiated on passive dendrites a hundred or more microns from the soma and observed at the soma can exhibit very slow rise times, perhaps as long as 5 to 10 msec.

PSPs can be either excitatory (EPSPs) or inhibitory (IPSPs) depending on whether they are positive or negative fluctuations in transmembrane potential (inside relative to outside). Unitary PSPs initiated in dendrites as observed at the soma are usually very small, of the order of 0.1 to 1 mV. Their peak values in dendritic trees may be considerably larger, perhaps approaching 50 or even more mV.

The potential recorded with an intracellular electrode in the soma regions of neurons is typically a graded continuous analog signal, one such manifestation of which is illustrated in Figure 2.1. Such a signal represents the confluent resultant of the interactions of all the input PSPs continually bombarding the neuron, and it may be referred to as the *generator potential.* To a first approximation one may picture this interaction as a simple algebraic summation of the positive and negative excitatory and inhibitory PSPs, although, as we will show in Chapter 4, the interactions are in fact, nonlinear. The magnitude of the generator potential varies between about 10 mV negative and some 10 to 25 mV positive. This composite signal can be called the *neuron generator potential*, because it generates the ongoing sequence of action potentials that constitute the output spike train of the neuron. The triggering of a single action potential by a generator potential is illustrated in Figure 2.1.

Individual action potentials are triggered when the generator potential exceeds a critical value, known as the *threshold.* The threshold may be an approximately constant value in a given neuron or may vary depending on the time history of the generator potential. This last factor, known as *accommodation,* represents the desensitizing of neurons by maintained input activation. It results in transient responses to steady inputs known as on and off responses. When a neuron fires a single action potential, the neuron is thrown into a rather short period of absolute and then relative refractoriness, wherein it is impossible, then difficult to fire subsequent action potentials. Absolute refractoriness typically lasts less than a millisecond or so, and relative refractoriness may last of the order of 10 to 30 msec. Typical upper limits of neuronal firing in normal function are about 30 to 50 per second, determined in large part by these refractory mechanisms. Threshold values for vertebrate neurons are typically about 10 to 25 mV; they set the upper limits for somatic generator potentials. Output action potentials are typically triggered at the place where the axon emerges from the cell body, the axon hillock. When action potentials are generated at the axon hillock, they are propagated essentially without change to all the synaptic terminals of the neuron.

In summary, a surface-level view of the rudiments of neuroelectric signalling is quite simple: primary neuroelectric signals consist of PSPs and generator potentials at the input regions of neurons and action potentials and spike trains at the output regions. Action potentials are converted to PSPs at the synaptic junctions between neurons; generator potentials are converted to action potentials by means of threshold rules at the somas of neurons. Neural networks operate by feeding ongoing spike trains in large numbers of neurons to each other and to distal target neurons by virtue of vast numbers of synaptic interconnections determined by the particular anatomical structure of the system under consideration.

Biophysical Foundations of Neuroelectric Signals

All the primary neuroelectric signals just introduced are the directly observable reflections of fluxes of ions flowing in broad continuous loops through the local intra- and extracellular fluids and penetrating local regions of neuronal membranes through small pores or channels in the membrane. Indeed, both the activations of these primary signals and the resting energetic foundation from which they are triggered are rooted in a simple but elegant system of ionic balances and fluxes, whose resting nature is common to all but the most primitive living cells.

In brief, the resting state of this system is attained by ionic balancing processes wherein the neuronal membranes partition body fluids into intracellular and extracellular domains, which exhibit different concentrations of a few significant ion types, and a difference in electrical potential, manifested primarily immediately across the cell membrane. The ionic concentration differences are set up by metabolically driven molecular pumping agents in the neural membranes. The equilibrium conditions and properties are determined by balances of ionic pumping rates with the transmembrane diffusion and electrical forces that are resisted in a graded fashion by the semipermeable membranes.

The primary neuroelectric signals introduced previously reflect active modulations unique to neurons of this underlying general ionic system. Specifically, all the active neuroelectric signals are gated by highly selective modulations of membrane permeability highly localized in space and time to one or another set of these basic ion types in the intra- and extracellular fluids. The gatings allow particular ion types to flow downhill across the neural membrane on their concentration and electrical gradients at particular places at particular instants in time, thereby causing the closed loop composite ion flows that are measurable as the operative primary neuroelectric signals of the nervous system.

The nature of this ionic balance system is discussed more fully in Chapter 3.

The Active Permeability Gates for Neuroelectric Signals

PSPs are triggered or gated at individual synapses by brief, selective modulations in permeability localized in the subsynaptic membrane to particular sets of ionic types, weighted according to the transmitter–receptor combination of the synapse. The equilibrium potential of the synapse and therefore its functional influence—excitatory, inhibitory, and so forth—is determined by the particular set of weighting factors assigned by the synapse to its constitutive active conductance types. Action potentials in the presynaptic terminals cause the release of small packets of chemical transmitter from these terminals, which then diffuses across the synaptic gap to be taken up by molecular receptors on the postsynaptic membrane. These receptors, when activated alter their permeability to various ions as stated earlier. The general neurobiological variety of synapses is discussed in the next section of this chapter; their mathematical representation within the equivalent circuit approach presented here is discussed in Chapter 4.

Action potentials are gated by a coupling of very large brief excursions of permeability to two positive ion types. For action potentials generated and carried on axons, the initial permeability change is to sodium ions. For action potentials originating in the dendrites of pyramidal cells, the initial permeability change is to calcium ions. Both of these changes allow positive ions to rush into the cell causing the positive rise of the action potential. In both axonal and dendritic action potentials, the briefly delayed second permeability change is due to potassium. This change allows positive ions to rush out of the cell bringing the potential back down toward resting level and producing a postfiring refractoriness.

These permeability modulations are triggered by active gating molecules stored in the membrane along those regions of the neurons, which are thereby capable of generating action potentials. For outgoing action potentials this normally encompasses the entire axonal system, from its junction with the soma (hillock) out to its outgoing synaptic terminals. Their distribution in some dendrites may be similarly continuous or may be more sparsely scattered, corresponding to a set of "hot spots" within the dendritic tree. Typically, axonal action potentials are very brief, about a millisecond long with a refractory time constant of about 5 to 10 msec, whereas dendritic action potentials may last 5 to 15 msec with a refractory time constant of about 30 msec. The mathematical representation of action potentials and spike trains by these mechanisms is discussed in Chapter 6.

The Broader Context of Neuroelectric Signalling

Chemical synapses are a great and pervasive key to neurobiological organization. Synapses are critical elements in communication among neurons, in the adaptation of neural networks to their final functioning structure in higher representation, memory, and learning, and moreover may in their chemical individuality participate in the chemical dimension of systemic organization, which seems in some senses to transcend the neuroelectric signalling in which they participate.

The Variety of Synaptic Activations

The fundamental significance of chemical synapses in neuroelectric signalling is clear: these are the mediators of the triggering of PSP responses to presynaptic action potentials. The arrival of an action potential at a presynaptic terminal triggers the release of packets of chemical transmitter in the subsynaptic space between the two neurons; these packets diffuse through the fluid to the postsynaptic cell, where they are taken up chemically by particular molecular structures on the postsynaptic membrane that change their structure to allow the inward and outward fluxes of ions across the membrane in combinations determined by the particular transmitter–receptor pair.

These traditional chemical synapses exhibit great variety. There are likely some few dozen of neural transmitters in the brain of which a dozen or so are quite prominent. Acetylcholine, dopamine, epinephrine, norepinephrin, serotonin, GABA, and glycine, for example, are quite common and functionally significant transmitters. Chemical synapses can produce many shades of excitation and inhibition with various temporal properties, according to the mechanics of the particular transmitter–receptor types; they can be axodendritic, or axosomatic; they can further alter, presumably corresponding to different functional "modes", the proportions of ion channel gatings they trigger for the different types of ions, and thereby alter their relative excitatory or inhibitory effects. Such synapses may be excitatory in one mode and inhibitory in another.

There are, moreover, additional types of synapses.

Further, there are dendrodendritic synapses. These are chemical synapses between the dendrites in the input ends of neurons. Presumably, graded generator potentials in dendrites of one neuron initiate graded, continuous synaptic influences on the graded generator potentials of neighboring neurons to which it is connected by such synapses. The properties of such synapses and the overall functional significance is virtually unknown. Such synapses often entail bidirectional transmission.

Chemical synapses can be physiologically plastic over short time periods in the sense that the magnitudes of synaptic responses are influenced by the recent activation history of the synapse. Later members in a time series of action potentials may generate larger or smaller responses than earlier ones.

Such responses are called *facilitatory* or *adaptive*. Individual synapses tend to be more or less fixed in the degree to which they do or do not exhibit such behavior. This kind of plasticity depends only on the time history of activation of the synapse from the presynaptic side.

There are also synapses between neurons, called *electrical synapses*, that are not chemically mediated. These synapses are anatomically developed junctions that allow the neuroelectric currents of action potentials in the presynaptic neuron to flow across the membrane of the postsynaptic neuron, producing postsynaptic potentials.

The effects on neuroelectric signalling of all these types of synapses and their possible short-term plasticity properties can be represented in the equivalent circuit approach by the actions of postsynaptic modulations in membrane permeability (chemical synapses) or postsynaptic influxes of current (electrical synapses), triggered by activity in the presynaptic element.

Furthermore, there are chemical presynaptic synapses. These consist of a synapse on a synapse mediated by a presynaptic terminal falling on another presynaptic terminal. An action potential in the first terminal produces a depolarization in the second terminal that diminishes the number of transmitter packets released by an action potential in the second terminal. The effect of the first action potential is thereby to inhibit the effectiveness of the second or primary terminal. Such terminals are relatively common in the spinal cord. This effect can be readily included in the equivalent circuit approach by making the magnitude of conductance modulation at the given primary synapse susceptible to diminution by activity in the second presynaptic terminal.

Representation, Development, Synaptic Plasticity, Memory, Learning

Information is represented in the nervous system by coordinated patterns of activity involving large or vast numbers of interconnected neurons. There are some 100 billion or more neurons in the brain, with approximately 10 billion in the human cerebral cortex, and perhaps as many as 100 billion cerebellar Purkinje cells alone. Typical central neurons receive from a few thousand (spinal cord) input synapses to tens or even a hundred thousand (cerebral cortex) input synapses. The degree of interconnectivity and the corresponding amount of neuronal material attributed to thin, filigreed instruments of interconnection—dendrites and axons—is astounding. The axons from a single human brain, for example, if laid out end to end would stretch to the moon and back.

The operations of neural networks must be interpreted in terms of vastly numerous divergent–convergent synaptic junctions between vastly numerous populations. These junctions are divergent in the sense that any one projecting neuron activates typically thousands of receiving cells, and convergent in the sense that any one receiving neuron receives from typically thousands of sender cells. Typical central neurons must receive several hundred active PSPs instantaneously, or alternatively PSPs numbering into the low thousands spread over a time constant or two, to be driven across threshold to spike production. Then the firing of this particular cell will be significant only if its firings are correlated in time with firings of hundreds or thousands of other neurons projecting to common receiving cells. To properly grasp the operative dynamics of these systems one needs to develop visual images that represent this extensive parallel as well as serial integration of signals.

Synapses are centrally significant in this contemporary view that coordinated patterns of activity over vast numbers of interconnected neurons are the neuroelectric manifestations of meaning and representation—that neural networks, and not neurons, are the fundamental functional units of neuroelectric signalling in the brain. Specific individual patterns of this sort are seen to be served and maintained by corresponding specific sets of synapses whose strengths are appropriately adjusted to maintain the pattern and resist the incursions of alien patterns. The enormous numbers of synapses in the brain (about 10**14) are appropriate and perhaps necessary to allow for the embedding of very large numbers of distinct patterns of this sort. One theory of the structure and mechanics of such patterns is introduced in Part III of this book.

In this context, long-term synaptic plasticity is thought to be the primary anatomical substrate of learning, memory, and representation in the mammalian and human nervous system. Individual synapses are thought to be selectively increased or perhaps decreased in strength to enhance or diminish their participation as desired in the formation of various multineuronal dynamic firing patterns, which in turn constitute the neuroelectric representation of particular psychological elements. The neuroelectric firing patterns can be considered the realizations of such representations, and the synaptic specializations underlying them can be considered their anatomical beds.

A specific salient embodiment of this theory that changing synaptic effectiveness is the substrate of learning, memory, and representation is the suggestion that the stalks of synaptic spines under synapses on pyramidal cells in the cerebral cortex and hippocampus increase in size to allow larger responses to the same level of presynaptic activation by deceasing longitudinal resistance to synaptic currents.

That neural networks rather than neurons are the functional units of organization of the nervous system is underscored by recent principles found in the field of development. The nervous system overbuilds: considerably more neurons and more synaptic connections are originally laid down than are ultimately retained. The specific connectivities of mammalian and human neural nets are constructed under the general guidance of an individual's unique genetic inheritance, but the development is winnowed and carefully fine tuned according to one's unique life experience, primarily during the very critical prenatal and early postnatal learning periods. Large numbers of neurons and synaptic connections that are not actively used die and are weeded away during these periods. It seems likely that only those cells and networks involved in mediating useful systemic operations and interactions with the environment are maintained.

This process may reach some sort of climax in adolescence. It has been suggested that a particularly critical stage in adolescence occurs wherein as many as 80% of cortical synapses fall away. The picture is that early in adolescence local and regional cortical connections are close to all-to-all, whereas in late to postadolescence only some one in six or seven of these survive. If this is so, it would seem likely that such an event would reflect the end of long trial period in which one's basic representation of many features of the external world were being established. After this event, one's representations of many basic features of this world would be relatively fixed.

Extrasynaptic Communication in the Brain

In addition to the more or less traditional, highly localized synapses discussed previously, there is an entire second class of neuronal chemical transmitterlike substances called *regulators*. These regulators are released by neuronal activity at terminals that are similar to presynaptic terminals but are released broadly into extracellular fluids to diffuse widely to various target sites in sharp contrast to the absorption of transmitter at highly localized postsynaptic sites in conventional chemical synapses. These regulators may diffuse hundreds of microns or millimeters through nervous tissue. Some at least seem to be taken up by specific specialized molecular target receptors. They act over time periods of hundreds of milliseconds, seconds, minutes, and longer, rather than the very few milliseconds of conventional synaptic transmitters. They generally seem to serve more

TABLE 2.Ia: Primary Neurotransmitters

Cholines
Acetylcholine
Monoamines
Adrenaline
Dopamine
Noradrenaline
Serotonin
Tryptamine
Amino Acids
γ-Aminobutyrate
Aspartate
Glutamate
Glycine
Histamine
Taurine
Purines
Adsenosine
ADP
AMP
ATP
Peptides
Angiotensin
Bombesin
Carnosine
Cholecystokinin
Endorphins
Leutinizing hormone releasing hormone
Methione and leucine enkephalins
Motilin
Neuromedins
Neuropeptide Y
Neurotensin
Oxytocin
Somatostatin
Substance P
Thyroid hormone releasing hormone (TRH)
Vasoactive intestinal peptide (VIP)
Vasopressin

global and diffuse functions, such as mood, pain suppression, and the like. These regulators include the various neurohormones associated with control of the endocrine (hormonal) system through the pituitary gland by the hypothalmus. It has been only recently realized that many more such

Table 2.Ib: Molecular Regulators[2]

NEUROPEPTIDES	NEUROHORMONES
Angioptesins I, II, III	Adrenocorticotropic hormone
Bombesin	Corticotropin-releasing hormone
Bradykinin	β-Endorphin
Calcitonin	Glucagon
Carnosine	Gonadotropin-releasing hormone
Cholecystokinin	Growth hormone
Corticotropin-releasing factor	Hypothalamic-releasing hormone
Ependymin (β, γ)	Insulin
β-Endorphin	Luteinizing hormone-releasing hormone
Gastrin	Melanocyte-stimulating hormone
Glucagon	Melanocyte-stimulating hormone-releasing hormone
Gastrointestinal polupeptide	Neurohypophysial hormone
Insulin	Somatomedins
Leu- and Met-Enkaphin	Thyroid hormone
Melatonin	Thyroid stimulating hormone
α- and α-Melanocyte-stimulating hormone	Thyrotropin-releasing hormone
Motilin	Estrogens (female): estradiol, estrone, progesterone
Neurophysin	
Neuropeptide Y	Androgen (male): testosterone
Neurotensin	Glucocorticoids: cortisone
Oxytocin	Mineralocorticoids: aldosterone
Pancreatic polypeptides	
Physalaemin	
PItuitary peptides	
Proctolin	
Prolactin	
Secretin	
Somatostatin	
Substance P	
Thyrotropin	
Argininae vasopressin	
Lysine vasopressin	
Vasotocin	
Vasointestinal polypeptide	

[2] Reprinted with permission from F. O. Schmitt (1984), "Molecular regulators of brain function: a new view," *Neuroscience* 13. Copyright 1984, Pergamon Press, Ltd.

regulators are operative generally within the nervous system. It is possible that only a portion of this field has been thus far seen. Salient examples are the brain opiates, the endorphins and enkaphlins released in sustained periods of high stress. Other examples are given in Table 2.I, which summarizes the currently known neurotransmitters and neuroregulators.

In addition to the diffuse, global action of chemical regulators, neuroelectric activity may also sometimes exhibit less selective global effects. At many places in the brain and nervous system where extensive dense packing of dendrites, cell bodies, or axons occurs, there is electrical coupling between neighboring neurons in the absence of specialized synaptic connections. Here, the ionic currents subserving a neuroelectric signal in a given neuron pass in part through neighboring neurons because of the high electrical resistance of the densely packed extracellular fluid. Such effects can be particularly pronounced when large numbers of neurons are driven simultaneously so that large numbers of dendritic PSP currents are in synchrony. This is precisely what happens in the salient brain rhythms recorded by the EEG. Such composite extracellular currents constitute a global event that then can permeate and influence vast numbers of neurons in the region.

Such diffuse, extrasynaptic fields of influence and communication can be, in principle, incorporated into a mechanics of neuroelectric signalling of the type developed in this section. However, these effects begin to transcend the more fine-grained representational operations and functions toward which such a mechanics is primarily directed. These more global effects seem, indeed, to point to broader dimensions of influence and organization.

The Chemical Dimension

The chemical dimension of brain function manifests itself in terms of both the chemical synaptic transmitters and the chemical regulators discussed above. The transmitters, as we have seen, serve neuroelectric signalling directly by triggering fluctuations in postsynaptic membrane potentials. At this level of interpretation, the particular chemical transmitter used is irrelevant; only the resulting local conductance modulation and electrical effect is significant. However, in a somewhat different perspective, particular individual transmitter systems seem to have a coherence and systemic role of their own. The monamines, in particular—dopamine, epinephrine, norepinephrine, and serotonin—are all associated with particular functional qualities in the limbic system: reward, incentive, euphoria; arousal; sleep stages. GABA and glycine are universally used for local recurrent inhibitory systems, which are in turn critically important in the spatiotemporal sculpting of network neuroelectric patterns.

One primary manifestation for the regulator effects in the chemical dimension is the hypothalamus. The hypothalamus exerts pervasive governing influences on both the autonomic nervous system and the pituitary gland. The autonomic nervous system is the primordial peripheral nervous system for the basic life support and visceral functions of the body. The pituitary gland is the central governing controller of the endocrine or hormonal system of the body. Both of these systems operate with broadly released chemical regulator agents.

The central point in the present context is that the chemical regulators and the apparent systemic functions of the chemical transmitters seem to define a chemical dimension of organization that transcends, or at least supplements and enhances as well as contributes to, the neuroelectric level, which is the main focus of this book. For example, many common clinical nervous conditions, ranging from mild anxieties and depressions through Parkinson's and Alzheimer's diseases through severe depression, mania, and schizophrenia, are routinely treated by specific drugs that act on these various transmitters and regulators individually or in various combinations. Great progress has been made in this field of psychopharmacology over the last 25 years. Treatment is becoming highly successful, and remarkably fine tuned.

The Dimension of Molecular Neurobiology

The last quarter century has also seen a tremendous revolution and explosion of knowledge in molecular biology. This expansion will likely continue over at least the next 50 years and will include many fundamental contributions to our understanding of the principles of structural and functional organization of the brain. We see emerging here a staggering vision of systemic intelligence housed within the extensive operations of small number of primary types of molecules. We see this molecular intelligence guiding the construction and maintenance of very large-scale biological systems, including the brain, and of their physiological operations. We see further that this intelligence operates in a fundamental sense from genetically prescribed foundations. Since molecular interactions are at root mediated by intramolecular electromagnetic forces, one cannot help but wonder in amazement if these alone are the principal organizational factors of this intelligence.

In the coming half century we will likely see some of the molecular secrets underlying the significant structural and physiological features of the brain, including many of those central to the systemic organization of neuroelectric signalling: things like fabrications of particular neuronal shapes in particular regions, particular distributions of particular cell classes in different regions, particular classes of synaptic junctions, particular patterns of neuronal connectivity, particular transmitters, particular distributions of particular transmitters, structural foundations of memory embedding and of short term plasticity—the list can be multiplied endlessly.

It has been recently estimated that each human being has on the order of 30,000 genes specific to the brain. The amount of specificity available to this many genes can be recognized as enormous. Since many of these genes can be anticipated to be unique to particular individuals, families, and types of people, we can anticipate that one factor in this development will be the discovery and recognition of significant individual, familial, and type differences in brain structural and operational propensities. We should get used to accepting this eventuality in the same way that we accept differences in size, hair color, and athletic ability, and get on with the broader purposes of understanding the constraints, contexts, and real significances of these differences.

General Milieu of the Brain

The average size of a male human brain is about 1375 cc; that of a human female brain is about 1275 cc. Intelligence is only loosely correlated to brain size, Lord Byron's brain was 2200 cc while Anatole France's brain was 1100 cc. More significant is the ratio of brain weight to body weight. The brain to body ratio for mammals (who exhibit significant learning) ranges from about 10 to 100 times that of reptiles (who do not); that for higher primates is about 2 to 20 that for mammals, with human the highest. Dolphins are second highest.

It may be that the quantity of glial cells in the brain are highly significant in intelligence. Einstein's brain had a remarkably supranormal level of glial cells. The neurons of the brain are outnumbered by glial cells by about 10 to 1. These cells are involved in development and maintenance of structure, nutrition, and metabolism and perhaps maintain spatial arrangements of neurons and neuron groups. They seem not to be significantly involved in neuroelectric signalling.

The brain uses about 25% of a normal person's daily nutritional energy. Since the brain is something like 5% of the body weight, it thus uses energy about five times the rate of the rest of the body. This ratio might be higher in people who are particularly active mentally. We can imagine that much of this energy goes to pumping ions across membranes (certainly the density of cellular membranes is much higher in the brain than elsewhere), operating and restoring the molecular apparatus associated with supra-astronomical numbers of molecular permeability gatings for all the action potentials and synaptic activations generated in a day and for the synaptic structural changes associated with the learning of new neuroelectric patterns.

Concluding Remarks on a Mechanics of Neuroelectric Signalling

In the light of the preceding brief survey one can begin to feel many of the limitations of even a successful mechanics of neuroelectric signalling, as well as the extreme difficulty of attaining such a goal. The mechanics in this book is presented with humility before the many limitations that constrain the work, and the many large mysteries its broader contexts suggest.

Chapter 3 Physical Foundations of Neuroelectrical Signalling: Transmembrane Ionic Balances and Fluxes

This chapter describes the biophysical transmembrane ionic system with particular reference to its resting state, which provides the energetic foundations for neuroelectric signalling.

The first and main part of the chapter provides a general mathematical formulation of the physical constraints governing the ionic fluxes in the system and applies this formulation to attain mathematical expressions of interest for the resting state. The derivations are presented rigorously from first principles, beginning with the fundamental mechanical law embodied in the Nernst–Planck constitutive equation. This approach allows one to see clearly the conditions and constraints to which the subsequent derivative rules, such as the Nernst equilibrium equation and the Goldman equation, are subject.

We emphasize that, in the context of neuroelectric signalling, the equations presented in this chapter are subsidiary to the membrane equation, which is taken up in conjunction with the equivalent circuit concept introduced in the next chapter.

The second part of this chapter contains a brief qualitative introductory overview of this transmembrane ionic system for beginning students. Advanced students and professionals may skip this second part. Beginning students should read it first and supplement this reading with a more detailed description from one of the neurobiology texts cited in the appendix.

Mathematical Description of Ionic Currents and Potentials

General Three-Dimensional Formulation

Figure 3.1 illustrates the essential features of the ionic foundations of neuroelectric signalling. Part A illustrates the stable resting base, while part B illustrates the spatial currents associated with a neuroelectric signal around its source in a momentary locus of altered membrane permeability.

Quantitative relations among the ionic currents, potentials, and various main parameters are obtained by considering the physical principles that control their interactions. Three such principles control the flow of ions in both the resting and active states: (a) the constitutive equation relating the fluxes of ions of a given type driven by the electrical force, the concentration gradient, and the molecular pumping mechanism; (b) the law of conservation of mass for a given ionic species, and (c) the law of conservation of charge for the fluid as a whole.

These laws are expressed mathematically in Eq. (3.1) for three-dimensional flow of ions.

$$
\begin{aligned}
\overrightarrow{J_i} &= -qz_i\mu_i c_i \overrightarrow{\nabla V} - kT\mu_i \overrightarrow{\nabla c_i} + \overrightarrow{P_i} \\
&= \left[-qz_i\mu_i c_i \frac{\partial V}{\partial x} - kT\mu_i \frac{\partial c_i}{\partial x} + P_{ix}\right]\overrightarrow{i} \\
&\quad + \left[-qz_i\mu_i c_i \frac{\partial V}{\partial y} - kT\mu_i \frac{\partial c_i}{\partial y} + P_{iy}\right]\overrightarrow{j} \\
&\quad + \left[-qz_i\mu_i c_i \frac{\partial V}{\partial z} - kT\mu_i \frac{\partial c_i}{\partial z} + P_{iz}\right]\overrightarrow{k}
\end{aligned}
\tag{3.1a}
$$

$$
\frac{\partial c_i}{\partial t} = -\nabla^* \overrightarrow{J_i} = -\left\{\frac{\partial J_{ix}}{\partial x} + \frac{\partial J_{iy}}{\partial y} + \frac{J_{iz}}{\partial z}\right\} \tag{3.1b}
$$

$$
\frac{\partial \rho}{\partial t} = -\nabla^* \overrightarrow{J} = -\left\{\frac{\partial J_x}{\partial x} + \frac{\partial J_y}{\partial y} + \frac{\partial J_z}{\partial z}\right\} \tag{3.1c}
$$

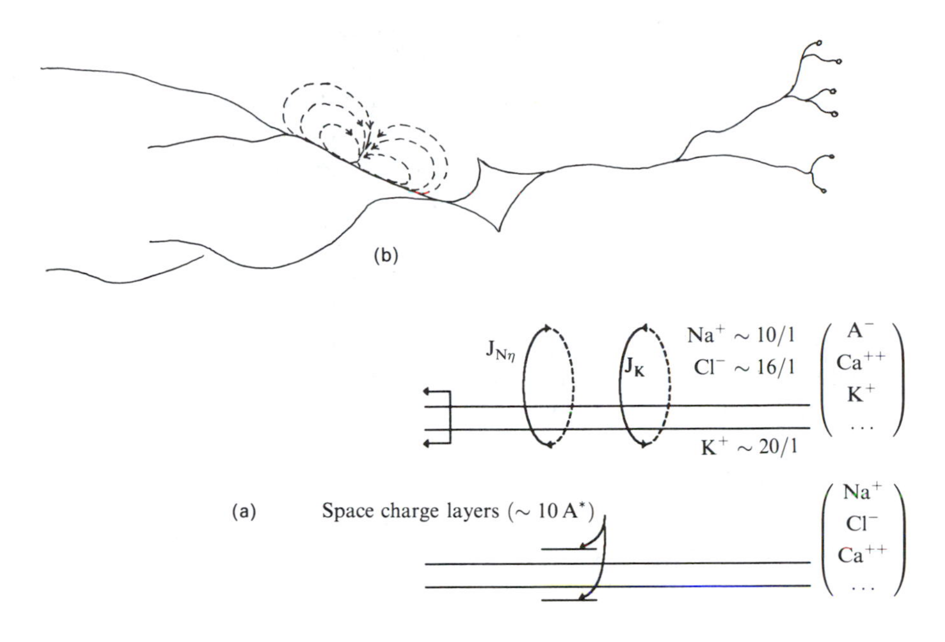

FIGURE 3.1 Essential ionic foundations of neuroelectric signalling: a. resting state; b. current loops around a source of activation.

These equations apply to the entire spatial region of the ionic system, including the intra- and extracellular fluids, and the membrane. The representative parameters in the equations will be expected to have different values in these different regions. Moreover, these equations do not explicitly recognize a temporal capacitative effect of the cell membrane, which must be considered in time-varying applications. Nonetheless, these equations are conceptually instructive, can be generalized for the time varying case, and are valid for steady-state fluxes.

Equation (3.1a) is the constitutive equation for ion fluxes for ions of type i; it is known as the *Nernst–Planck constitutive equation*. In this expression, J_i represents the number of particles crossing a unit area normal to the flow per unit time. In three-dimensional flow it is a vector with three components corresponding to the components of flow in each of the three coordinate directions. This equation is valid at any point in space at any time; that is, any of the variables may have different values at different places in space and time. There is a separate Nernst–Planck equation for each ion type involved in the flow.

The first group of terms on the right side of the equation is simply Ohm's law for the driving effect on ions of the electrical field. The driving factor is the gradient of the electrical field represented by three components for the three directions: $\partial V/\partial x$, $\partial V/\partial y$, and $\partial V/\partial z$; or, equivalently, in compact notation by $\overrightarrow{\nabla V}$. The term is negative because a positive ion will be driven in the negative direction by a field that increases in the positive direction. The constants of proportionality in this term are the unitary electronic charge, q; the valence of the ion, z_i (+1 for sodium and potassium, −1 for chlorine, +2 for calcium, etc.); a measure of the permeability or inverse resistance of the medium known as the mobility, μ_i; and the concentration of ions in particles per volume, c_i.

The second group of terms in the equation represent Fick's law for the driving effect of concentration gradients on ion motion. The driving factor here is the gradient of the concentration, $\overrightarrow{\nabla c_i}$, with its corresponding three scalar components: $\partial c_i/\partial x$, $\partial c_i/\partial y$, and $\partial c_i/\partial z$. Again, this term has a negative sign because an ion will be driven in the negative direction by a concentration gradient increasing in the positive direction. The constants of proportionality in this term are Boltzman's constant, k; the absolute temperature, T; and the mobility, μ_i.

The final term, P_i, represents possible pumping action in particles per area per time; this term is applicable to sodium and potassium only, and is zero everywhere except at cell membranes.

Equation (3.1b) is the equation of conservation of mass for the ith ionic species. This equation states simply that for the mass of ions of type i to be conserved at a point in space, it is necessary that the total amount of such mass flowing into the point (represented by the right side of the equation) must show up as an increase in the amount of mass at the point (represented by the left side of the equation). This equation is derived by considering a differentially small cubicle element of space and adding up influx and outflux terms. One may find such a derivation in standard textbooks on fluid mechanics.

Equation (3.1c) is the equation of conservation of electric charge. This equation, parallel to (3.1b), states that the amount of electric charge at a given point must vary according to the net electrical current flowing into the point. This last equation is a global equation containing contributions from each of the ion types. Its terms, the electric charge density, ρ, and the total electric current, J, are related to the concentrations and fluxes of individual ion types by Eq. (3.2):

$$\rho = \sum z_i q c_i \tag{3.2a}$$

$$\vec{J} = \sum z_i q \vec{J_i} \tag{3.2b}$$

When properly applied, Eq. (3.1) can be used to describe the three-dimensional flow of ions in neuroelectric signalling, and it can be reduced to describe the properties and constraints of one-dimensional flow directly across membranes. Strictly speaking the application of these equations to three-dimensional flow is a difficult and tedious engineering task. Four dimensions must be simultaneously integrated; one must properly adjust the various parameter values for the different regions of the flow—external, internal, membrane, and so on. Moreover, properties of the membrane can be described completely by these equations only for the steady state; as we will see in the next section, the membrane produces temporal variations in signals beyond those described by Eq. (3.1). Nonetheless, when properly applied, Eq. (3.1) form the foundations for studying extensions to nonsteady signalling at the ionic and spatial levels and can be used as they stand to provide instructive insight into properties of steady-state signalling in both spatial and one-dimensional contexts.

Ions are intermingled in the external and internal fluids to the extent that these fluids are electrically neutral over distances larger than that of several ions, except for very small layers called *space-charge layers* on the immediate outer and inner surfaces of neural membranes. These layers relate to charged lipoprotein molecules in the membranes. Surpluses of negative and positive ions in the intra- and extracellular fluids, respectively, are associated with positive and negative polarizations of the ends of the membrane lipoprotein molecules. These latter polarizations form the foundation of the membrane capacitive effect as discussed in the next chapter.

In the steady state, both the total electric current, J, and the individual ionic currents, J_i, flow in closed loops. We can see this physically from the fact that, since in the steady state no charge or ion type may accumulate at any one point, the amount of current going into any point must also go out. If there are no sources of current, the current lines must close into themselves to form loops. This can also be seen mathematically from Eqs (3.1b) and (3.1c), which show that the total electric current, J, and the individual ion flux currents, J_i, must have zero divergences if there is no change in concentrations with time. This is an instructive insight and a very general constraint on the geometry of neuroelectric currents in three dimensions: their steady flow must involve closed loops of ions and total currents.

In the time-varying case, currents also move in closed loops when one takes into account the effective "displacement current" associated with structural displacements and tensions in the charged lipoprotein molecules of the cell membrane discussed in the next chapter.

Application to One-Dimensional Flow Across Membrane

Equations (3.1) can be reduced for application to one-dimensional flow across a membrane, as illustrated in Figure 3.2, by reducing the three-dimensional spatial terms to one dimension as shown in Eqs. (3.3):

$$J_i = -qz_i \mu_i c_i \frac{\partial V}{\partial x} - kT\mu_i \frac{\partial c_i}{\partial x} + P_i \tag{3.3a}$$

$$\frac{\partial c_i}{\partial t} = -\frac{\partial J_i}{\partial x} \tag{3.3b}$$

$$\frac{\partial \rho}{\partial t} = -\frac{\partial J}{\partial x} \tag{3.3c}$$

This formulation implicitly assumes that the effect of the membrane on the ion fluxes can be represented by a mobility term (as in the external and internal fluids). This is valid for steady fluxes, but not for time-varying fluxes.

In this formulation one is picturing a small control volume of membrane that houses a sufficiently large number of channels so that the idea of a concentration of ions has meaning, even though ions must pass through individual channels in series. This is parallel to the assumption of continuous media used in deriving governing equations in fluid mechanics.

Further notice that the concentrations of the ions inside the membrane channels are higher than in the external fluid because the channels constitute only a portion of the membrane. This jump in concentration values across these space–charge layers is accounted for mathematically in terms of a partition factor. That is, at the appropriate place in some equations, the external concentrations are taken as $c_{2_i} = f^* c_{out_i}$, and $c_{1_i} = f^* c_{1_i}$, where f is the partition factor.

We shall now apply Eq. (3.3) to identify two characteristic values of the resting state of this governing system for ionic fluxes; namely, the equilibrium potential for individual ionic types and the membrane resting potential.

Equilibrium Potential for Single-Ion Types

It is instructive to consider the constitutive equation (3.3a) that defines the transmembrane driving forces on any given single type of ion. If we consider the case where the ion flux approaches zero, then Eqs (3.4) apply.

$$\frac{\partial V}{\partial x} = -\frac{kT}{qz_i c_i}\frac{\partial c_i}{\partial x} \tag{3.4a}$$

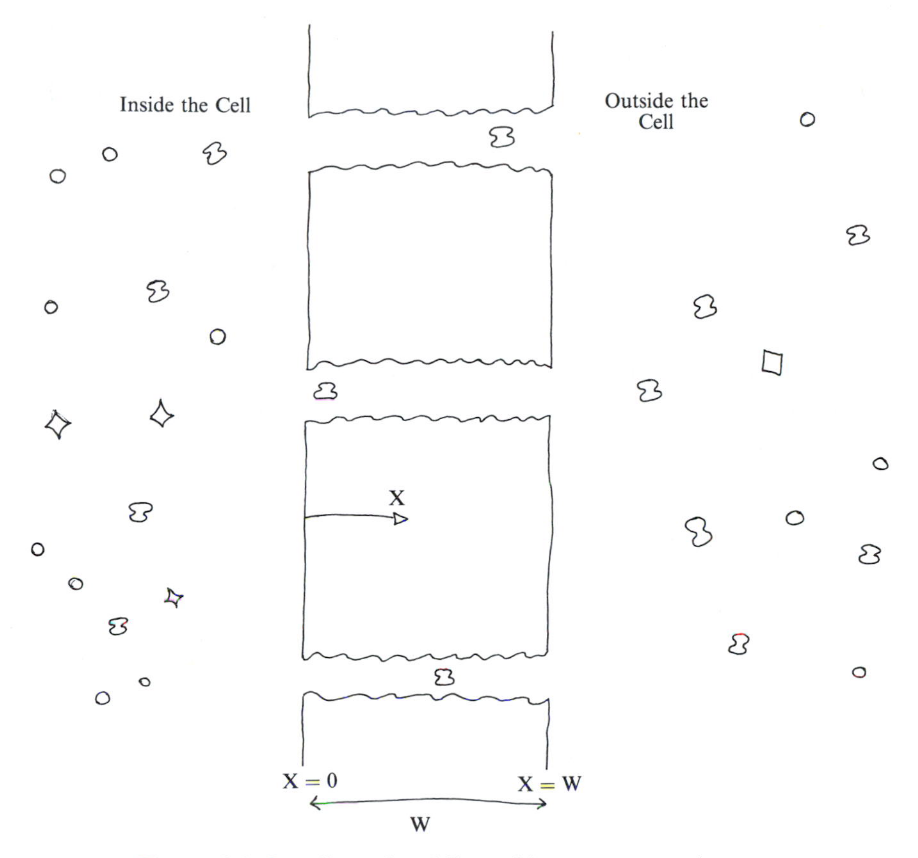

FIGURE 3.2 One-dimensional flow of ions across membrane.

$$V_2 - V_1 = -\frac{kT}{qz_i}\ln\left(\frac{c_2}{c_1}\right) \tag{3.4b}$$

$$J_i = -\mu_i\left\{qz_ic_i\frac{\partial V}{\partial x} + kT\frac{\partial c_i}{\partial x}\right\} + P_i \tag{3.4c}$$

Equation (3.4a) follows from (3.3a) when J_i and P_j are zero. Equation (3.4b) results from integrating (3.4a) from the inside ($x = 0$, $c = c_1$) to the outside ($x = w$, $c = c_2$) of the membrane. This equation represents the equilibrium potential of the ionic species. Physically, it relates the membrane electrical potential to the ratio of external to internal ionic concentrations for the situation in which the diffusion force exactly balances the electric force. One can see moreover, by casting Eq. (3.3a) in the equivalent form shown in Eq. (3.4c), the essential physical significance of this quantity. If we imagine a case where the mobility of the membrane to the ith species, μ_i, is increased without limit, then so that the current, J_i, stays finite, it is necessary that the term multiplying μ_i in Eq. (3.4c) go to zero. This, however, results exactly in Eq. (3.4a) which is the condition for the equilibrium potential shown in Eq. (3.4b). That is, if one increases the membrane permeability to the ith species without bound, the potential of the membrane must approach the equilibrium potential of that ionic species, or else the flux of that ion type will become infinitely large.

One may further see, with the help of Eqs (3.3a) and (3.4), that if the potential across the membrane is slightly larger than the equilibrium potential, (positive) ions will flow inward to reduce it; whereas if the potential is slightly less than the equilibrium potential, (positive) ions will flow out to increase it. That is, both effects tend to drive the potential toward the equilibrium potential.

This equilibrium potential is called the *Nernst equilibrium potential.* It is valuable in interpreting the comparative driving propensities of the various single-ion channels and the composite-ion channels associated with synaptic and action potentials.

Application to Membrane Resting Potential

A second quantity of interest is the membrane potential that occurs when the total current across the membrane is constrained to be zero in a steady state. This is the resting membrane potential. In this situation individual ions are continually flowing (indicated by nonzero values of J_i), but the net electrical current, J, determined by Eq. (3.2b) is zero.

The following four steps are instructive in considering the resting potential. First, one can show that if the ions involved are all monovalent (for example, sodium, chlorine, potassium, but not calcium) and the membrane is assumed to be homogeneous, then the electrical field across the membrane is constant; that is, $\partial V/\partial x$ is equal to $\Delta V/w$ at all points in the membrane. (This is proved on pages 44 and 45 of NBM.)

Second, if the field across the membrane is constant, then the fluxes, J_i, are given by Eq. (3.5a), and the concentrations are given by Eq. (3.5b). These are seen as follows:

$$J_i = \frac{z_i q \mu_i}{w}\left\{\frac{c_{2i} - c_{1i}\exp[D]}{1-\exp[D]}\right\}\Delta V + P_i \tag{3.5a}$$

$$c_i = \frac{c_{1i}\{\exp[Dx/w] - \exp[D]\} + c_{2i}\{1-\exp[Dx/w]}{1-\exp[D]} \tag{3.5b}$$

$$\frac{dc_i}{dx} = \frac{Dc_i}{w} - \frac{J_i - P_i}{kT\mu_i} \tag{3.5c}$$

$$c_i = a_i = b_i \exp[Dx/w],$$

$$c_i(x = 0) = c_{1i} = a_i + b_i, \tag{3.5d}$$

$$c_i(x = w) = c_{2i} = a_i + b_i \exp[D]$$

$$a_i = \frac{c_{2i}c_{1i}\exp[D]}{1 - \exp[D]}, \qquad b_i = \frac{c_{1i} - c_{2i}}{1 - \exp[D]} \tag{3.5e}$$

$$D = \frac{qz_i\Delta V}{kT} \tag{3.5f}$$

Equations (3.3b) and (3.3c) tell us that, for the steady-state condition, both the J_i and J are constant; they depend on neither time nor space. Then Eq. (3.3a) can be written as a differential equation for c_i, as in Eq. (3.5c). The solution of this equation for c_i in terms of integration constants a and b is given in Eq. (3.5d). By insisting that solution (3.5d) reduce to the boundary conditions ($c_i = c_1$ at $x = 0$, and $c_i = c_2$ at $x = w$) the expressions given in Eq. (3.5e) for a and b are found. Substitution of these values for a and b into Eq. (3.5c) produces the expression for c_i given in Eq. (3.5b). Equation (3.5a) for the flux J_i is obtained by using this expression for c_i in Eq. (3.3a). These equations tell us that the concentrations, c_i, vary with position in the membrane channels between c_1 and c_2, whereas the electric field and ionic fluxes, J_i, are constant. They, moreover, give the dependence of the fluxes, J_i, on concentrations, potential, and parameters as in (3.5a). (Note in passing that if J_i and P_i are zero for a nonzero V, then the term in brackets in Eq. (3.5a) must be zero. This again gives the Nernst potential for the equilibrium potential of the species.)

Third, if the individual ion fluxes are given by Eq. (3.5a), then Eq. (3.6b) expresses the constraint that the total electric current across the membrane be zero.

$$J = \sum z_i q J_i = 0 \tag{3.6a}$$

$$\sum z_i \left\{ \frac{q^2\mu_i}{w} \left[\frac{c_{2i} - c_{1i}\exp[D]}{1 - \exp[D]} \right] \Delta V + qP_i \right\} = 0 \tag{3.6b}$$

This follows directly from using Eqs (3.2b) and (3.5a) in this constraining condition. Equation (3.6b) is significant because we can solve it for the membrane potential, E, which exists at these conditions, in terms of the parameters, c_{1i}, c_{2i}, μ_i, and P_i. That is, we can solve (3.6b) for the resting membrane potential. When nonmonovalent ions (such as calcium) are present, the electric field gradient varies within the membrane, and then Eq. (3.6b) does not apply.

Fourth, if the ions in Eq. (3.6b) are all monovalent (e.g., sodium, chlorine, potassium, but no calcium) and if we neglect the pumping action, P_i, then the resting potential can be found explicitly from Eq. (3.6b) as shown in Eq. (3.7).

$$\Delta V_{\text{rest}} = kT \ln\left\{\frac{\sum \mu_i\{c_{2i} \text{ for } + \text{ ions}; \quad c_{1i} \text{ for } - \text{ ions}\}}{\sum \mu_i\{c_{1i} \text{ for } = \text{ ions}; \quad c_{2i} \text{ for } - \text{ ions}\}}\right\} \tag{3.7a}$$

$$c_{\text{out}i} = \xi_i c_{2i}, \qquad c_{\text{in}i} = \xi_i c_{1i}, \qquad \text{permeability} = \xi_i \mu_i \tag{3.7b}$$

This last expression for the resting potential is known as *Goldman's equation.*

The membrane resting potential can thereby determined explicitly in terms of the external and internal ionic concentrations and membrane permeabilities from Eq. (3.7) when only monovalent ions are present. Observed and predicted values are in agreement at values of about 70 mV, ± 10 or 20 mV, inside negative with respect to outside. From the point of view of neuroelectric signalling, this potential drop is a base level on which neuroelectric signals appear as modulations.

General Quantitative Values of the Resting State

The concepts of the ionic equilibrium potentials and the membrane resting potential allow us to survey in passing various relevant quantitative and physical information which is contained in Table 3.I. The ratios of sodium and chlorine ion concentrations outside to inside are typically about 10 and 16, respectively, while the ratio of potassium ion concentration inside to outside is about 20. The resting membrane permeabilities to chlorine, sodium, and potassium ions, respectively, are .052, 2, and 100 all in 10^{-8} cm per sec. These values reflect the size of the ion types, and the idea that they must move through relatively small pores in the membrane. The equilibrium potentials for these ions are respectively, -10, $+100$, and -10, in mV measured with respect to resting level. The equilibrium potential for calcium is about 55 mV above the resting level.

Potassium ions are driven inward by the resting membrane potential drop but outward by its concentration gradient. There is a slight preponderance of the outward diffusion force. Sodium ions are driven inward by both the

TABLE 3.I: Ionic Balances in the Resting State.

Relative Concentrations:

Extracellular fluid: $Na^+ \sim 10/1$, $Cl^- \sim 16/1$
Intracellular fluid: $K^+ \sim 20/1$

Permeabilities ($k^* T^* \mu/w$ in 10^{-8} cm/sec)
K^+: 100
Na^+ : 2
Cl^-: 0.52

Force Balances:
K^+: [diffusion out, electric in] = slight tendency out * high permeability ~ Na^+ flux
Na^+: [diffusion in, electric in] = strong tendency in * low permeability ~ K^+ flux
Cl^-: [diffusion in, electric in] = strong tendency out * very low permeality ~ very small leakage

resting membrane potential and its concentration gradient. The weaker outward driving force on potassium and its much higher membrane permeability combine to produce an outward leakage of potassium ions about equal in magnitude to the inward leakage of sodium, which results from a much stronger combined inward driving force with a 50-fold weaker membrane permeability. Chlorine ions are driven inward by diffusion forces and outward by the resting electrical potential; the small difference in these forces together with the very low permeability to chlorine results in a very low degree of chlorine migration in the resting state. Other negative ions, which are necessarily present inside the cell to maintain electrical neutrality, are strongly pressed outward by both diffusion and electrical potential, but do not diffuse because of essentially zero membrane permeability.

Qualitative View of the Physical Character of the Ionic System

The system discussed here provides the elemental physical mechanisms by which neuronal signals are generated. It consists of membrane pumping and resting permeability properties that partition the fluids into distinct intra- and extracellular compartments with different ionic compositions and electrical potentials. The mechanism of generating neuroelectric signals consists of localized selective modulations of membrane permeability that allow selective sets of ions to flow down their concentration and electrical gradients across the membrane and in closed loops through the extra- and intracellular fluids.

The entire system is a pristine example of simple but elegant engineering design based on volumetric ionic concentrations, balances, gatings, and fluxes. It may be seen as a global systemic battery that stores energy in the concentration and electrical differences of the extra and intracellular fluids. Metabolic energy is used steadily and continually to maintain the functional integrity of this systemic battery by molecular pumping of ions (sodium and potassium) to their uphill poles. Actual signalling, effected by microscopically placed permeability modulations selective in space, time, and ion type allow individual ions to flow downhill across the membrane utilizing this energy as needed. The signalling is something like power steering or passive Eastern philosophy: the actuating signal mechanism (gatings) require small amounts of energy, highly localized and highly selective, but bring about and make use of larger (and highly directed) energy fluxes from the ambient environment.

The resting state attained by this system is universal in all living cells. The nervous system develops specialized permeability gating mechanisms to draw on it as a source for neuroelectric signalling.

The Resting State

The ionic resting state is determined by the action of metabolically active molecular "pumping" agents in the membranes of cells, together with natural physical laws by which electrical and concentration forces tend to drive ions across a membrane that is semipermeable. Here *semipermeable* means simply that the membrane exhibits selective and partial resistance to the flows of ions through its structure, larger forces leading to larger fluxes.

In the absence of additional activations, these factors interact to produce a resting state wherein the extracellular and intracellular fluids are composed of distinctly different but balanced ionic composition. The extracellular fluid is very much like seawater, consisting of high concentrations of sodium (Na^+) and chloride (Cl^-) ions, while the intracellular fluid is high in potassium (K^+) and other variable and nonactive negative ions.

The ionic pumps carry out sodium ions and carry in potassium ions. These transfers are going on continually at low rates in many portions of all cells. The number of ions transferred at any one short time period at a given locale is small, but accumulates significantly over space and time. Clearly, ionic pumping alone would force the external fluid to become increasingly saturated with sodium ions and the internal fluid with potassium ions. As this happens, however, a compensating diffusion of sodium into the cell and potassium out of the cell occurs because of diffusion. The larger is the concentration difference between external and internal sodium concentration, the larger will be the compensatory inward flux of sodium. Clearly, at some value of external sodium concentration the inward diffusion of sodium ions will exactly balance the outward pumping of sodium ions; the system at this point would be in equilibrium, continual but equal outward pumping and inward leakage of sodium occurs but the concentrations remain constant. This happens similarly for potassium.

The situation is slightly more complicated because transmembrane electrical forces also involve the interactions of the ions with the lipoprotein molecules of the cell membrane. Nonetheless the principle is clear: at some levels of external and internal concentrations of ions, the transmembrane fluxes of individual ion types due to diffusion and electrical forces will exactly balance with the active pumping rates for each type, thereby establishing an equilibrium state wherein sodium is elevated in the extracellular fluid and potassium in the intracellular fluid because this is the direction the pumps produce.

This understanding of the mechanics of ions fluxes also allows one to ascertain mathematically two relevant quantities of interest and their dependencies on the physiological and anatomical parameters of the membrane and fluids. The first of these is the value of the transmembrane electrical potential at which the combined leakage of ions is equal to zero; that is, the value at which the leakage due to the electrical potential is exactly equal and opposite to that due to diffusion. This value is of particular interest because it can be shown that, if the membrane permeability to this ion type is progressively increased, then the transmembrane potential must approach this value. These particular potential values are called *equilibrium potentials*; the equation for the equilibrium potentials of single ions is called the *Nernst equilibrium equation*.

This concept of the equilibrium potential provides a very useful way to picture the resultant effect of changing membrane permeability—increasing permeability for a given ionic species tends to drive the potential toward the equilibrium potential of that species. We will show in the next chapter that this foundational result can be easily generalized to synapses that use constant relative weightings of certain combinations of ion types.

The second value of membrane potential that is of interest is the value at which the net electrical current across the membrane is zero. This value is the resting potential of the neuron because the transmembrane potential necessarily is constant if there is no current flowing the membrane. The total current across the membrane in the resting state is simply the sum of the individual ionic currents. The resulting expression for the resting potential in the simple representative case where only monovalent ions are involved (Na^+, K^+, Cl^-) is called the *Goldman equation*. This equation is useful because it shows how the anatomical and physiological parameters of the membrane and fluids influence the resting transmembrane potential.

The Active States

The generic active neuroelectric signals, post-synaptic potentials and action potentials, are mediated by selective changes in permeability which allow selective sets of ions to flow down their concentration and electrical gradients.

When action potentials arrive at presynaptic terminals, packets of molecular transmitters are released into the subsynaptic region to diffuse to the subsynaptic membrane of the receiving neuron. The transmitters there are taken up by particular molecular receptors in the postsynaptic membrane. Depending on the highly individualized molecular chemistry of the particular transmitter–receptor combination, selective modulations to the various ionic types are effected in the receptor. These modulations are localized to the synaptic region; they typically exhibit a transient duration about the same as that of the action potential, about a half-millisecond, but may be longer. The permeability modulations induce ions to flow in or out of the cell through the subsynaptic membrane. The ionic fluxes close in loops by crossing the membrane again at some distance away from the synapse.

The precise ionic makeup and underlying molecular chemistry is not yet explicitly known for most receptors. A useful, but oversimplified, generalization is that synapses are functionally excitatory or inhibitory. Eccles pointed out 40 years ago that excitatory synapses could operate by opening channels to all ionic species; this would drive the local potential toward a depolarized state, meaning the intracellular potential would rise (toward threshold). Eccles also noted that inhibitory synapses could operate by opening smaller channels—about 23 Å: large enough for all ions except sodium ions. The ionic current in this case would be dominated by negative inward, positive outward fluxes thereby pulling the intracellular potential away from threshold and causing an inhibitory functional effect.

Action potentials are triggered by particular types of excitable gating molecules, which are discretely distributed over the surface of excitable axons from their junctions with the soma out to their synaptic terminals. These gating molecules trigger and govern action potentials by changing their configuration to produce transmembrane ionic channels in response to the neuron generator potential at their location. First, gating molecules are activated that open membrane channels for sodium ions. Sodium ions rush into sodium-sparse intracelluar fluid from the sodium-rich extracellular fluid driving the intracellular potential up; this is the rise of the action potential. After a short delay (some tenths of a millisecond) these channels close and other gating molecules open channels to potassium. Potassium ions now rush out from the potassium-rich intracellular fluid into the potassium-poor extracellular fluid, bringing the potential back down toward or past its resting level—the downside of the action potential. The potassium channels close to recover their resting values over a longer time period, usually from about 5 to 15 msec.

Conductance of action potentials along the axon is achieved by interrelated repetitions of these same events; much like the propagation of a flame along the fuse of a firecracker by many distinct small local explosions. The currents generated locally by such a gating package pass longitudinally along the neuron and traverse its membrane in contiguous regions thereby causing the generator potential in neighboring regions to attain critical levels and activate the gating molecules in this second region; and so on. Backpropagation is prevented by the phenomenon of refractoriness. For a short time after the firing of an action potential a particular region is much less prone to fire another action potential. This is governed in part by the shunting conductance to potassium (relative refractoriness, about 5 to 15 msec) and in part by more obscure molecular mechanisms (absolute refractoriness, about 1 msec).

Thus, neurons exert active molecular control over their own output signals in terms of triggering and conducting series of large, fast action potentials to downstream cells according to the character of their generator potentials at their cell bodies; and they effect this control by way of selective modulations in membrane conductance to salient ions of the surrounding fluids.

The broader "pseudo-action potentials" occuring in the dendritic regions of larger and more elaborate neurons of higher brain regions (for example, pyramidal cells of cerebral cortex and hippocampus) are mediated by very similar excitable gating processes. Here, though, the excitable gating is for calcium ions; the recovery gating is again for potassium ions; the peak potential is smaller and broader, the refractory shutdown is much more pronounced and much slower. It has been suggested that the molecular gating molecules for these dendritic action potentials may be limited to certain local regions ("hot spots") rather than distributed continuously for rapid conduction as in axons.

Chapter 4 The Membrane Equation and its Equivalent Circuit as the Central Governing Equation for Neuroelectric Signals

The purpose of this chapter is to present the equivalent circuit for representing neuroelectric signals most directly in terms of their underlying causal and constraining physical agents and to derive from this representation their central governing equation, the membrane equation.

The chapter discusses the physical components of the equivalent circuit, including both the fluxes of individual ion types and the capacitive displacement currents associated with structural tensions and displacement of the membrane lipoprotein molecules. This latter feature is what allows neurons to capture fleeting extraneous stimulations briefly in time, and thereby is the root of its ability to represent information neuroelectrically.

The membrane equation is derived from the equivalent circuit by imposing a current balance across its constituent elements. The equation is manipulated and normalized into a dynamic engineering system form. It is shown that the terms representing individual ion types can be transformed into groups representing individual active processes, such as single synapses or action potential generators.

The final canonical form of the membrane equation expresses

trans-membrane potential in a given local region in terms of two active dynamic variables, the total transmembrane current and the local active permeability modulations, and one explicit constituent parameter, the membrane time constant. The total transmembrane current is determined by the spatial geometry of the neuron and its surroundings as these have some impact on the local region—these are taken up in Chapter 5. The local active permeability modulations represent the local generation of neuroelectric signals, synaptic responses and action potentials—these are taken up in Chapter 6. The single explicit parameter is the membrane time constant. The other local parameters (area of the local region, resting conductance) simply scale the effectiveness of the active dynamic variables. Other important physiological parameters will be manifest in the expressions for these activating variables as is shown in Chapters 5 and 6.

The Equivalent Circuit Representation of Neural Membrane

The preceding chapters have shown that the primary neuroelectric signals, synaptic responses, and action potentials are reflective of local transmembrane ionic currents, triggered by brief modulations in membrane permeability restricted to localized regions, represented by potential drops restricted very largely to drops directly across the membrane itself; and propagated longitudinally along the surface of the membrane by current loops through the adjacent intra- and extracellular fluids.

It follows that the most direct means of describing the physical processes and constraints that control these signals will be a representation of the physical processes and properties of the local membrane that control its current–potential relationships and its current–potential relationships with neighboring regions. Such a representation will allow one to express the potential of a given local region in terms of the passive and active states of membrane components, and relate this potential to the potentials in neighboring regions in terms of the portions of current loops it shares with them. The equivalent circuit representation is precisely such a means.

In this context it should be very clear that this equivalent circuit device is not merely an analogy, or model, in a weak sense of that word. It is a shorthand device for representing the actual physical process of the neurobiology of the system.

The Equivalent Circuit

Figure 4.1 displays the foundational equivalent circuit in two manifestations.

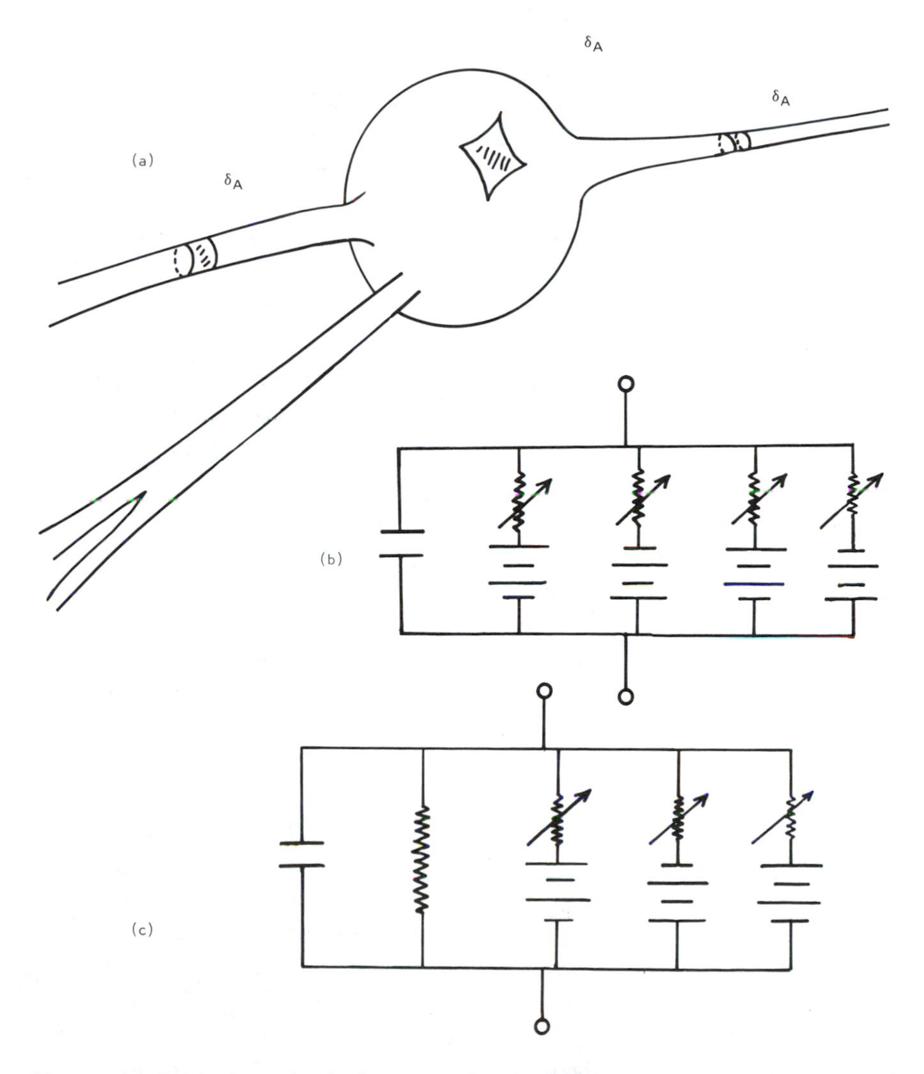

FIGURE 4.1 Equivalent circuit for neuroelectric signals: a. representative areas of circuit; b. circuit components for individual types of ions; c. circuit components for resting and active states. Reprinted with permission from *Neural and Brain Modeling*, R. MacGregor, Copyright 1987, Academic Press.

The form most readily mapped onto single ion fluxes is shown in Figure 4.1b.

The circuit represents the electrical properties of an area of neuronal membrane. Strictly speaking, the model becomes accurate only for areas of membrane sufficiently small that the transmembrane electrical potential is the same at all points on it. Many cells have soma for which this is approximately true. The diameters of these soma are so large that the resistance to current flow inside the soma is very small resulting in an isopotential inner surface. When this occurrence is coupled with a local extracellular space free of electrical obstructions, the external surface is also isopotential. On dendrites and axons, however, the internal radii are generally small enough that longitudinal electrical resistance is large. This implies that large transmembrane potential gradients may occur in these regions. In these cases, the equivalent circuit model is accurate only for areas with very small longitudinal dimensions. As indicated in Figure 4.1a, one may accurately represent these cases by taking the reference area as a differentially small annular ring. The ways of using equivalent circuits to represent entire neurons and networks of neurons are discussed in Chapter 5. For now, it is sufficient to emphasize that the equivalent circuit always has reference to a particular area of neuronal membrane, over which the transmembrane potential has been assumed to be uniform.

The nodes of the circuit represent points in the extracellular and intracellular fluids immediately outside the space-charge layers. The electrical potential between these nodes is therefore the transmembrane potential. The various branches in the circuit represent the current–potential relations mediated by the ionic fluxes introduced in Chapter 3.

Passive and Active Permeabilities as Conductances

Each of the several branches containing a battery in series with a variable conductance represents the current–potential properties of membrane channels for an individual ionic species. The battery represents the effective driving force of the general neuronal systemic battery on this particular species; its value is the Nernst equilibrium potential for that species. The variable conductance represents the possibly time-varying permeability to that species. We note that, if this conductance is allowed to be infinitely high, the potential across the membrane is then clamped at this equilibrium potential, as we know from the last section it must be. In general, if the transmembrane potential is instantaneously at a value different from this equilibrium potential, current will flow through the branch in proportion to the difference between these two potentials; the constant of proportionality is the conductance. The conductance value in these branches is the composite from all the channels to this species in the area of membrane represented by the circuit. The conductance value can be considered the sum of two components: the resting (nonactive) value that contributes to the resting properties of the neuron, and an active component that represents the molecularly mediated permeability fluctuations at synapses and action potential sites, and perhaps for other less common active neuroelectric signals.

Membrane Capacitance: The Capturing of Electrical Signals over Time

The branch containing the capacitance represents structural displacements and tensions in the membrane lipoprotein molecules in response to changes in the transmembrane potential, and the associated effect on the charges in the space–charge layers. These displacements and tensions are the direct result of changing electrical forces across the cell membrane on the electrical charges of these molecules. They include minute transverse motions of these molecular charges, bringing about corresponding tensions and minute compressions or distentions of the molecules. They include minute transverse motions of these molecular charges, bringing about corresponding tensions and minute compressions or distentions of the molecules. They also include minute rotations of the molecules driven by corresponding rotations of the charged heads of the molecules.

These structural displacements and tensions and their associated interactions with altering charge concentrations in the space–charge layer bring about the particular short-term temporal sluggishness in the electrical response of neurons. This sluggishness is conveniently subsumed by the mathematically based concept of a time constant, which is introduced later. This elemental temporal sluggishness is the root of the nervous system's ability to capture signals in time and therefore to represent information. Without it, neuroelectrical signals would be merely instantaneous reflections of instantaneous external stimulation fields.

It is instructive to consider at this point an example that illustrates most clearly the biophysical foundation for the capacitance branch of the equivalent circuit. The example is illustrated in Fig. 4.2. The figure and discussion are presented in terms of the contribution of the rotation of the membrane lipoprotein molecules because it is simplest to illustrate graphically. The figure and discussion apply also directly and equally well to the transverse displacement and tension effect.

The lipoprotein molecules are electrically polarized: one end is positive, and the other end is negative. (There are two such molecules in series across the membrane.) The positive end of the molecule is at the inner surface of the membrane, and the negative end at the outer surface. Negative ions accumulate along the inner surface and positive ions along the outer surface (inner and outer space–charge layers, respectively), corresponding to the

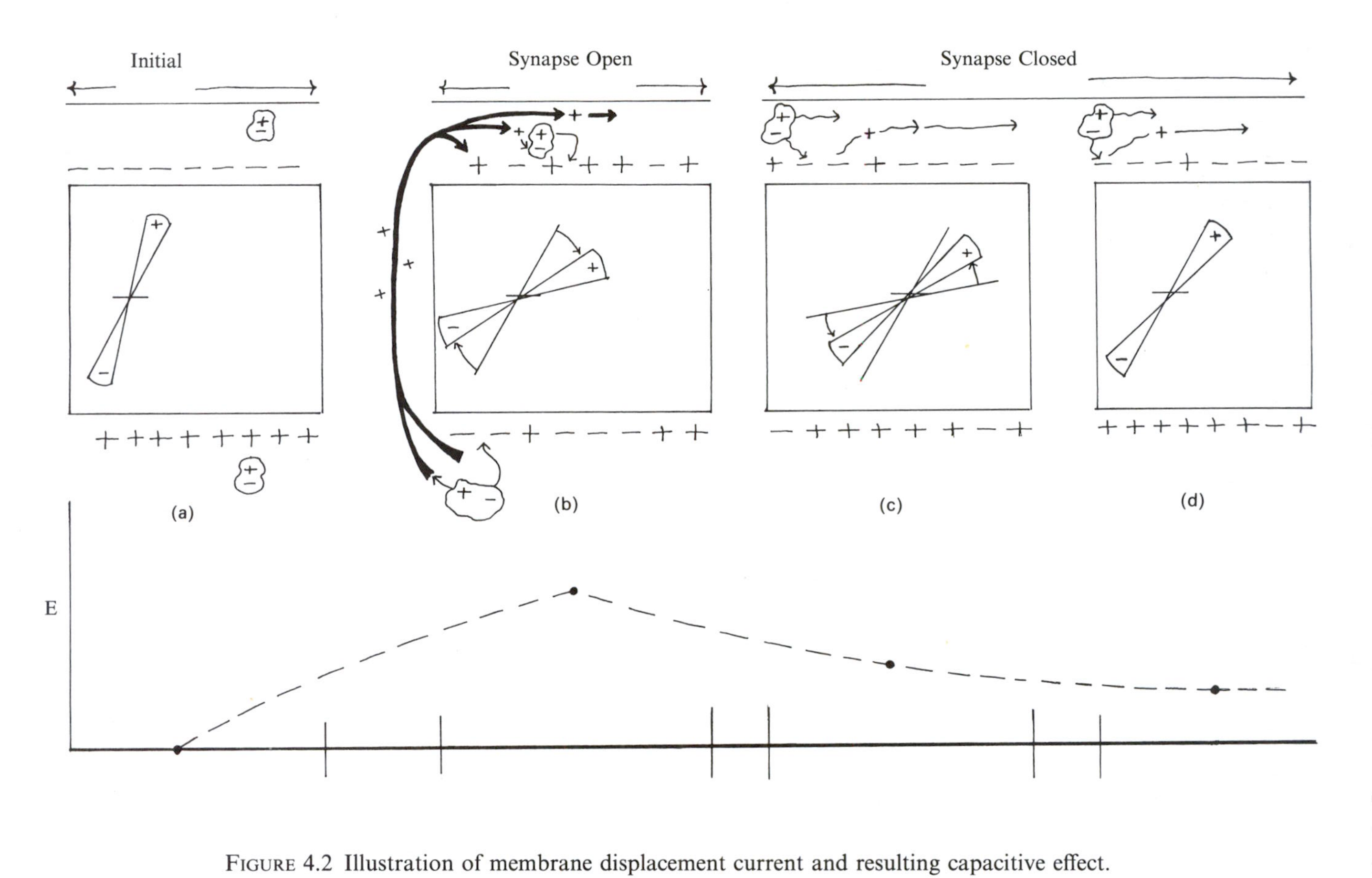

FIGURE 4.2 Illustration of membrane displacement current and resulting capacitive effect.

inner negativity of the resting potential. This resting state is indicated in Figure 4.2a. When the transmembrane potential varies to increase the internal potential (depolarization), the charged ends of the lipoprotein molecules are repulsed by the altered potentials at either end and tend to rotate toward the center, as shown in Figure 4.2b. There is a mechanical resistance to this effect from neighboring structures in the membrane, so that the molecule rotates a certain increment depending on the magnitude of the potential change, to a new equilibrium position. Corresponding to the decreased electrical effects of the charged ends of these molecules on the space–charge layers, some of the negative ions on the inner surface are replaced by positive ions, and vice versa on the outer surface. These changes are also illustrated in Figure 4.2b. If now the transmembrane potential returns to its original resting value, the polarized molecules rotate back to their equilibrium positions, and the positive ions on the inner surface (and negative ions on the outer surface) are released to move back into the intra- (and extra)cellular fluid. This is illustrated in Figures 4.2c and 4.2d.

Suppose that the case illustrated in Figure 4.2 represents an area of membrane immediately adjacent to an active excitatory synapse; that Figure 4.2a is the state before activation, 4.2b is the state when the synapse is active, 4.2c and 4.2d are states shortly after the synapse is again quiet. In this case the hypothesized increase in transmembrane potential and the presence of positive ions on the inner surface (and negative ions on the outer surface) correspond to positive ions flowing in through the synapse.

It can be seen clearly from this consideration that the fact that the physical rotation of the lipoprotein molecules and the associated movement of ions in the surrounding fluids requires a nonnegligible time period imparts a sluggishness to the electrical response of the membrane. Specifically, since the capture and release of the charged particles in the space–charge layers is directly (and immediately) coupled to the position of the charged ends of these rotating and distending–compressing molecules, their motion (which makes up the responsive neuroelectric signal) lags the activating signal by the time associated with the rotation and compression–distention of these molecules and the corresponding accumulation of charge in the space–charge layer. Figures 4.2c and 4.2d, for example, illustrate clearly that the motion of at least some of these charges is subsequent to the termination of the synaptic permeability modulation that triggered the motion. In this case of an active synapse the positive ions freed from the inner surface in Figures 4.2c and 4.2d can be imagined to pass longitudinally along the neural membrane away from the synapse to initiate similar responses in the lipoprotein molecules and space–charge layers of adjacent regions.

This type of physical current–potential relationship is precisely what is known as a capacitance in electromagnetic theory. The motion of the charged ends of the molecules (and the linked motion of ions in the space–charge layers) constitutes a current. This current is what was labeled the *displacement current* by Maxwell. Its significant feature is that, since the molecular charges are structurally displaced only when the applied electrical field changes, the current is proportional to the rate of change of the potential rather than to its amplitude.

Physically, this effect can be seen as a continuity of effective current as ions in the space–charge layers vary during dynamic activity. For example, when a positive ion replaces a negative ion in the inner space–charge layer as shown in Figure 4.2b, the motion of that positive ion represents a positive current up to the inner surface of the membrane. The positive ion physically stops here momentarily. This represents a time change in potential and a local cessation of physical current. However, its presence drives the positive ends of the local lipoprotein molecules into motion, forming the displacement current. The magnitude of this displacement current is proportional to the positive ionic charge temporarily deposited at the inner space–charge layer, and therefore to the time rate of change of the transmembrane potential. The constant of proportionality depends on the physical links between the rate at which positive charge is deposited and the induced motion of the membrane molecules. This latter, in turn, depends on the inertial and resistive interactive properties of the molecules within the membrane. The constant of proportionality is what we call the *membrane capacitance*.

The explicit value of the membrane capacitance reflects the density of the membrane lipoprotein molecules. This density is very constant throughout the animal kingdom. The capacitance per unit area is about one microfard per square micron for all neurons, perhaps for all animal cells.

The essential ingredient for neuroelectric signalling is that the tendency for charged membrane molecules to move sluggishly in response to changing ion concentrations in the space–charge layers introduces a corresponding sluggishness in the ionic currents subserving neuroelectric signalling. This temporal effect on transmembrane potential must be represented in neuronal dynamics. It is accurately represented by the shorthand of the capacitive current–potential relationship. In a deep, systemic sense, the effect represents the engineering design by which the nervous system captures stimuli temporarily over time, to allow physical representation of them within electrical signalling.[1]

Manipulations of the Circuit

The equivalent circuit illustrated in Figure 4.1b, then, represents very succinctly and very directly the central biophysical driving forces and biophysical processes underlying the ionic foundations of neuroelectric signalling. In this figure the individual conductance branches represent individual ion types. One may add as many branches as there are types of ions in any particular application.

The circuit of Figure 4.1b can be transformed into the form shown in Figure 4.1c, which applies more directly to the variables and quantities of neuroelectric signals in normal functioning. To make this transformation, one combines the resting values of all the conductances into one branch to represent the total resting conductance of this area of membrane and measures potentials relative to resting level. One further combines those portions of the active conductances that act in parallel in a given active event (at for example a given type of synapse) into single branches wherein the constituent active conductance represents the total active conductance of the process, and the battery represents the effective equilibrium potential of the process. We will justify this transformation mathematically.

Mathematical Representation of Neuroelectric Signals by Means of the Equivalent Circuit

The circuits shown in Figure 4.1 succinctly express the action of the biophysical processes governing neuroelectric signals. This section will show

[1] Note that transmembrane potential fluctuations associated with conductance changes only would be instantaneous in ongoing neuroelectric signalling. Voltage changes for conductances depend only on the instantaneous current; they require only that ions be in motion; they do not require that ions traverse any particular distance nor accumulate anywhere. This is essentially action-at-a-distance: all the ions arranged serially in closed loops are thrown instantaneously in motion by a conductance change; the conduction is instantaneous. The charged lipoprotein molecules, however, tend to trap ions in the space–charge layer by their electrical charges. Changes in the electrical composition of the intra- and extracellular fluids trigger changes in the electrical forces on these space–charge layers and the lipoprotein molecules. These in turn bring on the structural displacements and tensions in the molecules referred to previously, and the corresponding accumulated change in ionic composition of the space–charge layers. These are the processes that require time for completion and introduce sluggishness into the neuronal responses.

how to represent those processes mathematically, based on these circuits. The resulting equations are shown as Eq. (4.1):

$$SC = CA\frac{dV}{dt} + \sum G_i A[V - V_i] + \sum g_i A[V - V_i] \tag{4.1a}$$

$$C\frac{dV}{dt} = -\sum G_i V + \sum G_i V_i + \sum g_i [V_i - V] + \frac{SC}{A} \tag{4.1b}$$

$$C\frac{dV}{dt} = -GV + \sum G_i V_i + \sum g_i [E_i - E] + \frac{SC}{A} \tag{4.1c}$$

$$G = \sum G_i, \quad V_r = \frac{\sum G_i V_i}{\sum G_i}, \quad E = V - V_r, \quad E_i = V_i - V_r \tag{4.1d}$$

$$\frac{dE}{dt} = \frac{-E + \sum (g_i/G)[E_i - E] + SC/(GA)}{C/G} \tag{4.1e}$$

$$\frac{dE}{dt} = \frac{-E + \sum g_i' [E_i - E] + SC'}{\tau} \tag{4.1f}$$

$$\tau = C/G, \quad g_i' = G_i/G, \quad SC' = SC/(GA) \tag{4.1g}$$

Basic Physical Constraint: Current Balance across the Membrane

Equation 4.1a is obtained by demanding a conservation of current through the membrane. We make the constraint that no charge accumulate within the membrane, so that the total current that crosses into the membrane over the area of the circuit must equal the sum of all those currents that pass through it. Therefore, the total transmembrane current through the region, *SC*, must equal the sum of all the ionic currents that flow through the individual ionic channels (represented by the conductance branches) plus those diverted to the space–charge layers in association with the moving lipoprotein molecules of the membrane.

The total transmembrane current, *SC*, is the dynamic variable that links the neuronal regions electrically with the neighboring regions and the surrounding environment. In the artificial case where stimulation is provided by an experimentally applied electrode, *SC* is the current supplied to the region by the electrode. In the normal case, *SC* links the potentials of adjacent compartments in terms of current–voltage relationships characteristic of the intra- and extracellular pathways connecting them. These are elaborated further in Chapter 5. In an isolated unipotential region, the *SC* is zero even during activation by local permeability modulations. This is because *SC* represents total transmembrane current for the region, and the total currents must flow in closed loops, therefore crossing the membrane of such a region twice, once in each direction, and canceling out to zero.

Interpretation of Terms

In this equation V is the transmembrane potential, inside minus outside; A is the area of the membrane of the circuit; the V_i are the equilibrium potentials of the ith ionic channels; C is the membrane capacitance per unit area; and the G_i and g_i are respectively the resting and active conductances per unit are for the ith ionic species.

The total transmembrane current, *SC*, is equal to the sum of the currents through the capacitive and conductance branches. The first term on the right side of Eq. 4.1a represents the current associated with the rotating lipoprotein molecules; this current is proportional to the time rate of change of V; the constant of proportionality is by definition the total capacitance of the branch. The second and third terms on the right represent the total current through the resting and active conductance branches, respectively. These terms are proportional to the differences between the membrane potential and their equilibrium potentials. The constants of proportionality are by definition, the total conductances of these channels.

Reorganization to Dynamic System Form

It is more convenient from an analytic point of view to rearrange Eq. 4.1a into the form shown in Eq. 4.1b. This form conveniently presents us with a first-order ordinary differential equation for the significant neuroelectric signal of interest, V. The terms involving the active conductances and *SC* show up as driving forces; the resting conductances and the capacitance show up as parameters. Eq. 4.1b is the first manifestation of the central governing equation for neuroelectric signals. We shall refer to it as the *membrane equation*. This form of the membrane equation is most directly related to underlying biophysical events. The individual ionic fluxes are individually identified.

Expression in Terms of Resting Values

It is more convenient, however, in studying ongoing neuroelectric signalling to further rearrange the equation to more explicitly represent dynamic variables of interest. We will do this according to the intervening variables defined in Eq. 4.1d.

First we define the total resting conductance, G, to be the sum of the individual resting conductances, as shown in Eq. 4.1d. Second, by applying Eq. 4.1b to the resting case where all the active conductances and SC are zero, and V does not change with time, we can see that the potential V, which for this case is the resting potential V_r, must approach the expression for V_r given in Eq. 4.1d. Now, we further define the relative transmembrane potential, E, to be the deviation of V from its resting level, V_r, as also defined in Eq. 4.1d. Therefore, E is zero when the membrane is in the resting state. E measures only those fluctuations in potential that form the neuroelectric signals most directly. If we make these substitutions in Eq. 4.1b, Eq. 4.1c is obtained.

Normalization of Terms

Equation 4.1c thus expresses the relative transmembrane potential, E, as driven by active processes, g_i and SC, in terms of resting properties of the membrane. It is very useful theoretically to normalize these latter quantities so as to explicitly and clearly identify the extent and nature of influence of the various constituent anatomical and physiological parameters that enter the systems.

If we divide Eq. 4.1c by the membrane capacitance, C, and further divide the top and bottom parts of the right side by the resting conductance, G, Eq. 4.1e is obtained. At this point it one gains the clearest view of the influence of the parameters by using the total resting conductance, G, and the regional area, A, to normalize the active conductance and SC terms, and, with the membrane capacitance, C, to define the membrane time constant, τ, all of which are shown in Eq 4.1g. Substituting these values into Eq. 4.1e produces 4.1f.

This normalization shows us that the value of the resting conductance influences neuroelectric behavior only as a scaling factor on the active conductances and by its influence on the time constant. This shows us that the universally significant unit of measure of given activations is their percentage of resting level conductance. Such a given percentage will evoke a given magnitude of potential response in any neuronal compartment to which it is applied, irrespective of the particular biophysical values at hand.

The normalization further shows that the membrane time constant is the only free parameter of neuroelectric signals at this generic level of development. It measures the effective time periods over which the neuronal membrane retains representations of its passing stimulation fields.

Grouping of Individual Ion Currents into Single Active Processes

The conductance terms in Eq. 4.1 represent conductances to individual ionic species. In considering the generation of some neuroelectric signals, such as synaptic activations, it is much more direct and convenient to express the activations directly in terms of the unitary activating processes, which usually involve sets of individual ion types activated in parallel.

Suppose that a single synapse affects active conductances to several ions in prescribed proportions, represented by a set of fractional numbers, a_i, as indicated in Eq. 4.2a.

$$g_1' = a_i g', \quad \sum a_i = 1 \tag{4.2a}$$

$$\sum g_i'[E_i - E] = \sum a_i g'[E_i - E] = g' \sum [a_i E_i - a_i E] = g' \left[\sum a_i E_i - 1E \right] \tag{4.2b}$$

$$\sum g_i'[E_i - E] = g'[E_{\text{syn}} - E], \quad E_{\text{syn}} = \sum a_i E_i \tag{4.2c}$$

The total active conductance for the synapse is g. The relevant activating terms for Eq. 4.1 are represented by in these terms as shown in Eq. 4.2b. These steps show clearly that the activating terms can be represented by one single equivalent expression involving the total synaptic conductance, g, acting at the effective equilibrium potential of the synapse, E_{syn}, as shown in Eq. 4.2d.

Final Canonical Form of the Membrane Equation and Concluding Comments

These last steps show that the activating conductance expressions in Eq. 4.1f can be interpreted in terms of single active synapses or other active processes with their effective equilibrium potentials as well as in terms of individual ionic types. This form, then, with this generalized interpretation constitutes the final canonical form of the membrane equation.

We should like to emphasize again that the equivalent circuit on which this equation is based is not merely an analogy, or model, in a weak sense of the word. It is a shorthand device for representing the actual physical process of the neurobiology of the system. When one looks at the circuit, one should envision passive ion pores and active ion channels under the shorthand symbols for transmembrane conductances; rotating and compressing–distending, charged lipoprotein molecules under the shorthand representation of the capacitance symbol; dynamic balances of diffusion and electrical forces on ions under the symbols for batteries.

This book considers that this equation is the analog of Newton's second law of motion in classical physics. It is the keystone by which the underlying operative biophysical processes can be most efficiently and succinctly understood by us to relate to the hierarchical organization of ongoing neuroelectrical signalling in normal function. The equation produces the time courses of the salient variable of interest—the deviation of transmembrane potential from resting level; it prescribes this variation in terms of the driving membrane permeability modulations; it contains implicitly the means to express the central underlying neurobiological processes explicitly in any given situation. The next two chapters show how this may be begun.

The canonical form of the membrane equation, Eq. 4.1f, is expressed in terms of two activation functions, g_i' and SC'. SC' is determined by the compartment's relation to contiguous compartments. It is through this function that we may effectively integrate the membrane equation over space in terms of the neuron's morphology. This topic is taken up in Chapter 5. The function g_i' represents the time-varying active conductance modulations (synaptic and intrinsic) within the compartment. It is primarily through this function that we may integrate the membrane equation over time. This topic is considered in Chapter 6.

Chapter 5 The Membrane Equation in Space: The Mechanics of Neurons with Dendrites

The purpose of this chapter is to introduce the principles and procedures for describing neuroelectric signalling in neurons with dendritic trees and, more generally, under any conditions wherein spatial interactions are significant. The primary governing approach here is to partition the neuron into individual compartments, in each of which the membrane potential is assumed to be uniform, and to constrain the total transmembrane current of each compartment according to the local spatial geometry of the compartments and their surroundings. The primary mathematical device for introducing these spatial constraints into the governing equations is the total membrane current for each compartment, *SC*, as represented in the canonical form of the membrane equation, Eq. 4.1f.

Spatially isolated neuronal compartments, such as a model for a neuronal soma which neglect current leakage into dendrites and axons, are necessarily constrained to have the total current, *SC*, equal to zero, except in the cases when this current may be applied by electrodes inserted into the cell or used to represent current flowing into the soma from dendritic regions.

This chapter introduces and discusses the general physical approach taken here, and both the method of finite compartmentalization, which is valuable for representing ongoing neuroelectric signalling, and the continuous representation by partial differential equations, which is useful for various mathematical characterizations. Computer representations of the finite compartmental approach is considered in Chapter 7.

Introduction

The central governing equation for neuroelectric signals, the membrane equation, derived in Chapter 4, is limited to single localized regions of neural membrane over which the transmembrane potential can be assumed to be uniform. To represent the general, and indeed typical, situation where potentials in a given neuron vary continuously in graded fashion across its dendritic trees, one simply partitions the neuron into a collection of compartments, such that the potential in each compartment can be considered to be uniform.

The potential in each compartment is represented by an equivalent circuit. The activity of the entire neuron is then accurately represented by the collection of compartments when the representations of current flows among the neighboring compartments are accurately reflective of the physical constraints imposed by the neuron and its surroundings. In this approach, each compartment, being represented by its individual equivalent circuit, is therefore represented by its individual transmembrane potential and its individual membrane equation.

The membrane equation (4.1f) shows us that the potential in each compartment is driven by both the conductance modulations within the compartment, G_i, and the total current passing through the membrane of the compartment, SC. These quantities, the SC for each compartment, then, are the mathematical means by which the spatial configurations and constraints of neurons are represented in this mechanics. Since we can equate the total current traversing a given compartment with the total current passing to and from that compartment through the extracellular and intracellular fluids, and since these latter currents are composed of ions flowing through volume conductors, it follows that SC terms can be expressed in terms of ohmic relationships. Therefore, the spatial influences on neuronal signals will be represented by effective resistances and conductances associated with the geometry of the neuron, and the geometry and composition of the local extracellular and intracellular space.

The mathematical description of the collection of compartments taken to represent a neuron can approach either in terms of a set of distinct, finite compartments, or in terms of a continuum wherein individual compartments become infinitesimally small. The first approach with n compartments defines n variables, consisting of the membrane potentials for each compartment, and n equations, consisting of the membrane equation for each compartment. The second approach produces a single partial differential equation for each nonbifurcating segment of neuron. Both approaches succeed in capturing the essential spatial and physical influences in mathematical terms and equations, and thereby provide the foundational theoretical mechanics needed to describe neuroelectric signalling in entire neurons.

The next two sections of this chapter take up these two approaches in turn.

Finite Compartmentalization: Representation of Neuroelectric Signalling

The finite compartmentalization procedure is straightforward. One simply partitions the neuron into a finite number, n, of compartments, represents each compartment with the equivalent circuit discussed in Chapter 4, expresses the current flow between contiguous compartments in terms of ohmic pathways reflecting the neural geometry, and uses current balances to obtain a system of n equations for the n unknown transmembrane potential values. The approach can be made as accurate as desired by taking the number of compartments increasingly large.

We will illustrate the procedure by a specific example. The generalization to any arbitrary configuration is direct and obvious. Indeed, once the procedure is understood, one may write the final equations directly from a succinct simple general rule that the example will provide for us. The example taken here uses a rather rough compartmentalization; the generalization to larger numbers of compartments is obvious.

Prototypical Example

Consider the neuron with two parallel dendrites, one of which does not bifurcate, and one of which bifurcates once, as illustrated in Figure 5.1a. Suppose we compartmentalize this neuron roughly as shown in Figure 5.1b. For the time being we will neglect current leakage into the axon and out of the small caps at the terminal ends of the dendrites. These terms can be added easily, are typically small, and would only clutter the present discussion.

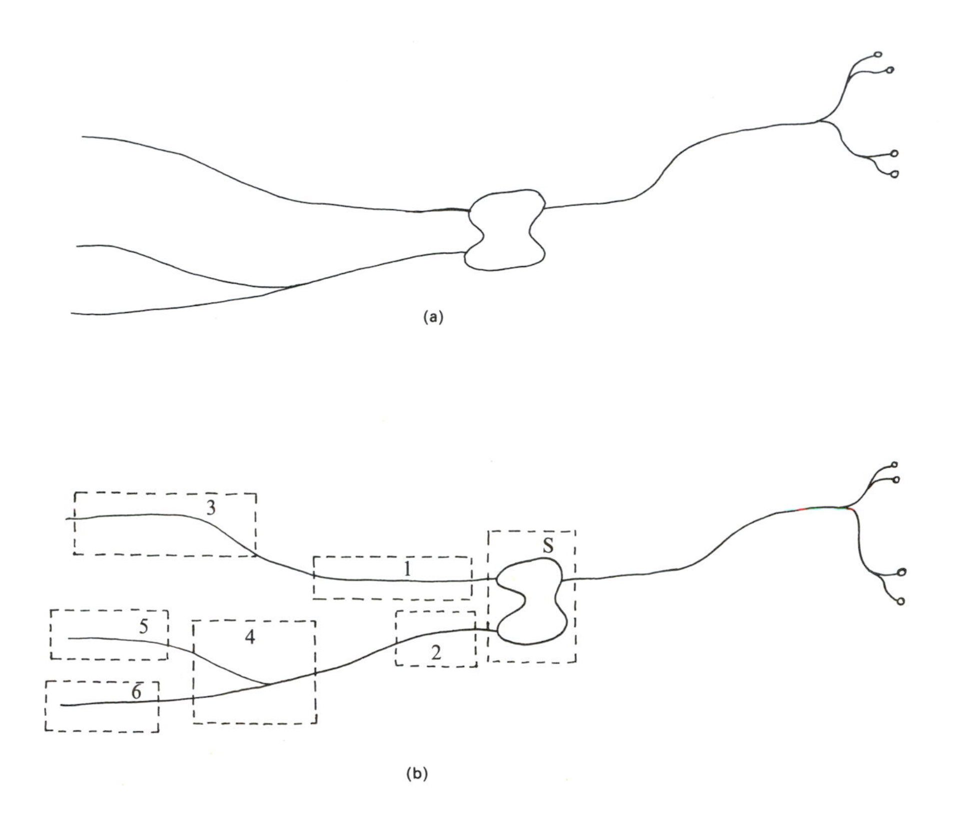

FIGURE 5.1 Example to illustrate compartmentalization of dendritic trees: a. example neuron; b. compartmentalization. (If compartments are always defined so that all points of branching can be taken as the effective centers of compartments, as indicated in part b, then any compartment exchanges current with at most only one other compartment at each of its ends.)

The next step is to represent each compartment with an equivalent circuit and interconnect the compartments by ohmic conductances; this is done in Figure 5.2a. In this figure the upper intercompartmental resistances represent current pathways through the extracellular fluid, R^e_{ij}, and the lower intercompartmental resistances represent current pathways through the intracellular fluid, R^i_{ij}. Each of the equivalent circuits has the form shown in the insert in the lower right of the figure. The details have been omitted from the general figure for clarity.

The intercompartmental current pathways are illustrated in Figure 5.2b. The resistances for these pathways can be expressed readily in terms of basic

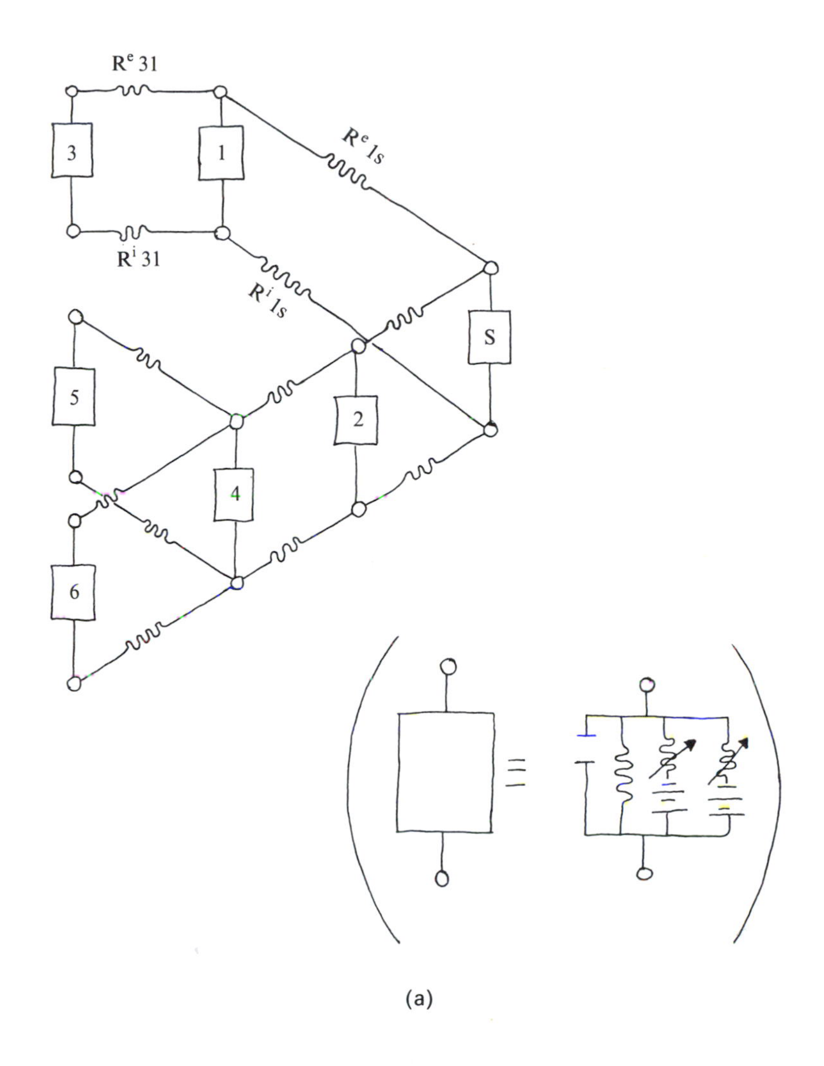

FIGURE 5.2 Representation of neuron with interconnected equivalent circuits: a. arrangement of circuits to represent dendritic tree; b. longitudinal current flow between contiguous compartments (generally, compartments will have different effective radii as indicated in part b; however, if compartments are always defined so that all points of branching can be taken as the effective centers of the compartments, then longitudinal pathways between contiguous compartments may typically be assigned a single effective radius); c. manipulation of circuit and definition of local current loops for mathematical analysis.

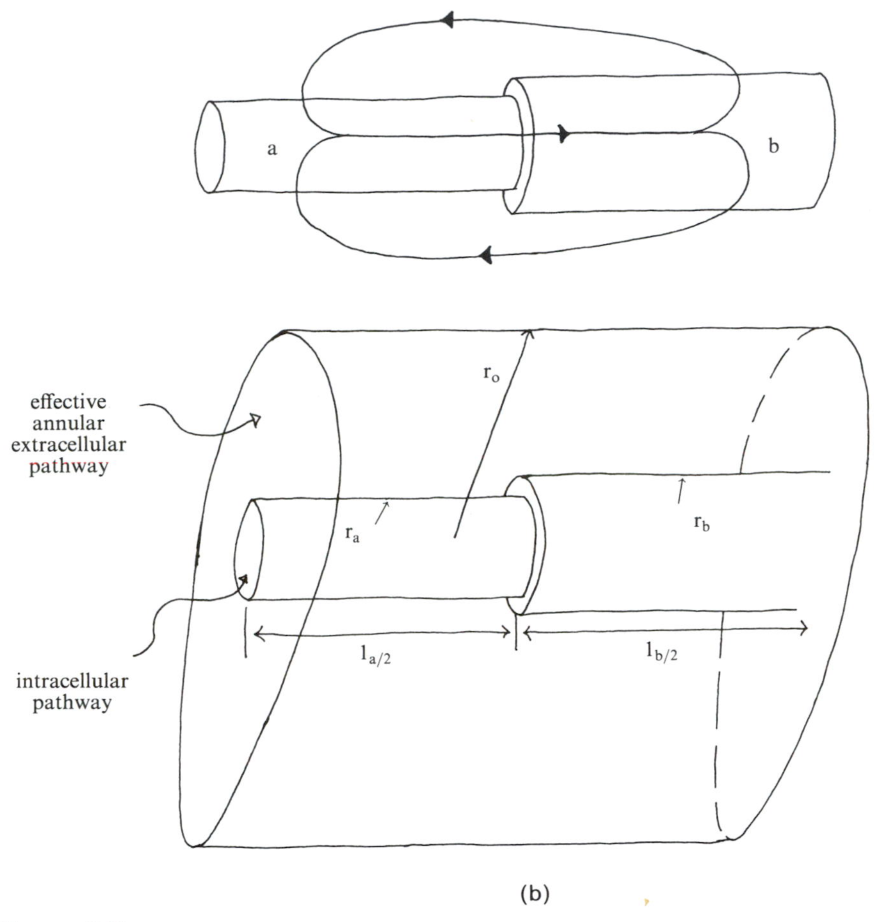

FIGURE 5.2b

expressions for volume conduction, as indicated in Eq. 5.1:

$$R = \frac{\rho l}{A} \tag{5.1a}$$

$$R^i_{ab} = \frac{\rho l_a/2}{\prod r_a^2} + \frac{\rho l_b/2}{\prod r_b^2} \tag{5.1b}$$

$$R^e_{ab} = \frac{\rho l_a/2}{\prod}\left\{\frac{1}{r_o^2 - r_a^2}\right\} + \frac{\rho l_b/2}{\prod}\left\{\frac{1}{r_o^2 - r_b^2}\right\} \tag{5.1c}$$

$$R_{ab} = R^i_{ab} + R^e_{ab}$$

$$= \frac{\rho l_a/2}{\prod r_a^2}\left\{1 + \frac{1}{(r_o/r_a)^2 - 1}\right\} + \frac{\rho l_b/2}{\prod r_b^2}\left\{1 + \frac{1}{(r_o/r_b)^2 - 1}\right\} \tag{5.1d}$$

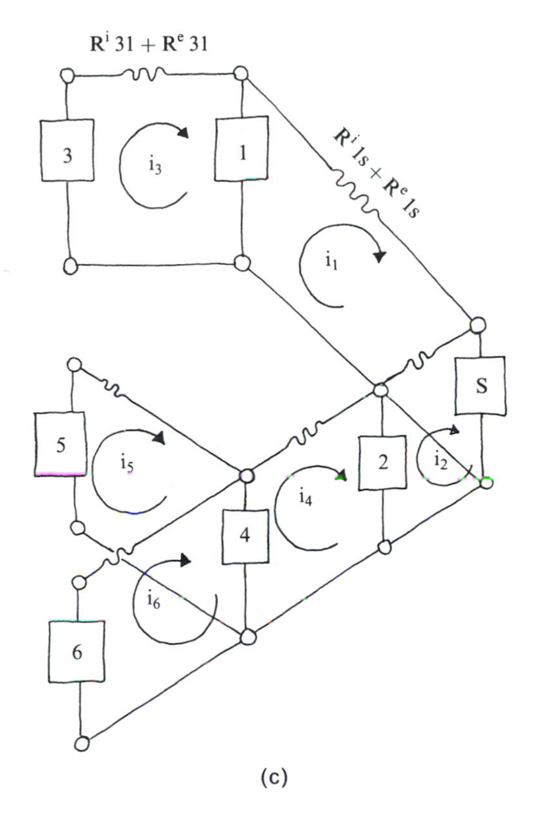

FIGURE 5.2c

For compartment *a* taken as soma: $r_a \approx r_s/2$. Note that if compartments are defined so that all points of branching are effective centers of compartments, as shown in Figure 5.1, then any pathway between contiguous compartments has a single radius, simplifying Equation 5.1.

Here the individual dendrites are taken as continuous circular cylinders with constant radii, the soma is taken as a sphere, and the extracellular pathways are represented by annular cylinders. One can adjust the effective outer radii of such annuli to reflect the morphology and degree of packing of neuronal material in the adjacent extracellular space. Of course, one may develop any desired degree of precision for these resistance values. These simple expressions are adequate for most uses and can be expected to show the correct trends of influence of the key geometric features.

The next step is to write the equations that govern the behavior of the compartment membrane potentials and their interrelations. To do this it is convenient to redraw the composite circuit slightly as shown in Figure 5.2c. One can readily justify mathematically the step of combining the intra- and extracellular pathways into one equivalent branch as we have done in this figure. The only restriction introduced is that one is now directly given only the transmembrane potentials and not their particular values or changes on each side. This is not generally important.

We have defined in Figure 5.2c unknown loop currents for each closed loop in the composite model. These are intermediate constructs to aid in the derivation of the circuit governing equations. In our example there are five such loop currents, i_1–i_5. The current i_1, for example, represents the current flowing in the pathway between the soma compartment and compartment 1. The total current flowing through the soma compartment is i_1 plus i_2. The total current flowing through compartment 1 is i_3 minus i_1. We are using the sign convention that currents flowing from the top down are positive, and those from bottom up are negative. These currents will be used to describe the total current flowing into each compartment for use in the membrane equations. They will then be expressed in terms of the potentials and eliminated from the equations.

The governing equations for the circuit shown in Figure 5.2c can now be written as in Eq. (5.2a) and (5.2b).

$$\begin{aligned}
\frac{dE_s}{dt} &= \frac{-E_s + \sum g_{sj}\,[E_j^* - E_s] + (i_1 + i_2)/(A_s G)}{\tau s} \\
\frac{dE_1}{dt} &= \frac{-E_1 + \sum g_{1j}\,[E_j^* - E_1] + (i_3 - i_1)/(A_1 G)}{\tau 1} \\
\frac{dE_2}{dt} &= \frac{-E_2 + \sum g_{2j}\,[E_j^* - E_2] + (i_4 - i_2)/(A_2 G)}{\tau 2} \\
\frac{dE_3}{dt} &= \frac{-E_3 + \sum g_{3j}\,[E_j^* - E_3] + (-i_3)/(A_3 G)}{\tau 3} \\
\frac{dE_4}{dt} &= \frac{-E_4 + \sum g_{4j}\,[E_j^* - E_4] + (i_5 + i_6 - i_4)/(A_4 G)}{\tau 4} \\
\frac{dE_5}{dt} &= \frac{-E_5 + \sum g_{5j}\,[E_j^* - E_5] + (-i_5)/(A_5 G)}{\tau 5} \\
\frac{dE_6}{dt} &= \frac{-E_s + \sum g_{6j}\,[E_j^* - E_6] + (-i_6)/(A_6 G)}{\tau 6}
\end{aligned} \tag{5.2a}$$

$$
\begin{aligned}
i_1 &= (E_1 - E_s)/R_{1s} & i_4 &= (E_4 - E_2)/R_{42} \\
i_2 &= (E_2 - E_s)/R_{2s} & i_5 &= (E_5 - E_4)/R_{54} \\
i_3 &= (E_3 - E_1)/R_{31} & i_6 &= (E_6 - E_4)/R_{64}
\end{aligned}
\tag{5.2b}
$$

$$
R_{1s} = \frac{\rho l_1/2}{\prod r_{12}}\left\{1 + \frac{1}{(r_0/r_1)^2 - 1}\right\} + \frac{\rho r_s}{\prod (r_s/2)^2}\left\{1 + \frac{1}{\left(\dfrac{r_0}{r_s/2}\right)^2 - 1}\right\}
$$

$$
R_{31} = \frac{\rho l_3/2}{\prod r_3^2}\left\{1 + \frac{1}{(r_0/r_3)^2 - 1}\right\} + \frac{\rho l_1/2}{\prod r_1^2}\left\{1 + \frac{1}{(r_0/r_1)^2 - 1}\right\}
\tag{5.2c}
$$

Equations (5.2a) represent the membrane equation applied to each of the six compartments of the example. The total currents through each of these compartments is expressed in terms of the unknown loop currents, i_1–i_5.

The loop currents can now be expressed in terms of the transmembrane potentials of the compartments as shown in Eq. (5.2b). The intercompartmental resistances are given in terms of the geometry of the neuron as in Eq. (5.2c).

The essential mathematical physics of the approach is contained in Figures 5.2 and Eq. (5.2). The remainder of the procedure is simply mathematical manipulation. Equations (5.2b) can be substituted into (5.2a) to eliminate the loop currents, leaving five equations for the five compartment membrane potentials. These results are shown in Eq. (5.3a):

$$
\tau_s \frac{\mathrm{d}E_s}{\mathrm{d}t} = -E_s + \sum g_{sj}[E_j^* - E_s] + \frac{E_1 - E_s}{R_{1s}A_sG} + \frac{E_2 - E_s}{R_{2s}A_sG}
$$

$$
a_1\tau_1 \frac{\mathrm{d}E_1}{\mathrm{d}t} = -a_1E_1 + \sum g_{1j}[E_j^* - E_1] + \frac{E_3 - E_1}{R_{31}A_sG} + \frac{E_1 - E_s}{R_{1s}A_sG}
$$

$$
a_2\tau_2 \frac{\mathrm{d}E_2}{\mathrm{d}t} = -a_2E_2 + \sum g_{2j}[E_j^* - E_2] + \frac{E_4 - E_2}{R_{42}A_sG} + \frac{E_2 - E_s}{R_{2s}A_sG}
$$

$$
a_3\tau_3 \frac{\mathrm{d}E_3}{\mathrm{d}t} = -a_3E_3 + \sum g_{3j}[E_j^* - E_3] - \frac{E_3 - E_1}{R_{31}A_sG}
$$

$$
a_4\tau_4 \frac{\mathrm{d}E_4}{\mathrm{d}t} = -a_4E_4 + \sum g_{4j}[E_j^* - E_4] + \frac{E_5 - E_4}{R_{42}A_sG} + \frac{E_6 - E_4}{R_{2s}A_sG} - \frac{E_4 - E_2}{R_{42}A_sG}
$$

$$
a_5\tau_5 \frac{\mathrm{d}E_5}{\mathrm{d}t} = -a_5E_5 + \sum g_{5j}[E_j^* - E_5] - \frac{E_5 - E_4}{R_{54}A_sG}
$$

$$a_6\tau_6 \frac{\mathrm{d}E_6}{\mathrm{d}t} = -a_6E_6 + \sum g_{6j}[E_j^* - E_6] - \frac{E_6 - E_4}{R_{64}A_sG}$$

$$a_j = A_j/A_s \tag{5.3a}$$

The terms are slightly rearranged for convenience in Eq. (5.3b):

$$\tau_s \frac{\mathrm{d}E_s}{\mathrm{d}t} = -E_s\left\{1 + \frac{1}{R_{1s}A_sG} + \frac{1}{R_{2s}A_sG}\right\} + \sum g_{sj}[E_j^* - E_s] + \frac{E_1}{R_{1s}A_sG} + \frac{E_2}{R_{2s}A_sG}$$

$$a_1\tau_1 \frac{\mathrm{d}E_1}{\mathrm{d}t} = -E_1\left\{a_1 + \frac{1}{R_{31}A_sG} + \frac{1}{R_{1s}A_sG}\right\} + \sum g_{1j}[E_j^* - E_1] + \frac{E_3}{R_{31}A_sG} + \frac{E_2}{R_{1s}A_sG}$$

$$a_2\tau_2 \frac{\mathrm{d}E_2}{\mathrm{d}t} = -E_2\left\{a_2 + \frac{1}{R_{42}A_sG} + \frac{1}{R_{2s}A_sG}\right\} + \sum g_{2j}[E_j^* - E_2] + \frac{E_4}{R_{42}A_sG} + \frac{E_s}{R_{2s}A_sG}$$

$$a_3\tau_3 \frac{\mathrm{d}E_3}{\mathrm{d}t} = -E_3\left\{a_3 + \frac{1}{R_{31}A_sG}\right\} + \sum g_{3j}[E_j^* - E_3] + \frac{E_1}{R_{31}A_sG}$$

$$a_4\tau_4 \frac{\mathrm{d}E_4}{\mathrm{d}t} = -E_4\left\{a_4 + \frac{1}{R_{54}A_sG} + \frac{1}{R_{64}A_sG} + \frac{1}{R_{42}A_sG}\right\} + \sum g_{4j}[E_j^* - E_4] + \frac{E_5}{R_{54}A_sG} + \frac{E_6}{R_{64}A_sG} + \frac{E_2}{R_{42}A_sG}$$

$$a_5\tau_5 \frac{\mathrm{d}E_5}{\mathrm{d}t} = -E_5\left\{a_5 + \frac{1}{R_{54}A_sG}\right\} + \sum g_{5j}[E_j^* - E_5] + \frac{E_4}{R_{54}A_sG}$$

$$a_6\tau_6 \frac{\mathrm{d}E_6}{\mathrm{d}t} = -E_6\left\{a_6 + \frac{1}{R_{64}A_sG}\right\} + \sum g_{6j}[E_j^* - E_6] + \frac{E_4}{R_{64}A_sG} \tag{5.3b}$$

Equations (5.3b) represent a clearly stated mathematical solution to the physical problem. One now is in the position to prescribe the various input activation functions, g_s, g_1, g_2, g_3, g_4, and g_5 as functions of time and mathematically evaluate the resulting compartmental potentials. Typically one is advised to use computer simulation techniques for these purposes. Various limiting cases and prototypical values can be obtained mathematically directly from the equation system, usually by neglecting the time varying terms.

The General Compartmentalization Equations

The finite compartmentalization procedure for the general case of arbitrary dendritic configurations can be succinctly exemplified by Figure 5.2 and Eq. (5.2). Further, once the procedure is grasped and the reduction of the neuron to a figure like Figure 5.2 is constructed, one may immediately write the resulting governing equation system as shown here in Eq. (5.4):

$$a_j \tau_j \frac{\mathrm{d}E_j}{\mathrm{d}t} = -\left\{ a_j + \sum_{\substack{\neq j \\ l=1}}^{n} \frac{1}{R'_{lj}} \right\} E_j + \sum g_j^{j'} [E_j^* - E_j] + \frac{\sum E_l}{R'_{lj}}$$

$$\begin{aligned} R'_{lj} &= R_{lj} A_s G \qquad \text{if } l \text{ and } j \text{ are connected} \\ &= 0 \qquad \text{if } l \text{ and } j \text{ are not connected } (R_{lj} \to \infty) \end{aligned} \qquad (5.4)$$

$$g_j^{j'} = \frac{g_j^j A_j}{G A_s}$$

This equation is the generalized form of the final solution of the example, Eq. (5.3b). Notice carefully how the terms in Eq. (5.4) have been normalized. This will be very useful in the dynamic similarity theory developed in Part II.

An effective computational technique for integrating the membrane equation and equation systems of the type involved in here is introduced in Chapter 7. These and related topics are discussed in NBM.

Continuous Mathematical Representation

The mathematical approach introduced here is useful for characterizing the influence of dendritic trees on the spatial propagations of signals. Widely studied over the last 30 years, it rests on a single second-order partial differential equation known as the *equation of electrotonus*. In this section we will derive this equation, discuss its properties and uses, and show by several examples how it may be applied.

Derivation of the Equation of Electrotonus

In this section we will derive the classical equation for electronic conduction in cylindrical neuronal elements. This is the fundamental differential equation that governs longitudinal conduction of transmembrane electrical signals in dendrites and axons.

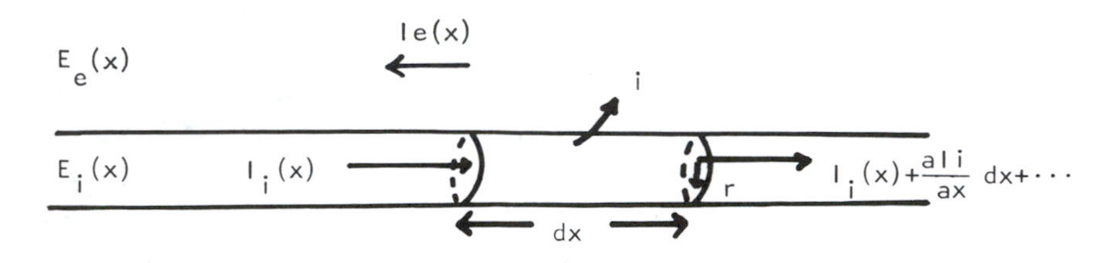

FIGURE 5.3 Differentially small section of dendrite or axon to illustrate current balance for equation of electrotonus. Reprinted with permission from *Neural and Brain Modeling*, R. MacGregor, Copyright 1987, Academic Press.

Figure 5.3 illustrates a section of a dendrite or axon. Equation (5.5) is obtained by writing a current balance for the control volume of length dx defined in the figure:

$$\begin{aligned} I_i &= i\,2\pi r\,\mathrm{d}x + I_i + \frac{\partial I_i}{\partial x}\mathrm{d}x + \cdots \\ \frac{\partial I_i}{\partial x} &= -i\,\pi r \end{aligned} \tag{5.5}$$

I_i is the total longitudinal current inside the dendrite as a function of position x. As dx is assumed to be differentially small and I_i is assumed to be continuous, its value at the right-hand side of the control volume, $x + \mathrm{d}x$, is expressible in a Taylor series expansion, as indicated in Figure 5.3. Here i is the current per unit area exiting the control volume through the membrane. Physically, Eq. (5.5) says that, since no charge accumulates in the control volume, the net change is current from one side of the control volume to the other is equal to the amount of current lost across the membrane.

Equations (5.6a) state the assumption that the intra- and extracellular fluids are ohmic conductors:

$$I_i = -\frac{1}{R_i}\frac{\partial E_i}{\partial x}, \qquad I_e = -\frac{1}{R_e}\frac{\partial E_e}{\partial x} \tag{5.6a}$$

$$I_i + I_e = 0, \qquad \frac{\partial I_i}{\partial x} + \frac{\partial I_e}{\partial x} = 0 \tag{5.6b}$$

$$-\frac{1}{R_i}\frac{\partial^2 E_i}{\partial x^2} - \frac{1}{R_e}\frac{\partial^2 E_e}{\partial x^2} = 0, \qquad \frac{\partial^2 E_e}{\partial x^2} = -\frac{R_e}{R_i}\frac{\partial^2 E_i}{\partial x^2} \tag{5.6c}$$

$$E \equiv E_i - E_e, \qquad \frac{\partial^2 E}{\partial x^2} = \frac{\partial^2 E_i}{\partial x^2} - \frac{\partial^2 E_e}{\partial x^2} \tag{5.6d}$$

$$\frac{\partial^2 E}{\partial x^2} = \frac{\partial^2 E_i}{\partial x^2}\left(1 + \frac{R_e}{R_i}\right) \tag{5.6e}$$

$$\frac{\partial^2 E}{\partial x^2} = -R_i \frac{\partial I_i}{\partial x}\left(1 + \frac{R_e}{R_i}\right) = \frac{\partial I_i}{\partial x}(R_i + R_e) \tag{5.6f}$$

That is, the currents are proportional to the gradients of the electric fields. R_i and R_e are the resistances per unit length in the internal and external fluids, respectively. These in turn are equal to the reciprocal of the conductivity of the fluid times the effective cross-sectional area. Equation (5.6b) expresses the law that currents must flow in closed loops when there are no sources and sinks of charge. For a neuronal dendrite or axon, the current passing longitudinally in one direction within the dendrite must be counterbalanced by an equal amount of total current passing in the other direction external to the dendrite at any instant in time. Equations (5.6c) result from differentiating (5.6a) and substituting them into (5.6b). The transmembrane potential E is defined as the difference between the internal and external potentials, as defined in Eq. (5.6d). Combining Eqs. (5.6c) and (5.6d) results in (5.6e). Substituting Eq. (5.6a) into (5.6e) results in (5.6f). Finally, substitution of Eq. (5.5) into (5.6f) results in (5.7):

$$\frac{1}{2\pi r(R_i + R_e)} \frac{\partial^2 E}{\partial x^2} = i \tag{5.7}$$

Equation (5.7) is the fundamental conduction equation that relates the transmembrane current at any point in space to the second derivative of the transmembrane potential at that point in terms of the radius of the dendrite and the external and internal resistances per unit length. In effecting the mathematical operations (5.6), it has been assumed that the resistances per unit length both internal and external to the dendrite do not vary with x. Therefore the derivations are not strictly applicable to dendrites or axons that taper or otherwise change radius with coordinate x. They are also not applicable directly to regions where a dense extracellular neuropile changes in density with position x. Generally, however, it is reasonable to model most dendritic and axonal fields by cylinders of constant radius, where the radius changes only at bifurcations. In this situation, Eq. (5.7) may be applied to each continuous segment between bifurcations.

Equation (5.7) then may be extended to the full electrotonic equation by expressing the transmembrane current i in terms of the current–voltage relations for the equivalent circuit model for the membrane as developed in

Chapter 4. This results in Eq. (5.8):

$$\frac{1}{2\pi r(R_i + R_e)} \frac{\partial^2 E}{\partial x^2} = C \frac{\partial E}{\partial t} + \sum (g_i + G_i)(E - E_i) \tag{5.8}$$

Here C is the membrane capacitance, g_i is the resting conductance of the membrane to ionic species i, and G_i is an increment in membrane conductance to ionic species i, which may reflect any neuronal process (e.g., synaptic activation, spike generation, or active conductance changes) that acts by modulating membrane conductances. Equations (5.9a) define the total resting conductance of the membrane G and express the convention used here that potentials are measured relative to the resting potential of the membrane, which is taken as zero.

$$G \equiv \sum g_i, \qquad \sum g_i E_i \equiv 0$$

$$\frac{1}{2\pi r(R_i + R_e)} \frac{\partial^2 E}{\partial x^2} = C \frac{\partial E}{\partial t} + G * E + \sum G_i(E - E_i) \tag{5.9a}$$

$$\frac{1}{2\pi r(R_i + R_e)G} \frac{\partial^2 E}{\partial x^2} = \frac{C}{G} \frac{\partial E}{\partial t} + E + \sum \frac{G_i}{G}(E - E_i) \tag{5.9b}$$

$$\lambda \equiv \frac{1}{\sqrt{2\pi r(R_i + R_e)G}}, \qquad \tau \equiv \frac{C}{G}, \qquad G'_i \equiv \frac{G'}{G} \tag{5.9c}$$

Equation (5.8) can then be rewritten as (5.9b). The length constant and the time constant are defined as in Eq. (5.9c). Moreover, it is convenient to normalize the active conductance modulations G_i, as indicated in Eq. (5.9a). The standard form for the classical equation of electrotonus that governs the dynamics and longitudinal conduction of electrical signals in circular dendrites and axons then is expressed in Eq. (5.10a). Finally, if one chooses to work with spatial coordinates normalized to the length constant, λ, and time units normalized to the time constant, τ, Eq. (5.10a) can be expressed as in Eq. (5.10b):

$$\lambda^2 \frac{\partial^2 E}{\partial x^2} = \tau \frac{\partial E}{\partial t} + E + \sum G'_i(E - E_i) \tag{5.10a}$$

$$\frac{\partial^2 E}{\partial x'^2} = \frac{\partial E}{\partial t'} + E + \sum G'_i(E - E_i) \tag{5.10b}$$

$$x' \equiv \frac{x}{\lambda}, \qquad t' \equiv \frac{t}{\tau}, \qquad G'_i \equiv \frac{G_i}{G}$$

Properties and Uses of the Continuous Approach

The equation of electrotonus is mathematical tool for describing the longitudinal propagation of signals in dendritic trees. It lays bare that the fundamental feature of signalling in dendrites is the coupling of longitudinal propagation with transmembrane leakage. It tells us further that the dynamics of single, nonbifurcating regions can be characterized by just two characteristic parameters, the time constant, τ, which we have seen in Chapter 4, and the length constant, λ, which characterizes this lateral propagation. The equation can show us readily that steady potentials in infinitely long dendrites must decay exponentially, and this length constant is the measure of that exponential decay. Thus the approach tells us how the geometry and resistive properties of the neuron and its surroundings influence the spatial decay of signals.

The approach informs us that neuroelectric signalling in entire neurons will involve the decremental propagation of signals longitudinally to and from local regions of neurons along their dendrites. The application of the continuous approach to neuroelectric signalling in entire neurons, however, is very cumbersome and fraught with mathematical detail. The equation usually must be integrated over single dendritic lengths and entire dendritic trees to give useful information. Moreover, ongoing neuroelectric signalling involves ubiquitous, localized active membrane permeability fluctuations, which are terribly inconvenient to the mathematical approach.

The mechanics of neuroelectric signalling in neurons are much better approached computationally than mathematically, and the natural language for this approach is the finite compartmentalization method rather than the mathematical continuous approach. The continuous approach, however, can be useful for mathematical characterizations of central influences of dendritic trees on signal propagation and in providing solutions to accompany certain experimental studies of particular neuronal processes and parameters. We will now consider how these uses are approached.

Mathematical Applications of the Equation of Electrotonus

The equation of electrotonus may be applied to single lengths of dendrite, or to entire neurons. If it is applied to entire neurons, one must either piece together collections of nonbifurcating dendritic compartments or define conceptually equivalent structures to which the equation can be applied. In this compartmentalization approach, one writes the equation of electrotonus for each compartment and specifies constraining boundary conditions at each junction: currents and potential values must be equal at the junctions. In any of these cases, mathematical solutions are made, using the fundamental techniques for solving partial differential equations: typically, transforms and series solutions, constrained by the boundary conditions.

Simple Illustration of Passive Decay of PSPs

Consider a rough but illustrative use of the equation of electrotonus to find an approximate mathematical expression for the character of spatial decay of PSPs in passive propagation along dendrites. Suppose a synapse on an otherwise quiet dendrite is activated at $x = 0$ and $t = 0$, by a brief excursion in permeability, which is represented by a Dirac delta function. (A Dirac delta function is a function whose width approaches 0 while its height approaches infinity so that the area it contains is 1.)

One can show with Fourier transforms that the solution to this problem is the function given in Eq. (5.11) and illustrated in Figure 5.4.

$$C\frac{\partial E}{\partial t} = \frac{1}{2\prod r(R_i + R_e)}\frac{\partial^2 E}{\partial x^2} - \sum g_i[E - E_i] \tag{5.11a}$$

$$\sum g_I = G + P\delta(x - 0)\delta(t - 0)$$

$$\sum g_i E_i = PE^*\delta(x - 0)\delta(t - 0) \tag{5.11b}$$

$$E(x, t) = [E^* - E(0, 0)]P\frac{\exp[-t - x^2/(4t)]}{t} \tag{5.11c}$$

Since synapses are active over a millisecond or so, this function cannot be a representative solution for small values of t (in fact, as t approaches 0, this function is one expression for a Dirac delta function localized at x). Nonetheless, as one can see in Figure 5.4, the function exhibits a typical PSP shape when evaluated at about $x = 1$, and it can therefore be used to estimate the subsequent spatial decay of such PSPs. The function shows the characteristic decrement and progressive delay of the PSP peak and the corresponding broadening of the signal, as seen at progressively distal points from the synapse.

This simple solution also shows the dependence of the response amplitude on the difference between the initial potential and the equilibrium potential. This is the characteristic effective driving force of the synaptic mechanism of changing membrane permeability. The solution further shows that the response amplitude is inversely proportional to the length constant of the

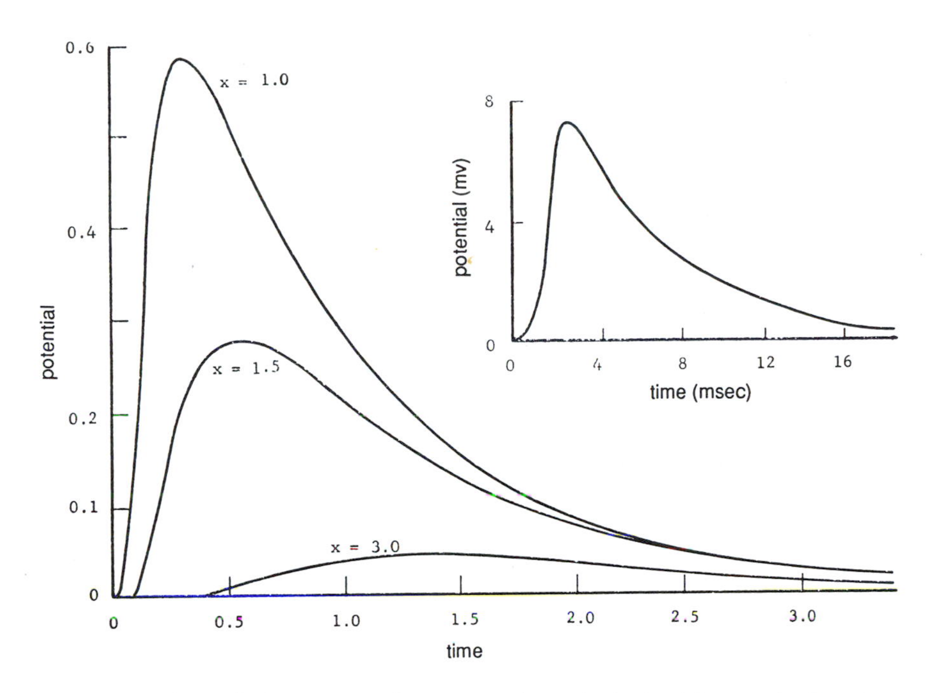

FIGURE 5.4 Theoretical postsynaptic response in a single dendrite: a. the function $\exp[-t - x^2/(4t)]/\sqrt{t}$ as a postsynaptic potential; b. experimentally observed postsynaptic potential. Reproduced from the *Biophysical Journal* 1968, vol. 8 pp. 307 by copyright permission of the Biophysical Society.

membrane. That is, the higher the longitudinal resistance of the dendrite, the larger is the proportion of the current that flows across the membrane locally; hence, the larger the local response and the larger the spatial decay.

The Equivalent Cylinder Model for Symmetric Dendritic Trees

Another simple but instructive use of the equation of electrotonus is that the passive decay properties of the symmetric dendritic tree shown in Figure 5.5, are given mathematically by those of an equivalent single cylinder when the branch diameters are appropriately constrained. The constraint is that the sum of the 3/2 power of the diameter of the two sibling branches at a bifurcation equal the 3/2 of the diameter of the parent branch at the bifurcation.

The value of the equivalent cylinder concept is that one may use it to estimate the spatial decay properties of signals in the corresponding dendritic trees. A PSP that can be shown to decay a certain amount in the

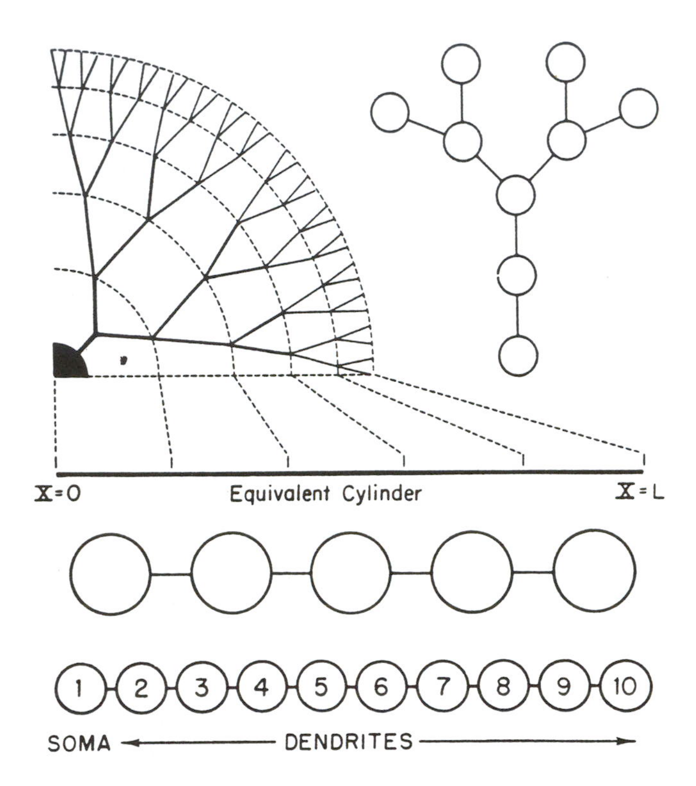

FIGURE 5.5 Equivalent cylinder representation of symmetrical dendritic tree (taken from W. Rall, "Core conductor theory and cable properties of neurons," *Handbook of Physiology: The Nervous System*, Wilfred Rall, copyright 1977, The American Physiological Society).

equivalent cylinder may be considered to decay the same amount in an equivalent symmetric passive dendritic tree, and so on.

Computing the Effective Resistance of an Arbitrary Dendritic Tree

The effective resistance as seen from the soma for any dendritic tree can be calculated relatively easily by the following set of rules. These rules, in turn, can be easily derived from the equation of electrotonus with the boundary conditions that currents and potentials must match at junctions.

The significant quantities here are the admittances, Y, seen when looking peripherally from the various places of the neuron as illustrated in Figure 5.6. The main rules of the calculus are shown in Eq. (5.12):

$$\bar{Y}_0 = \frac{1}{\lambda R}\tanh\left(\frac{l}{\lambda}\right) \tag{5.12a}$$

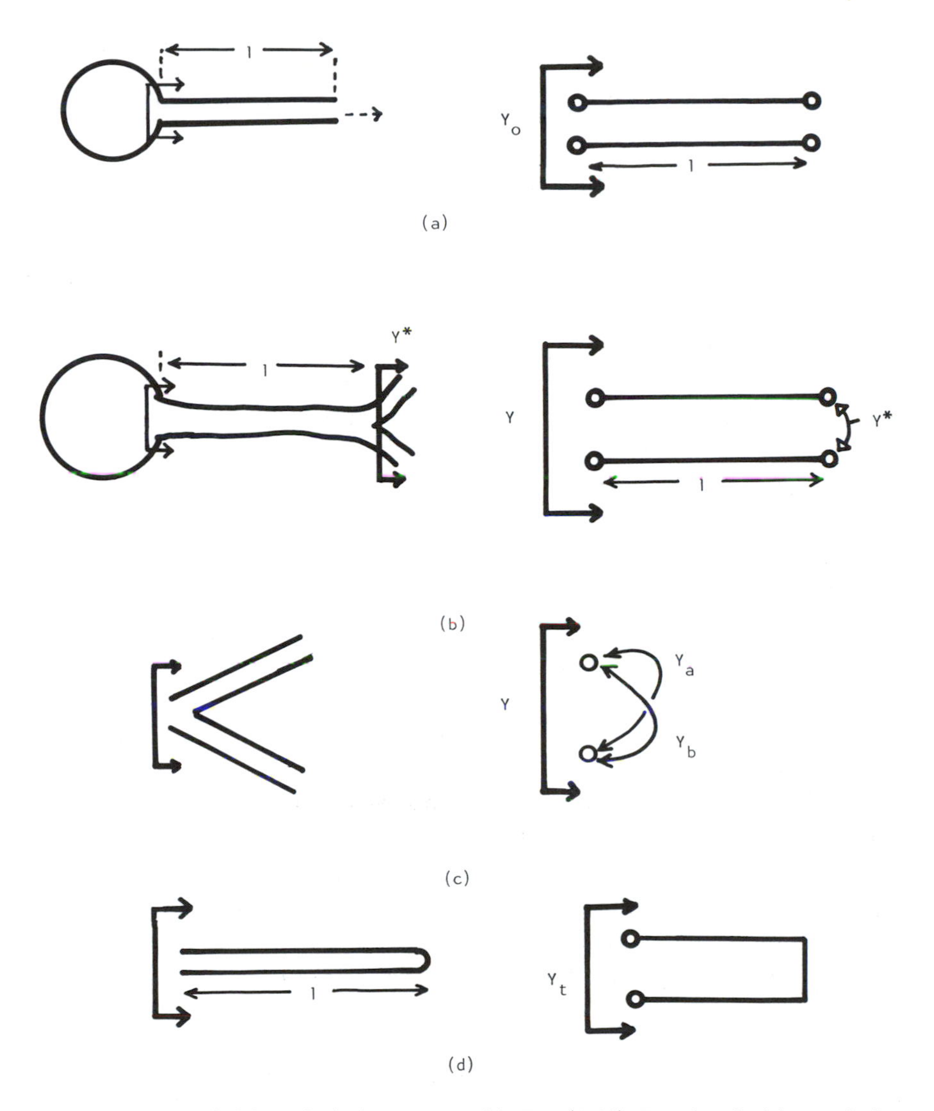

FIGURE 5.6 Definition of admittances used in Eq. (5.12). Reprinted with permission from *Neural and Brain Modeling*, R. MacGregor, Copyright 1987, Academic Press.

$$\bar{Y} = \frac{\bar{Y}_0 + \bar{Y}^*}{1 + \bar{Y}_0 \bar{Y}^* (\lambda R)^2} \tag{5.12b}$$

$$\bar{Y} = \bar{Y}_a + \bar{Y}_b \tag{5.12c}$$

$$\bar{Y}_i = \frac{1}{\lambda R} \left\{ \frac{\tanh(l/\lambda) + r/2\lambda}{1 + (r/2\lambda)\tanh(l/\lambda)} \right\} \tag{5.12d}$$

$$R = 1/\sigma\pi r^2 = \rho/\pi r^2, \qquad G\rho = r/2\lambda^2 \tag{5.12e}$$

An example and further discussion of this calculus is given in NBM.

Illustrative Continuous Compartmentalization Problem

The general mathematical methodology in applying the equation of electrotonus is particularly well illustrated by this example. The example also illustrates the usefulness of the general continuous approach to particular sharply defined experimental research. Indeed, this example contains the solutions to a number of significant and as yet unexplored specific questions. We will illustrate the general mathematical methods of approach by formulating and solving the general problem, giving explicit solutions for several particular questions of interest.

General Problem and Mathematical Formulation. Consider the experimental preparation illustrated in Figure 5.7, wherein experimentally applied stimulating and recording electrodes are applied in either or both of two positions in a neuron: one in the soma, and one in a dendrite a given distance, Δ, from the soma.

Under the following assumptions this geometry can be represented mathematically by the four compartmental scheme shown in part b of the figure. These assumptions are there are no synaptic or action potentials, the membrane time constant is uniform throughout the cell, the soma is isopotential, all k dendrites are identical, current leakage into the axon is negligible compared to that into the dendrites, and the potential is initially at a resting level everywhere in the neuron.

The general governing equations applicable to this system are given in Eq. (5.13):

$$\frac{\partial E}{\partial t} = -E + \lambda^2 \frac{\partial^2 E}{\partial x^2} + \left\{ \frac{SC}{G} - \frac{GE}{G}[E - E_e] - \frac{GI}{G}[E - E_i] \right\} \tag{5.13a}$$

$$I(x, t) = -\sigma A \frac{\partial E(x, t)}{\partial x} \tag{5.13b}$$

Equation (5.13a) is simply the equation of electrotonus expanded to include the possibility of externally applied stimulating current, SC, at particular values of x. Equation (5.13b) is the Ohm's law relating longitudinal current to the spatial gradient of the electrical potential.

The general governing equation, (5.13a), reduces to the form shown in Eq. (5.14a) for each of the three compartments indicated by x_1, x_2, and x_3 in

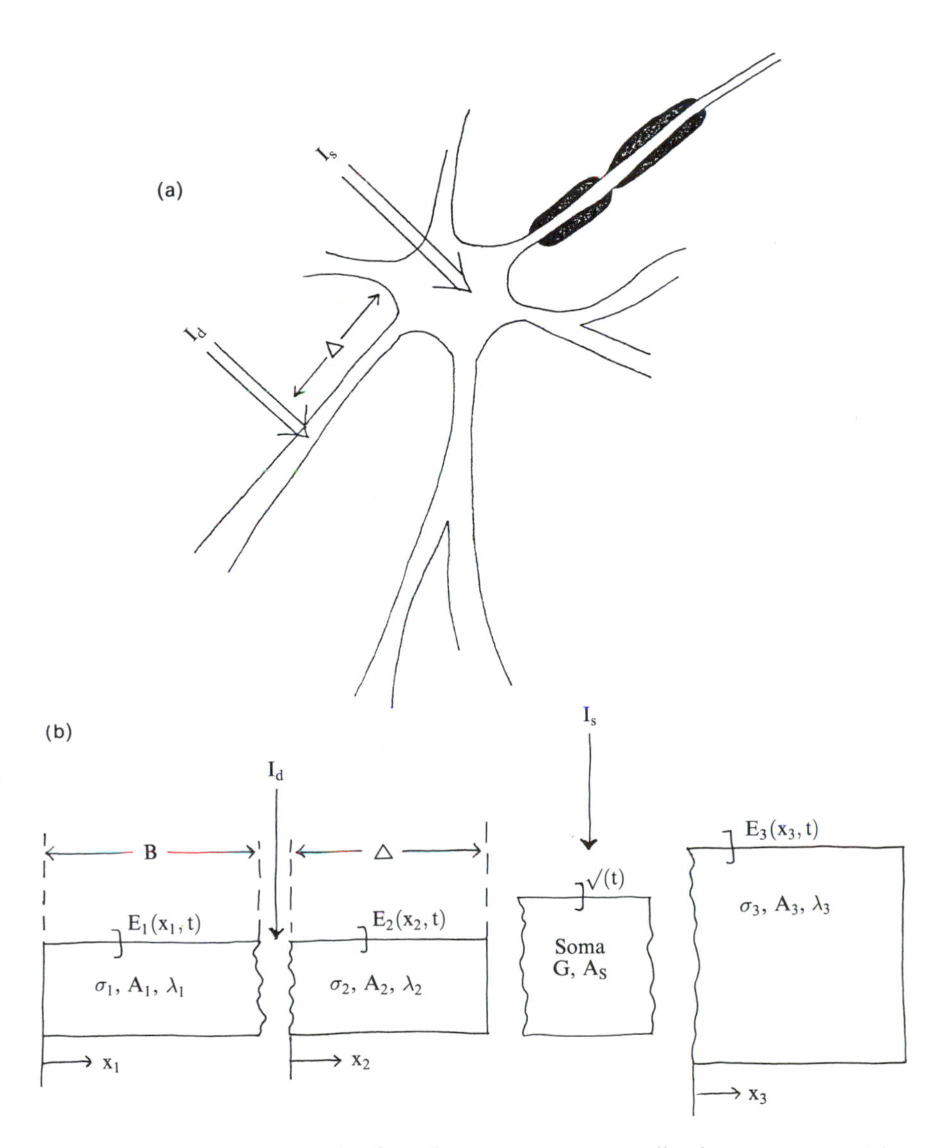

FIGURE 5.7 Illustrative example of continuous compartmentalization: a. neuron with two experimental electrodes; b. mathematical characterization of problem.

Figure 5-7b.

$$\lambda_i^2 \frac{\partial^2 E_i(x_i, t)}{\partial x_i^2} - E_i(x_i t) - \frac{\partial E_i(x_i, t)}{\partial t} = 0, \qquad i = 1, 2, 3 \tag{5.14a}$$

$$\lambda_i^2 \frac{\partial^2 \bar{E}_i(x_i, s)}{\partial x_i^2} - (s + 1)\bar{E}_i(x_i, s) = 0 \tag{5.14b}$$

$$\bar{E}_i(x_i, s) = C_i(s) \sinh\left(\frac{\sqrt{s+1}}{\lambda_i} x_i\right) + Di \cosh\left(\frac{\sqrt{s+1}}{\lambda_i} x_i\right) \tag{5.14c}$$

$$\frac{\partial \bar{E}_i(x_i, s)}{\partial x_i} = \frac{\sqrt{s+1}}{\lambda_i} C_i(s) \cosh\left(\frac{\sqrt{s+1}}{\lambda_i} x_i\right) + D_i \sinh\left(\frac{\sqrt{s+1}}{\lambda_i} x_i\right) \tag{5.14d}$$

These regions are free of any applied or active stimulation and characterized by simple passive decay of signals. Equation (5.14a) is three equations corresponding to $i = 1$, 2, and 3. If we take the Laplace transform of (5.14a), (5.14b) results. This equation in turn can be regarded as an ordinary differential equation in x, with the Laplace transform variable, s, as a parameter. Equations (5.14b) can then be integrated over x to give the solutions shown in (5.14c). To express currents it will be convenient to have the expression for the spatial gradients of E in these branches. These gradients are given in (5.14d).

Equations (5.14c) give the solutions for the potentials in the three dendritic compartments in terms of the Laplace transform variable, s. These solutions, however, are expressed in terms of the six unknown "constants" of integration, C_i and D_i. These terms are, in fact, dependent on the parameter s, and therefore on time. Moreover, we are also interested in the transmembrane potential of the soma, V, which is also an unknown function of time.

Boundary Conditions. We can solve for the seven unknown functions, C_i, D_i, and V, by applying seven boundary conditions. These boundary conditions will represent constraints on the values of currents and potentials at junctions in the model. They thereby represent the constraints imposed on the general mathematical solution by the physical geometry of the neuron.

These boundary conditions are given in Eq. (5.15):

$$\sigma_1 A_1 \frac{\partial \bar{E}_1(0, s)}{\partial x_1} = 0 \tag{5.15a}$$

$$\overline{I_d}(s) = \sigma_1 A_1 \frac{\partial E_1}{\partial x_1}(B, s) - \sigma_2 A_2 \frac{\partial E_2(0, s)}{\partial x_2} \tag{5.15b}$$

$$E_1(B, s) = E_1(0, s) \tag{5.15c}$$

$$E_2(\Delta, s) = V(s) \tag{5.15d}$$

$$E_3(0, s) = V(s) \tag{5.15e}$$

$$I_s(s) = GA_s\{sV(s) + V(s)\} + \sigma_2 A_2 \frac{\partial E_2(\Delta, s)}{\partial x_2} - \sigma_3 A_3 \frac{\partial E_3(0, s)}{\partial x_3} \tag{5.15f}$$

$$\sigma_3 A_3 \frac{\partial E_3(L, s)}{\partial x_3} = 0 \tag{5.15g}$$

Equations (5.15a) and (5.15g) express the constraint that current flowing out the ends of the dendrites represented by compartments 1 and 3 be negligibly small compared to that flowing out the cell membrane. Equation (5.15b) expresses the constraint that the total current flow be conserved in the small region where the dendritic electrode is placed. Equations (5.15c), (5.15d) and (5.15e) are obtained by equating expressions for potentials at the left and right sides of each of the three junctions in the model. Here we have assumed the soma is isopotential and that the region occupied by the dendritic electrode is negligibly small. Equation (5.15f) expresses the constraint that current be conserved in the soma. Here we have used the capacitive and leakage terms of the membrane equation to represent the current leaking out the soma membrane.

General Solutions. We can now solve for C_i, D_i, and V, by using Equations (5.14c) and (5.14d) in Eq. (5.15). In doing this we will also use the relations among parameters shown in Eq. (5.16):

$$\lambda_1 = \lambda_2 = \lambda_3 = \lambda$$
$$B + \Delta = L \tag{5.16}$$
$$\frac{\sigma_3 A_3}{\lambda_3} = (k-1)\frac{\sigma_1 A_1}{\lambda_1}$$

These relations are implied by the general assumptions stated earlier.

The general solutions for the potentials under the recording electrodes are produced by this procedure after a great deal of algebraic manipulation. These general solutions are shown in Eq. (5.17):

$$\overline{E_2}(\Delta, s) = \frac{\overline{Id}(s)\exp[-H\Delta/\lambda]\{1 + \exp[-2(L-\Delta)H/\lambda] + I_s(s)\{1 + \exp[-2LH/\lambda]\}}{DEN} \tag{5.17a}$$

$$\overline{E_2}(0, s) = \frac{\{1 + \exp[-2(L-\Delta)H/\lambda]\}\{Is(s)\exp[-H\Delta/\lambda] + Id(s)NUM/2\}}{DEN} \tag{5.17b}$$

$$NUM = 1 + (k-1)\tanh(HL/\lambda) + \frac{GAs\lambda H}{A} + \left[1 - (k-1)\tanh(HL/\lambda) - \frac{GAs\lambda H}{A}\right]\exp[-2\Delta H/\lambda] \quad (5.17c)$$

$$DEN = GAs(s+1)\{1 + \exp[-2HL/\lambda\} + \frac{k\sigma A}{\lambda}H\{1 - \exp[-2HL/\lambda]\} \quad (5.17d)$$

$$H = s + 1 \quad (5.17e)$$

These general solutions can be reduced to a number of particular experiments, as indicated in the following.

Stimulating and Recording Electrode at the Soma: Step Current. For the special case where the applied current, *Id*, is zero, the dendrites are very long, and a step current of amplitude, *I*, is applied to the soma, the resulting potential under the stimulating electrode is given by Eq. (5.18):

$$\overline{E_2}(\Delta, t) = \frac{|I|}{GA_s(\rho+1)(\rho-1)}\{\rho\,\mathrm{erf}(\sqrt{t}) - 1 + (\rho^2-1)t\,\mathrm{erfc}(\rho\sqrt{t})\} \quad (5.18a)$$

$$\rho = \frac{k\sigma A}{GA_s} \quad (5.18b)$$

This expression is characterized by one parameter, the quantity, ρ, which measures the ratio of current leakage to the dendrites to current leakage through the soma membrane. The procedure can be used to study the effective communication of dendrites with soma in particular neurons, including their length constant.

Stimulating and Recording Electrode at the Soma: Ramp Current. The solution for the same case when a ramp current is applied is given in Eq. (5.19):

$$I_s = mt \quad (5.19a)$$

$$E_2(\Delta, t) = \int_0^t \frac{m(t-\tau)}{GA_s}\exp[(\rho^2-1)]\,\mathrm{erfc}(\rho\sqrt{\tau})\,\mathrm{d}\tau \quad (5.19b)$$

$$E_2(\Delta, t) = \frac{m}{GA_s}\left\{\frac{\exp[rt] - rt - 1}{r^2} - \sum_{i=1}^{\infty} a_i t^{(i+3/2)}\right\}$$

$$a_i = \frac{2}{\sqrt{\Pi}}\frac{\sum_{l=0}^{i-1}\left\{\frac{(-1)^{(i-l+1)}(2(i-l)-1)r^l}{(i-l-1)!\,l![2(i-l)-1]}\right\}}{(i+1/2)(1+3/2)}, \qquad r = \rho^2 - 1 \quad (5.19c)$$

$$\text{if} = 1: \quad E_2(\Delta, t) = \frac{m}{GA_s}\left\{t^2/2 - \frac{2}{\sqrt{\pi}}\sum_{i=1}^{\infty}\frac{(-1)^{(i+1)}t^{(i+3/2)}}{(i-1)!(2i-1)(i+1/2)(i+3/2)}\right\}$$

Equation (5.19b) uses the convolution of the response to the step, and (5.19c) uses a series expansion for the error function. The explicit solution in Eq. (5.19c) can be evaluated computationally for any value of ρ.

Accommodation Curves with Leakage into Dendrites. The process of neuronal accommodation, which is manifested in such phenomena as "on" and "off" transient responses and postinhibitory rebound is studied in individual neurons by finding the latency to firing for applied ramp currents of various slopes. The solutions given here can show the influence of leakage of current into dendrites on this behavior. The governing equations are given as Eq. (5.20):

$$\eta\frac{d\theta}{dt} = -(\theta - \theta_0) + cE_2(\Delta, t) \tag{5.20a}$$

$$\theta(t) = \theta_0 + \frac{cm}{GA_s}\left\{\frac{\exp[rt] - \exp[-t/\eta]}{r^2\eta(r+1/\eta)} - \frac{\lambda(\exp[-t/\eta] - 1 + t/\eta)}{r} + \frac{1-\exp[-t/\eta]}{r^2} - \sum_{i=1}^{\infty} b_i t^{(i+5/2)}\right\} \tag{5.20b}$$

$$b_1 = 2a_1/(7\eta)$$

$$b_i = \frac{a_i - b_{i-1}}{(i+5/2)}, \qquad i = 2, 3, \ldots$$

$$\rho = \frac{k\sigma A}{Ga_s\lambda}, \qquad r = \rho^2 - 1, \qquad a_i \text{ given in Eq. (5.19)} \tag{5.20c}$$

Equation (5.20a) shows the general process by which accommodation occurs: the cell threshold, θ, rises (accommodates) to elevated levels of cell potential, E, according to a sensitivity parameter, c, and at a time constant, η. Equation (5.20b) is obtained by integrating (5.20a) while making use of (5.19c). To obtain the relevant latency data one needs simply to numerically expand Eqs. (5.19c) and (5.20b) simultaneously and find the value of t at which they are equal, for different combinations of the governing parameters, c, η, and ρ. The phenomenon of accommodation is discussed more thoroughly in terms of a point neuron in Chapter 6.

Solutions for an Electrode Placed in the Dendrites. The general solution shown in Eq. (5.17) can also be developed along similar lines for the case of a stimulating and recording electrode placed in a dendrite, some distance, Δ, from the soma. The possible distortions of data from misplacing an electrode in a large dendrite of a motoneuron for example can be studied. This same general solution can be applied to step and ramp currents and to accommodation.

An Action Potential Backpropagated into Dendrites. One may also show from this general solution that the shape of an action potential triggered at the soma and passively propagated into a dendrite is seen at a distance, Δ, from the soma as given in Eq. (5.21a):

$$E_2(0,t) = \frac{|I|}{GA_s}\int_0^t \exp\left[-\tau\right]\left\{\frac{\exp[-\Delta^2/(4\tau)}{\sqrt{\Pi\tau}}\right.$$

$$\left. -\rho\exp\left[\rho^2\tau + \rho\sqrt{\tau}\right]\operatorname{erfc}\left(\rho\sqrt{\tau} + \frac{\Delta}{2\sqrt{\tau}}\right)\mathrm{d}\tau\right\} \quad (5.21a)$$

$$I_s(t) \approx \operatorname{erf}\sqrt{t} + \frac{\exp\left[-t\right]}{\sqrt{\Pi t}} \quad (5.21b)$$

This makes use of the result that the current at the soma corresponding to a square wave pulse of duration, τ, is given by Eq. (5.21b). This solution can be useful in indicating the extent to which backpropagation of action potentials into dendrites might diminish the amplitude of incoming PSPs.

This chapter has discussed the integration of the membrane equation over space, as mediated by the geometrical constraints placed on its term for total local transmembrane current, SC. The next chapter discusses its integration over time, as mediated by its terms for active conductance modulations, g_i'.

Chapter 6 The Membrane Equation in Time: The Generation of Synaptic and Action Potentials

This chapter discusses the representation of the mechanisms of generation of the active neuroelectric signals—synaptic potentials and action potentials. As shown in Chapter 3, the direct physical causal agents for these generations are localized and selective modulations in membrane permeability to the various individual ionic species. As shown in Chapter 4, the mathematical representation of these agents is the active membrane conductance modulation terms, g_i', in the membrane equation, (4.4f).

The causal structure of these mechanisms should be seen clearly as follows: the changes in permeability are the causal agents for the neuroelectric signals; to correctly express the mechanics of neuroelectric signaling, it is sufficient to prescribe these underlying changes in permeability. To go to deeper levels is generally counterproductive in the broad development of the mechanics because it introduces too much subordinate detail, which obscures one's vision of the higher order processing.

However, it would be instructive to have a clear view of the deeper level physical processes that underly the permeability changes because this could conceivably provide a generic relation or relations for their temporal form and parameter dependence on the local anatomy, chemistry, and physiology. The permeability modulations, however, are governed by molecular gating processes currently known only in outline. As and when these processes become more clearly known, they may produce such general generic expressions for time course and parameter dependence for the active permeability modulations. When available, such expressions can be easily interjected in their appropriate place in the higher level mechanics of neuroelectric signalling.

The unavailability of such relations is, in fact, not very significant for the prescription and fulfillment of the theoretical mechanics for neuroelectric signalling. In the first place, one may rely on experimental descriptions of the generative conductance modulations and characterize these with direct, easily identifiable parameters (e.g., amplitude, decay time constant).

Alternatively, from a practical operational point of view, one may prescribe a generic functional form for the permeability modulations and adjust these forms and their parameters to match the desired potential responses, according to the particular situation being studied. The processes of short-term synaptic plasticity, facilitation, and adaptation, for example, can be readily handled this way.

Again, either of these approaches provides an adequate representation of the causal processes and peculiarities underlying neuroelectric signal generation.

Second, the fine-grained structure of the individual time courses is often not terribly significant as regards the broader nature of neuroelectric signalling. A main reason for this in synaptic responses is that the duration of many conductance changes is about 0.1 to 0.2 of the membrane time constant. This means that responses are sluggish and their time courses depend more on the membrane time constant than on that of the much shorter conductance modulation. The responses depend more on the amplitude and rough duration of the conductance modulation than on its fine grained temporal structure.

The conductance changes underlying action potentials can be accurately described by experimentally based equation systems. However, from the viewpoint of neuroelectric signalling, one is usually concerned with if and when action potentials occur and not very much with their individual shapes. Again, the general mechanics developed here allows one to insert at this point as much realistic neurobiological detail as desired or, alternatively, to simplify that detail to essential features, as may be equally desirous or advantageous.

This chapter describes, first, the time courses of the post synaptic conductance changes underlying synaptic electrical responses and then discusses those underlying action potentials. The latter section includes the description of a state-variable model for the generation of repetitive firings of action potentials in spike trains, which includes clear parameter control over the fundamental temporal processes of accommodation and relative refractoriness in ongoing activity. The chapter concludes with a model for the generation of calcium-mediated action potentials in dendrites.

The Generation of Synaptic Potentials

Postsynaptic permeability modulations are brought about by a complex of events triggered normally by the arrival of an action potential at the presynaptic terminal. Packets of transmitter molecules are released into the synaptic cleft across which they diffuse to be taken up by the receptor molecules in the subsynaptic membrane. Many factors in both the pre- and postsynaptic portions of this process contribute to the character of the resultant permeability change.

Although there are exceptions, the duration of the postsynaptic conductance change does not usually significantly outlast the duration of the triggering action potential. Figure 6.1 illustrates a typical PSP (part a) and an experimentally based and widely applicable generic representation for postsynaptic conductance modulations (part b). This form can be represented by Eq. (6.1):

$$g(t) = Pt \exp(-\alpha t) \tag{6.1}$$

In this expression, two parameters, P and α, control respectively the amplitude and exponential decay rate of the conductance change.

Much development of the mechanics of neuroelectric signalling is undertaken using discrete representations of variables over successive short-time windows. Therefore, one highly simplified but representative approach is to represent the conductance change by a simple square pulse of amplitude P over however many time units make up the duration of the action potential or desired duration of the conductance change.

A refinement of this simple block approach involves using a series of discrete values that decay exponentially, as illustrated in Figure 6.1c. This form is a discrete approximation similar to the generic form shown in Eq. (6.1) and Figure 6.1b. It is also governed by the two characteristic parameters, amplitude and exponential decay constant.

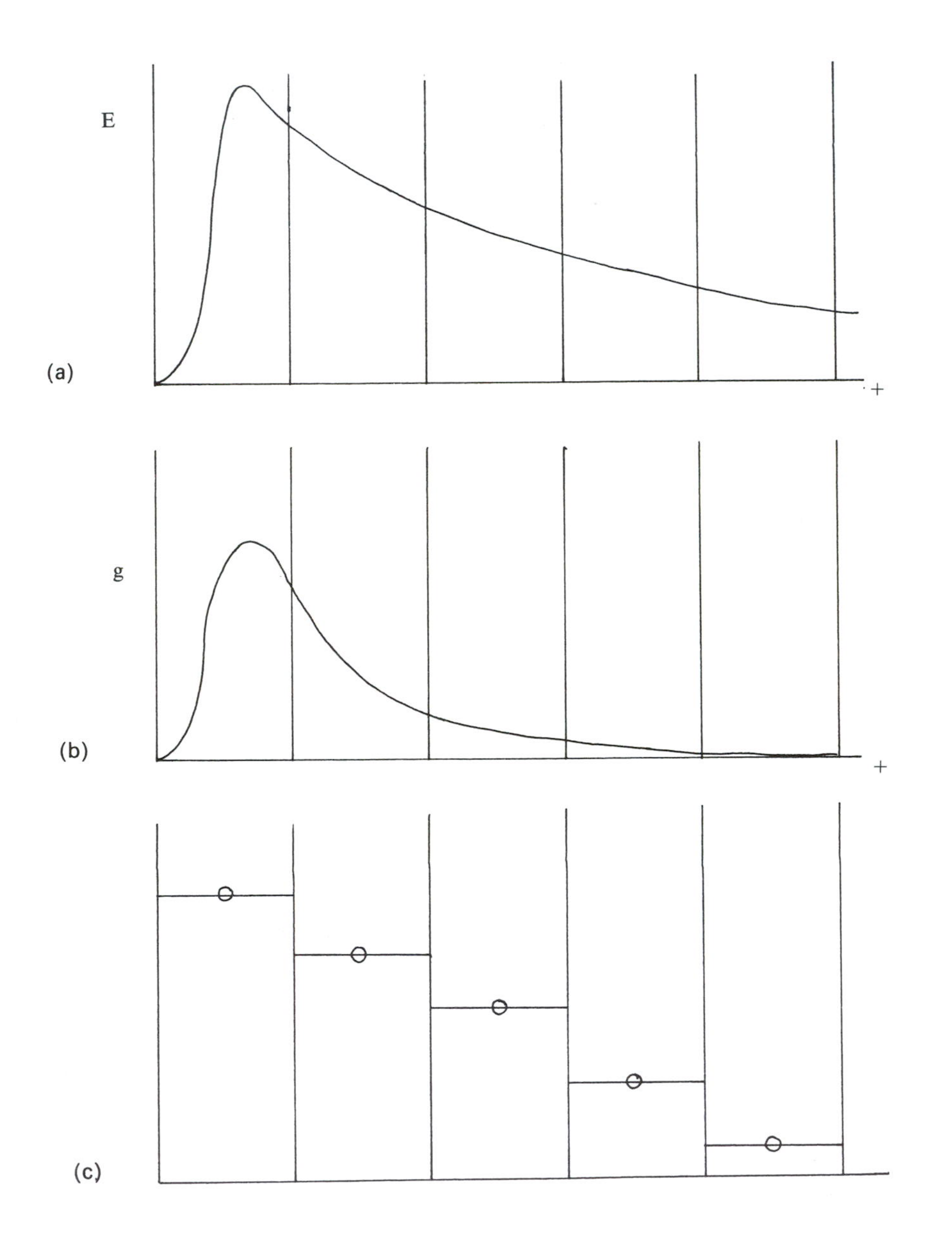

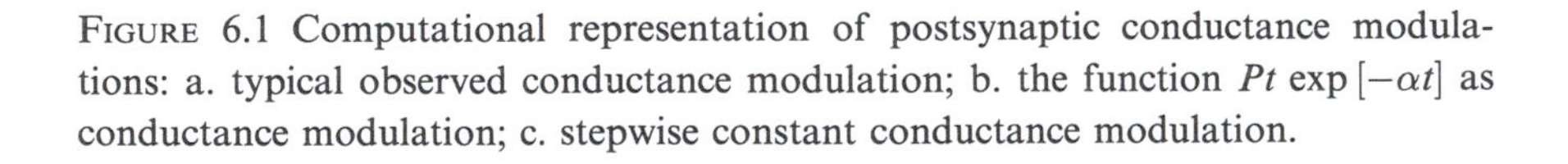

FIGURE 6.1 Computational representation of postsynaptic conductance modulations: a. typical observed conductance modulation; b. the function $Pt \exp[-\alpha t]$ as conductance modulation; c. stepwise constant conductance modulation.

These forms are most useful for large-scale, general purpose computer simulation programs. Any higher degree of precision can be substituted for these rougher but useful forms at will.

The Generation of Single-Action Potentials

The Hodgkin–Huxley Model for Action Potentials in the Squid

In 1952 Hodgkin and Huxley published the following equation system for the generation of single-action potential in squid giant axons:

$$C\frac{\mathrm{d}E}{\mathrm{d}t} = GNO * m^3h(ENA - E) + GKO * n^4(EK - E) + GL * (EL - E) + SC) \tag{6.2a}$$

$$\frac{\mathrm{d}n}{\mathrm{d}t} = \alpha_n(1-n) - \beta_n n$$

$$\frac{\mathrm{d}m}{\mathrm{d}t} = \alpha_m(1-m) - \beta_m m \tag{6.2b}$$

$$\frac{\mathrm{d}h}{\mathrm{d}t} = \alpha_h(1-h) - \beta_h h$$

$$\alpha_n = \frac{0.01(E+10)}{e^{(E+10)/10} - 1}, \qquad \alpha_m = \frac{0.1(E+25)}{e^{(E+25)/10} - 1}, \qquad \alpha_h = 0.07e^{E/20}$$

$$\beta_n = 0.125e^{E/80}, \qquad \beta_m = 4e^{E/18}, \qquad \beta_h = \frac{1}{e^{(E+30)/10} + 1}$$

$$C = 1\mu\mathrm{F/cm^2}, \qquad GNO = 120\frac{\mathrm{m \cdot mho}}{\mathrm{cm^2}}, \tag{6.2c}$$

$$GKO = 36\frac{\mathrm{m \cdot mho}}{\mathrm{cm^2}}, \qquad GL = 0.3\frac{\mathrm{m \cdot mho}}{\mathrm{cm^2}},$$

$$ENA = -115\,\mathrm{mV}, \qquad \mathrm{EK} = +12\,\mathrm{mV}, \qquad \mathrm{EL} = -10.6\,\mathrm{mV}$$

$$n_0 = 0.31767691, \qquad m_0 = 0.05293249; \qquad h_0 = 0.59612075 \tag{6.2d}$$

In this model the fundamental equation for the transmembrane potential E is Eq. (6.2a), which describes E in terms of the resting and active membrane conductances in accordance with the equivalent circuit model discussed in Chapter 4. The active conductances for sodium, G_{Na}, and potassium, G_{K}, are defined in terms of hypothetical continuous functions n, *m,* and *h* as shown in Eq. (6.2b); *n, m,* and *h* in turn are given by rate equations guided by α and β parameters as shown in Eq. (6.2c). Finally, these rate-governing α and β parameters, are represented by experimentally determined functions of E as shown in Eq. (6.2d). In conjunction with this model, Hodgkin and Huxley also described a large number of carefully controlled experiments on transmembrane currents, conductances, and potentials that quantitatively matched the predictions of the equation system. The work is significant because (1) it established that large, brief excursions in conductances to sodium and potassium are the fundamental physical processes underlying action potentials in neurons; (2) it helped firmly establish the ionic picture of the basis of transmembrane potentials in neurons; and (3) it held out by example the hope that significant neuroelectric events could be precisely and quantitatively modeled by mathematical and engineering techniques.

The Hodgkin–Huxley model has at the same time strong and elegant features and fundamental weaknesses. On the one hand, and most significantly, it accurately describes the transmembrane potential in the course of single-action potentials in terms underlying the generation of action potentials. On the other hand, the model does not embody a satisfactory picture of the biophysical basis of the membrane conductance modulations in terms of more fundamental molecular events. For example, one hypothetical interpretation of the Hodgkin–Huxley equation system, as presented by Hodgkin and Huxley themselves, would be that the functions *n, m,* and *h* represent the relative concentrations of three "gating" molecules in the intra- or extracellular fluid adjacent to the membrane. Then, $1 - n$, $1 - m$, and $1 - h$ would represent the concentrations of these molecules on the other side of the membrane. The rate equations for *n, m,* and *h* would be considered as representing the diffusion of these molecules across the membrane. The values of α then would represent the momentary ease of influx of these molecules and the values of β would represent the momentary ease of efflux (or vice versa). Finally, the form of the governing equations for the conductances G_{Na} and G_K in terms of *n, m,* and *h* might reflect that opening a channel for a sodium ion would require the confluence of one *m* molecule and three *h* molecules on one side of the membrane, whereas gating a potassium channel would require the confluence of four *n* molecules. That is, the likelihood of gating a channel would be proportional to the probability of finding the right combination of gating molecules at hand, which in turn would be proportional to the product of all the relevant concentrations.

Hodgkin and Huxley presented this equation system with the full understanding that it was merely a conceptual guide and probably did not accurately represent the underlying processes. Again, the fundamental contribution of the model is that it relates the level of electrical signals to the level of membrane conductances, not that it describes the molecular events at still lower levels. However, the precise form of the equation system is, in fact, quite arbitrary. The fundamental equation for the transmembrane potential E is based on the equivalent circuit model and is therefore reliable, but the functional forms of the other equations are somewhat suspect. Specifically, Eq. (6.2a), which relate G_{Na} and G_K to n, m, and h, and Eq. (6.2b), the rate equations for n, m, and h themselves, are without satisfactory physical foundation. Equations (6.2c) for the values of α and β are experimentally reliable since they are based on curve-fitting experimental data from squid giant axons, but they are only shorthand representations of data—that is, they represent no physical principles in themselves. Moreover the α and β themselves are hypothetical constructs defined in the weakly based equations for n, m, and h. Therefore, although the equation system does give a satisfactory quantitative description of the generation of action potentials and the time courses of the sodium and potassium conductances, one does not feel totally confident in the functional form of the equations or their biophysical basis. Therefore, one cannot reliably predict with the model how fundamental changes at the molecular and biochemical level will influence the behavior of the system, or how a neuronal tissue other than the squid giant axon may differ in behavior on the basis of known biophysical differences.

The model also has strengths and weaknesses as a model of information processing operations in neurons. It is accurate in terms of describing events leading up to the generation of single-action potentials and the shapes of single-action potentials in the squid axon. It exhibits accommodation to activating currents in ways consistent with detailed studies of spinal motoneurons (see discussion later in this chapter). On the other hand, the model does not accurately describe the generation of two or more action potentials even in the squid axon. Within an information processing orientation, one is most interested in ongoing repetitive activity in neurons. The mechanisms of interest are primarily those that determine the modulations of the ongoing graded generator potential, the threshold, and their interaction, rather than the detailed shape of the action potential once triggering has occurred. Therefore, this is a significant limitation of the model.

The Hodgkin–Huxley approach has been adapted to describe action potentials in frogs, the contraction of heart muscles, to the firing of two and three repetitive action potentials and, most recently, to the generation of dendritic calcium-mediated action potentials in the hippocampus. All these approaches are based on representing the time courses of the two main conductance changes by functions like *n, m,* and *h* governed by rate equations and their two parameters. This approach can be used to good purpose in studies dependent on fine-grained representation of action potentials.

It is instructive to realize, however, that the *n, m,* and *h* functions and their counterparts likely have no real physical foundation, and the approach is fundamentally as heuristic as that embodied in Eq. (6.1) for synaptic conductance modulations. Moreover, the approach has three central problems as regards the broader purposes of a mechanics of neuroelectric signalling. First, as mentioned previously, the Hodgkin–Huxley model does not describe repetitive firing accurately; this is essential with respect to neuroelectric signalling.

Second, the fundamental physical processes such as accommodation are obscurely represented in the parameters of the representation. One can not readily and directly adjust, for example, the very important accommodative parameters of the model. Third, the approach involves a great deal of computation to give what is considerably more detail than is usually required in neuroelectric signalling—for example, the detailed shaped of the individual action potential, and all the constituent *n, m,* and *h* functions with their parameters.

For these reasons it is desirable to provide a representation of the ongoing generation of multiple-action potentials in spike trains that describes the essential operative causal agents clearly, expresses their governing parameters directly, and is computationally efficient. The next section of this chapter describes such a model.

The Generation of Neuronal Spike Trains: A State-Variable Representation[1]

The model presented here describes the generative processes that produce action potentials and their postfiring effects, but makes no attempt to describe the action potentials themselves. The guiding idea here is that in the broader purposes of neuroelectric signalling, one may usually assume that these action potentials will show up with an appropriate time delay at the neuron's axonal terminals acting to trigger postsynaptic conductance changes in the cells to which they connect. Thus, in simulating or otherwise

[1] This state-variable model was first described in *Biological Cybernetics* (then known as *Kybernetik*) **16** (1974), 53–64.

developing the mechanics of neuroelectric signalling, one needs only to determine the times at which these pulses occur and represent the corresponding downstream synaptic conductance changes in the correct receiving cells at the correct times.

Specifically, this model effectively replaces the active spike-producing conductance, sodium conductance, with a generally time-varying state variable called *threshold.* Action potentials are generated when the transmembrane potential exceeds the threshold. This representation includes a representation of neuronal "accommodation," wherein the threshold varies according to recent input. This latter effect is related to the inactivation of the triggering sodium channels by membrane potential.

Similarly, the very large excursion of the potassium conductance, which pulls the action potential back down toward threshold and resting levels, is replaced by a postfiring potassium conductance, which rises abruptly at firing and then decays exponentially, producing a relatively long-lasting and low-valued tail that contributes to postfiring refractoriness. These tails from separate action potentials will sum when action potentials occur sufficiently close together in time.

This overall theoretical formulation by me is based on previous formulations and extensive experimental studies of the accommodative (prefiring) and adaptive (postfiring) properties of spinal motoneurons by Hill, Bradley, and Somjen and Kernell. The central behavioral properties of the model are closely matched to the relevant experimental results.

The model is described in terms of four operational state variables, three independent input functions, and four central parameters, each of which has a clear and direct functional interpretation and behavioral implications. Different types of neurons can be easily represented at will according to the values used for these four parameters.

The following sections provide a basic description of the governing processes and equations of this model, a qualitative description of its behavior, and then a more detailed mathematical examination of its behavior and its comparison to the relevant experimental studies.

The State-Variable Model for Repetitive Firing

The model considered here is shown in Eq. (6.3):

$$\frac{dE}{dt} = -E - g_K(E - E_K) + [SC - GE(E - E_{ex}) - GI(E - E_{in})] \tag{6.3a}$$

$$\frac{d\theta}{dt} = -\frac{(\theta - 1)}{\eta} + \frac{cE}{\eta} \tag{6.3b}$$

$$\frac{dg_K}{dt} = \frac{g_K}{\tau_K} + bS \tag{6.3c}$$

$$S = \begin{cases} 1, & \text{if } E \geqq \theta \\ 0, & \text{otherwise} \end{cases} \tag{6.3d}$$

These equations describe the ongoing dynamics of the neuron triggering section in terms of four state variables: the transmembrane potential, E; the threshold, θ; the post-firing potassium conductance above its resting level, g_K; and a spiking variable, S. All four of these state variables are functions of time. Equations (6.3) should be seen as the equations that determine these four variables.

The equations also indicate that these state variable are driven by several input functions, namely, any synaptic conductance modulations and the total transmembrane current, SC. The synaptic currents are represented in Eq. (6.3a) by a single excitatory and a single inhibitory representative, G_e and G_i, respectively.

There are four parameters of the system, c, η, b, and τ_k. These parameters characterize the prefiring (c and η), and postfiring (b and τ_j) behavior of the spike-generating mechanisms of any particular neuron. They are values that can be determined experimentally for any given cell. In a broader, electrical-signalling view, they are values by which a given neuron expresses some individuality in its spike-generating properties: transient or steady firing, sensitive or sluggish firing, and so on. From the standpoint of a theorist or modeller, they are values one can adjust to create various individual firing propensities and types.

The remaining terms, E_e, E_i, and E_K, are the equilibrium potentials of the active conductances, G_e, G_i, and g_K, respectively.

Viewed functionally, Equations 6.3 prescribe how the state variables, E, θ, g_k, and S, respond to various ongoing input driving functions, G_e, G_i, and SC, under the modulation of their constituent parameters, c, η, b, and τ_k.

The mechanisms of this model contribute several significant dynamic characteristics to neuroelectric signalling. The first of these is that neurons may exhibit either transient or steady responses, or both, to steadily maintained input. Such transient responses are generally referred to as *phasic* activity and include so-called on or off responses. Steady responses are referred to as *tonic* activity. We will see that these properties are determined by the accommodative effects and can be controlled by the parameters c and η.

A second characteristic feature is the tendency of neurons to be less sensitive (refractory) to input for a short time following generation of an action potential. This is represented in Eq. (6.3) by the postfiring potassium conductance, g_K. The effect increases in strength with b, and in duration with τ_k.

A third and related effect is that the curve of output firing rate versus input intensity levels off at to a straight line when output firing intervals become comparable to the refractory time constant, τ_K. This effect is influenced by both b and τ_K; it will become significant when we study the character of neuronal firing levels in Part II.

Qualitative Behavior of the Model

Equation (6.3a) is the membrane equation. Equation (6.3b) is the equation for the firing threshold, θ. It represents the physiological observation that many cells exhibit firing thresholds driven upward by prior transmembrane depolarization, E. This phenomenon is called *accommodation.* Equation (6.3b) allows one to represent varying degrees of accommodation according to the parameter c. If c is zero, for example, the cell does not accommodate. The displaced threshold tends to decay back toward a resting level, θ_0, with the time constant, η. Equation (6.3b) shows that if c or E is fixed at 0, the threshold will be fixed at θ_0. Equation (6.3b) is an empirical fit to extensive experimental studies of the phenomenon of accommodation in many neurons.

Equation 6-3-c defines the spiking variable, S. If the transmembrane potential, E, is greater than threshold, θ, then S is set equal to 1, indicating that an action potential has been triggered. In this representation we are implicitly assuming that sampling of this equation is done at regular time intervals of a given size (usually about a millisecond), and that an action potential when it occurs occupies some short finite time period (some fraction of a millisecond, but the exact length is not explicit in the model). When the cell does not fire an action potential, S is set equal to 0.

Equation (6.3d) determines the postfiring potassium excursion. When E is greater than θ, and S is correspondingly set to 1, g_K is driven upward in proportion to a parameter, b, as determined by this equation. There is a tendency for g_K to decay back to its resting value of zero; the time constant of this decay is a parameter, τ_K. If the cell fires multiple spikes within, say, two or three τ_K of each other, there will be significant multiple contributions

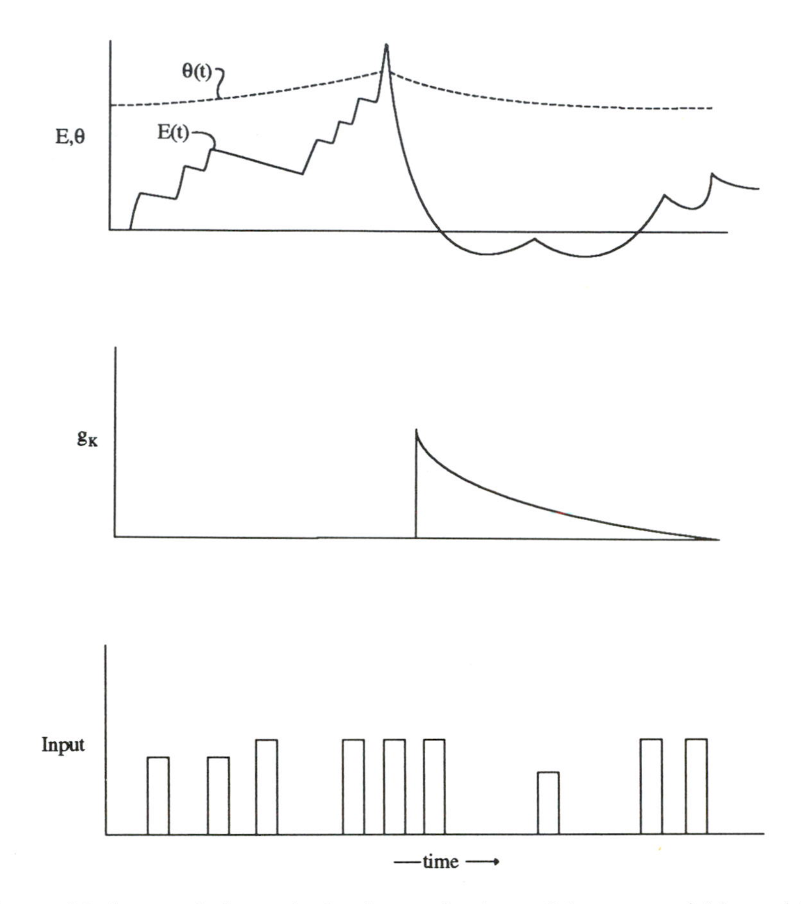

FIGURE 6.2 Accomodative and adaptive mechanisms of the state-variable model for repetitive firing.

to g_K by this equation, corresponding to what can be called an *accumulation effect*. These phenomena involving the postfiring potassium conductance have been labeled *adaptation*. They have been extensively studied experimentally, primarily in spinal motoneurons. Equation (6.3) is an empirical fit to these experimental studies. Further studies have focused on fine tuning the quantitative form of Eq. (6-3d). These modifications can be adapted at will. For most perspectives, however, Eq. (6.3d) is sufficiently accurate.

Figure 6.2 illustrates the qualitative behavior of this model. Suppose there is continuous activity in the excitatory input synapse on the soma triggering section, represented here by G_e. These tend to drive the potential, E, upward according to Eq. (6.3a). Then, according to (6.3b), the threshold, θ, is driven upward in proportion to the product of c and E. Equations (6.3a) and (6.3b), determine that both E and θ tend to decay back to their resting levels in lulls of G_e and E, respectively.

Usually c is less than 1 so that E goes up more than θ, and τ is less than η, so that E goes up faster than θ. If E thus ever becomes greater than θ, Eq. (6.3c) prescribes that S equals 1, indicating that an action potential occurs. Equation (6.3d) then immediately prescribes a jump increase in potassium conductance, g_K. This increase in g_K acts in the membrane equation (6.3a), to pull the membrane potential back down toward the equilibrium potential of potassium, E_k (typically, about 3 mV below the resting value).

Gradually, g_K reduces back towards 0. The system then returns to its initial state of responding to ongoing G_e according to Eq. (6.3a), (6.3b), and (6.3c).

In keeping with the philosophy expressed in Chapters 2 and 3, equation system (6.3), and particularly (6.3b) and (6.3d), represent succinctly the essential neurobiological mechanisms relevant to repetitive firing. The biology is kept in appropriate perspective by relatively simple and succinct expressions. Nonetheless, one may develop and elaborate more detailed or more exact descriptions of this biology in any particular application, as desired. The equation system provides a good foundation for any such extension, and shows how any further extensions can be integrated into the broader scheme in which it may play a part.

Equations (6.3), then represent an elaboration of the membrane equation in terms of a particular representation of the generation of action potentials. Where and when action potentials are not generated, Eq. (6.3b), (6.3c), and (6.3d) become irrelevant (θ is θ_0, g_K and S are 0), and the entire equation system reduces to Eq. (6.3a), which is simply the membrane equation with active synaptic or current inputs.

The next section of this chapter spells out the mathematical properties of this equation system, with particular reference to their relations to experimental studies and to the overall input–output behavior they produce.

Mathematical Behavior of the Model

The main ingredients in determining the repetitive firing characteristics in this model are its mechanisms of accommodation and adaption and their combination.

Accommodation. The basic picture of accommodation presented here is that the threshold mechanisms that bring about spike activation are affected by the transmembrane electrical potential. If the potential goes up, then the threshold goes up according to the first-order mechanism defined by Eq. (6.3b) and, thus, accommodation occurs. Moreover, if the potential goes down, the threshold goes down. If the membrane is sufficiently hyperpolarized for some time and suddenly released, the potential can come up quicker than the threshold returns and a spike may be initiated. This behavior has been sometimes called a *rebound* or *anode break*. If the threshold rises more slowly than does the potential, phasic firing patterns may result. The parameter, η, in Eq. (6.3b) is the time constant of the threshold fluctuations. A given constant value of SC such that E is always below threshold leads to $\boldsymbol{E} = \boldsymbol{SC}$, and $\theta = \mathbf{1} + \boldsymbol{c} \cdot \boldsymbol{E}$. In Hill's original model, c is taken equal to 1 so that $\theta - \boldsymbol{E}$ is always equal to 1 in the steady state. That is, in Hill's model after the cell has accommodated to a particular state, it is exactly as easy to excite as it was in the initial state. Our model thus differs from Hill's in assuming that cell patches may accommodate to various degrees which are determined by the parameter c. If c is zero, the cell does not accommodate at all. If c equals 1, the threshold rises just as much as the potential.

Figure 6.3a shows an example of elementary accommodation in this model. Our model, when subjected to a linearly rising current of slope, m, will result in strength–duration curves governed by Eq. (6.4):

$$I^* = \frac{t^*}{(1-c)t^* - \left\{(1-e^{-t^*}) + c\left[-(\eta+1) + \frac{\eta^2}{\eta-1}e^{-t^*/\eta} - \frac{1}{\eta-1}e^{-t^*}\right]\right\}} \tag{6.4}$$

This expression is obtained from Eq. (6.3a) and (6.3b) by supposing that $I = mt$ and demanding that $E(t*) = \theta(t*)$. The predicted family of strength-duration curves is shown for various values of c in Figure 6.3b.

If c is equal to 0, the monotonically decreasing function corresponding to no accommodation obtains, whereas if c is equal to 1, the minimal current gradient behavior of Hill's original model occurs. That is, if the current slope is below a certain value in this latter case, the cell will completely accommodate and no spike will be fired even as the current and the corresponding potential becomes increasingly large without bound. Moreover, in this case the slope approaches the value $1/\eta$. For example, for very large time constants of accommodation, η, one gets very small slopes for

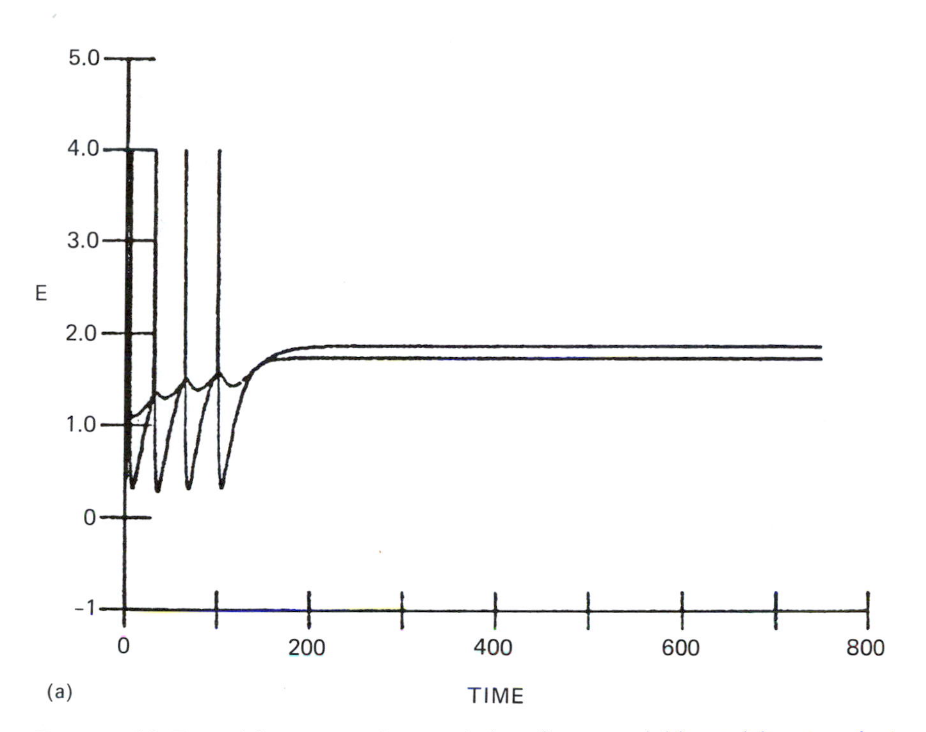

FIGURE 6.3 Essential response characteristics of state variable model: a. transient on response; b. intensity-latency dependence on model parameters; c. comparison of model to experimental intensity–latency curves (reprinted with permission from R. J. MacGregor & R. Oliver, A model for Representative Firing in Neurons, *Kybernetic* (Biological Cybernetics), Springer-Verlag, 1974, **16,** 53–64).

minimal current gradients. If c is between 0 and 1, a very striking characteristic pattern occurs, where there is a trough followed by a rise and a leveling off at some ceiling level. For c greater than 1, the curves follow the early portions of these curves but then shoot off to infinity at some finite time.

The expression given in Eq. (6.4) matches the available strength duration data reasonably well. Figure 6.3 shows a set of data obtained from cat motoneurons by Bradley and Somjen, and it is match with our Equation (6.4). Figure 6.3c shows data from a "high-ceiling" cell, which is matched by taking $c = 0.508$ and $\eta = 6.48$. Data from an only slightly accommodating "low-ceiling" cell are matched by taking $c = 0.116$ and $\xi = 21.68$. These values for c and η were obtained by using a least squares error method in matching the theoretical curve to the given data points.

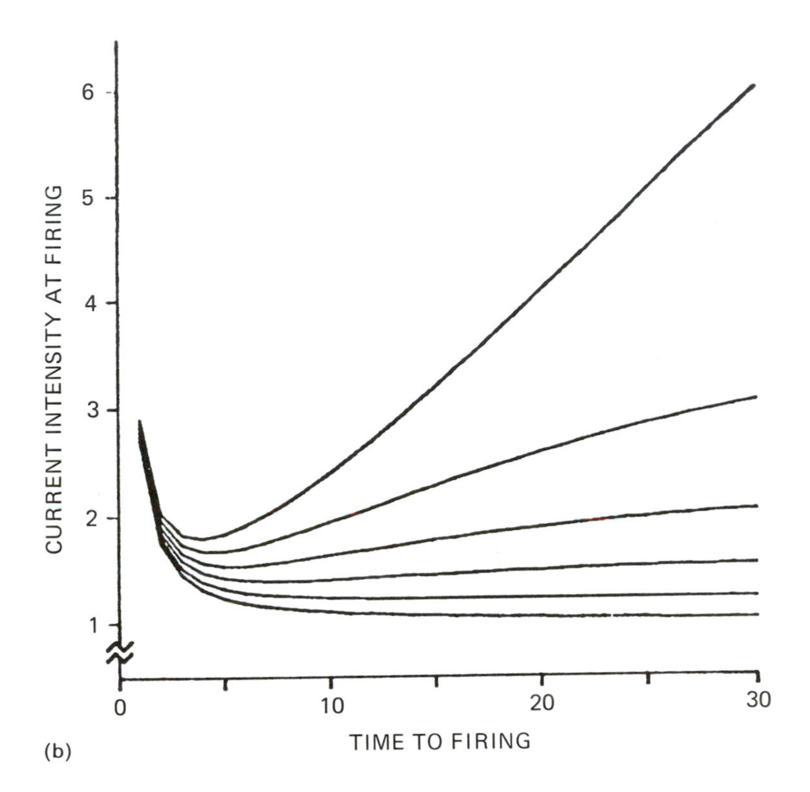

FIGURE 6.3b Reprinted with permission from R. J. MacGregor & R. Oliver, A model for Representative Firing in Neurons, *Kybernetik* (Biological Cybernetics), 1974, Springer-Verlag.

The essence of this extended Hill model is in assumption that the threshold can rise. This prediction is made more precise by "threshold-duration" curve given in Eq. (6.5), derived from this model:

$$\theta^* = \frac{t^* - 1 + e^{-t^*}}{[1-c]t^* - \left\{(1-e^{-t^*}) + c\left[-(\eta+1) + \frac{\eta^2}{\eta-1}e^{-t^*/\eta} - \frac{1}{\eta-1}e^{-t^*}\right]\right\}} \tag{6.5}$$

These results apply to the case of linearly rising ramp currents. It is convenient to define the "observed rheobase" current as the minimum amplitude of step current that will excite a pulse in the cell. One can find the expression given in Eq. (6.6) for observed rheobase and for the latency to

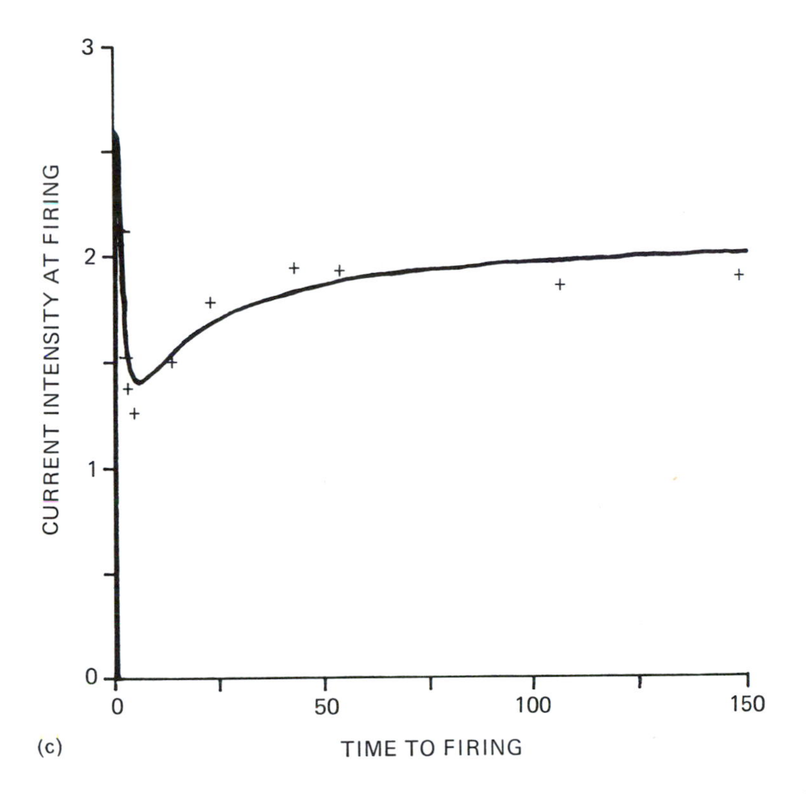

FIGURE 6.3c Reprinted with permission from R. J. MacGregor & R. Oliver, A Model for Representative Firing in Neurons, *Kybernetik* (Biological Cybernetics), 1974, Springer-Verlag.

firing at observed rheobase from Eq. (6.3).

$$I^* = \frac{1}{(1-c) + c\left[\dfrac{1}{\dfrac{\eta - 1}{c}} + 1\right]^{1/\eta - 1}}$$

$$t^* = \frac{\ln\left\{\dfrac{\eta - 1}{c} + 1\right\}}{1 - 1/\eta}. \tag{6.6}$$

These expressions are obtained by solving Eq. (6.3a) and (6.3b) for a step current and demanding that, at the rheobase, both $E(t*) = \theta(t^*)$, and $dE(t^*)/dt = d\theta(t^*)/dt$. One finds that the strength-duration curve for step

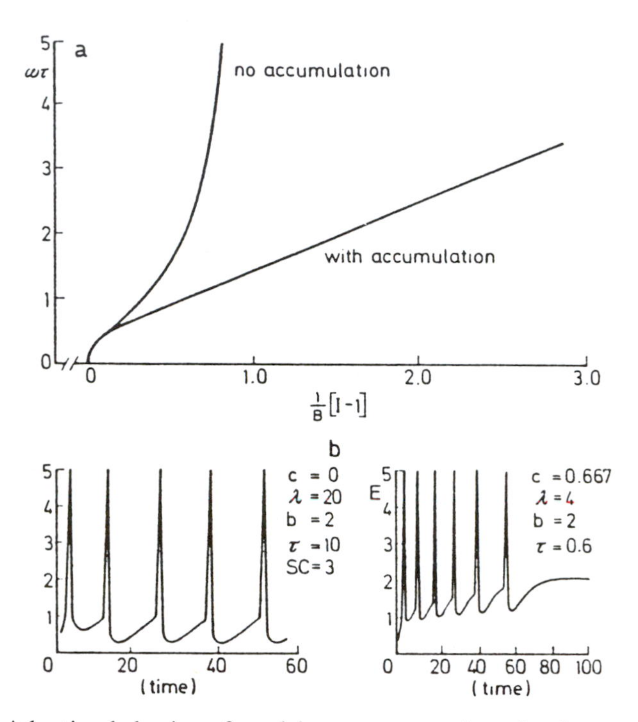

FIGURE 6.4 Adaptive behavior of model: a. representative after-hyperpolarization curves; b. steady firing rate with and without accumulation of after-hyperpolarization conductance (reprinted with permission from R. J. MacGregor & R. Oliver, A model for Representative Firing in Neurons, *Kybernetic* (Biological Cybernetics), 1974, Springer-Verlag, **16,** 53–64).

currents is given by Eq. (6.7):

$$I^* = \begin{cases} \dfrac{1}{(1-c) - \left[1 + \dfrac{c}{\eta - 1}\right] e^{-t^*} + \dfrac{c\eta}{\eta - 1} e^{-t^*/\eta}}, & t^* < \dfrac{\ln\left[\dfrac{\eta - 1}{c} + 1\right]}{1 - 1/\eta} \\[2em] \dfrac{1}{(1-c) + c\left[\dfrac{1}{\dfrac{\eta - 1}{c} + 1}\right]^{1/\eta - 1}}, & t^* > \dfrac{\ln\left[\dfrac{\eta - 1}{c} + 1\right]}{1 - 1/\eta} \end{cases} \tag{6.7}$$

Adaptation. The essence of this model for adaption is extremely simple. It is assumed that refractoriness, after hyperpolarization, and adaption correspond to a sudden elevation of potassium conductance during the emission of a spike potential, followed by an exponential decay of the potassium conductance to its resting level, and that such potassium conductance deviations are cumulative. The model as it stands has two parameters: b, which measures the amount of potassium conductance increase, and τ_K, which measures the time course of it subsequent exponential decay. The accumulation of $\boldsymbol{g_K}$ with repetitive firing and its influence on the membrane potential is illustrated in Figure 6.4a.

If we consider the case where some input pattern has generated a spike at time, 0, then been suddenly shut off, and if in this case we suppose that the cell membrane is very fast relative to the time constant of the conductance change (that is, $\tau \ll 1$), then for this case we will have the after-hyperpolarization given approximately by Eq. (6.8):

$$V(t) = \frac{b\,e^{-t/\tau}}{1 + be^{-t/\tau}} \cdot V_k \tag{6.8}$$

Indeed, one might think of the system in this case as always striving toward the potential given in Eq. (6.8) and going toward that value from any given position with time constant equal to $1/(1 + g_K(t))$.

This model for adaptation implies a particular curve that can be easily obtained under laboratory conditions. This is a curve of steady-state firing rate versus the amplitude of applied step current. Therefore, where we neglect accumulation of the potassium conductance changes corresponding to different spikes, we obtain the expression given in Eq. (6.9):

$$e^{-1/w\tau} = \frac{1/\theta - 1}{b\left[1 + \frac{(-V_k)}{\theta}\right]} \tag{6.9}$$

This expression with some readjusting is equivalent to one presented by Kernell. If we include accumulation of the conductance changes from different spikes, it is easy to show that the predicted transfer relation is given by Eq. (6.10):

$$\frac{e^{-1/w\tau}}{1 - e^{-1/w\tau}} = \frac{1/\theta - 1}{b\left[1 + \left(\frac{-V_k}{\theta}\right)\right]} \tag{6.10}$$

These relations are shown in Figure 6.4a. For small firing frequencies accumulation of g_K is not important and the two curves coincide as one

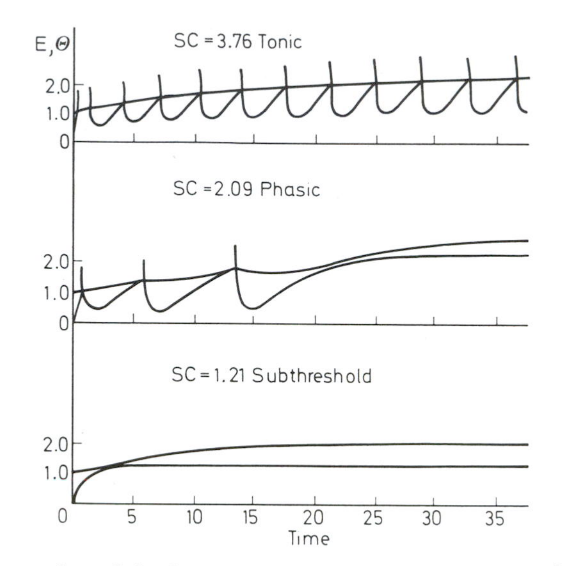

FIGURE 6.5 Examples of basic response type to step current (reprinted with permission from R. J. MacGregor & R. Oliver, A model for Representative Firing in Neurons, *Kybernetic* (Biological Cybernetics), 1974, Springer-Verlag, **16,** 53–64).

would expect. On the other hand, for higher firing rates the actual firing rate approaches a linear relationship with slope equal to $I/(b\tau)$. That is, at high firing rates, there is decreased gain corresponding to accumulation of after-hyperpolarization conductance change, as one would expect. This factor has the decided effect of increasing the dynamic range over which the cell can respond.

Responses to Step Currents. This model exhibits three modes of response to a step current depending on current intensity. For SC less than some critical lower limit, $\boldsymbol{SC_l}$ there is no activation; for SC greater than $\boldsymbol{SC_l}$ but below another critical limit, $\boldsymbol{SC_u}$ there is phasic activity only; for SC greater than $\boldsymbol{SC_u}$ there is a more or less pronounced burst at the onset of stimulation followed by steady sustained firing at a lower rate. Examples of these basic patterns are shown in Figure 6.5.

The values of SC_l and SC_u and their differences depend on system parameters c, η, b, and t. In particular, SC_l is exactly the "observed rheobase" given in Eq. (6.6), whereas SC_u is somewhat less than $1/(1-c)$.

Figure 6.6a illustrates how the no response, phasic, tonic, and (when applicable) inactivated regions occupy the parameter space of the model.

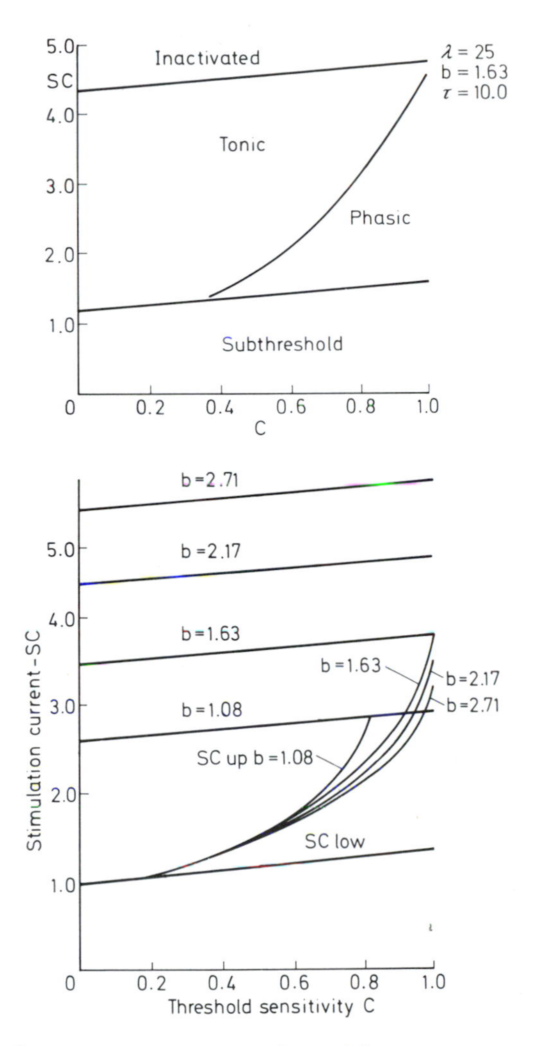

FIGURE 6.6 Map of response to step current in model parameter space: a. simulation results; b. qualitative map (reprinted with permission from R. J. MacGregor & R. Oliver, A model for Representative Firing in Neurons, *Kybernetic* (Biological Cybernetics), 1974, Springer-Verlag, **16,** 53–64).

Figure 6.6b shows more precise demarcation of these ranges for various parameter values obtained from the electronic realization of the model. The number of spikes in a burst increases with current intensity within the phasic range. This is illustrated in Figure 6.7.

Generally, the duration of individual bursts also increases with current intensity in this range. However, within a current range producing a given

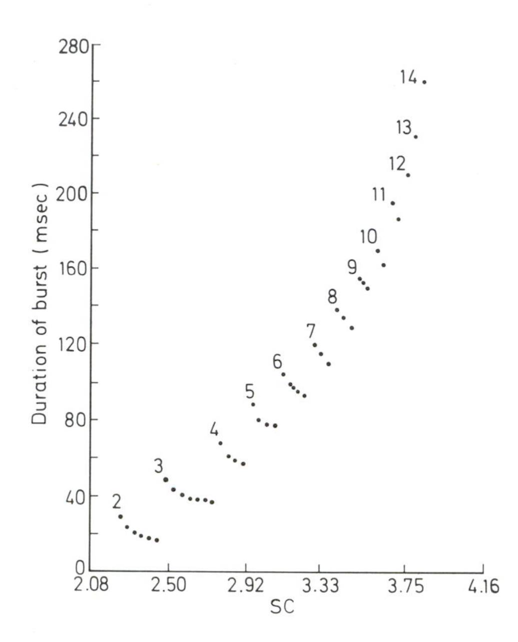

FIGURE 6.7 Duration of burst and number of spikes for phasic responses to step current (reprinted with permission from R. J. MacGregor & R. Oliver, A model for Representative Firing in Neurons, *Kybernetic* (Biological Cybernetics), 1974, Springer-Verlag, **16,** 53–64).

integral number of spikes per burst, the duration of the burst decreases with increasing current intensity. Moreover, the given range of intensity over which a given integral number of spikes per burst is obtained decreases as the number of spikes in the burst increases. These points are all indicated in Figure 6.

Calcium-Mediated Action Potentials in Dendrites

Action potentials mediated by calcium and potassium, much in the same way that output action potentials are mediated by sodium and potassium, are found in the dendritic trees of large pyramidal cells of the hippocampus and cortex. These action potentials are broad (on the order of 10 msec) and exhibit pronounced, long-lasting (30 msec) after hyperpolarizations. When they propagate to the soma, they generate short bursts of three to five typical output spikes over a five to ten msec interval. It seems that these dendritic action potentials signal integrated information over local dendritic regions, in much the same way that normal soma-generated action potentials signal integrated information over entire neurons.

Equations (6.11) define a simple and direct model for generating these potentials with calcium and potassium conductance functions:

$$\frac{dES}{dt} = \frac{-ES + \{SCS + GDS * (ED - ES) + GKS * (EK - ES)}{TS}$$

$$\frac{dTHS}{dt} = \frac{-(THS - TH0) + C * ES}{TTH}$$

$$S = \begin{cases} 0 & \text{if } ES < THS \\ 1 & \text{if } ES \geqslant THS \end{cases}$$

$$\frac{dGKS}{dt} = \frac{-GKS + B * S}{\tau GK}, \qquad PS = ES + S * (50 - ES)$$

$$\frac{dED}{dt} = \{-ED + SCD + GSD * (PS - ED) + GCA * (ECA - ED) + GKD * (EK - ED)\}/TD \tag{6.11}$$

$$\frac{dGCA}{dt} = \begin{cases} \dfrac{-GCA}{\tau GCA} & \text{if } ED < THD \\ \dfrac{-GCA + D * (ED - THD)}{\tau GCA} & \text{if } ED \geqslant THD \end{cases}$$

$$\frac{dCA}{dt} = \frac{-CA + A * GCA}{\tau CA}$$

$$\frac{dGKD}{dt} = \begin{cases} \dfrac{-GKD}{\tau GKD} & \text{if } CA < CA0 \\ \dfrac{-GKD + BD}{\tau GKD} & \text{if } CA \geqslant CA0 \end{cases}$$

Parameters of this model are defined in a glossary in the appendix. In this model, membrane conductance to calcium (GCA) is triggered at a local "hot spot" if the membrane potential there exceeds a given threshold (THD). This in turn causes a influx of calcium (CA). The calcium influx causes the membrane potential to rise, constituting the rise phase of the dendritic action potential. If the internal calcium concentration exceeds a certain level (CA_0) the potassium conductance (GKD) is triggered, bringing on the down phase of the dendritic action potential. Traub has presented a more elaborate model for this process in the style of the Hodgkin–Huxley formulation.

Concluding Remarks

The next chapter will show how the representations of the temporal factors underlying activation discussed in this chapter and the neuronal geometries discussed in the previous chapter may be represented in computer simulations, which are held here to be indispensable for higher order development of the mechanics of neuroelectric signalling.

Chapter 7 Computational Representation of Neuroelectric Signalling

The purpose of this chapter is to show how to construct neurobiologically realistic computational representations of the central governing mechanisms of neuroelectric signalling in neurons and in large-scale, highly interconnected local and composite neural networks, with particular reference to the hierarchical organizational principles of neuroelectric signalling discussed in the previous chapters.

The Value of Computational Representation in Neural Networks

Neurobiologically founded computational representations of the mechanics of neuroelectric signalling as presented in this book are essential in neural network studies for at least three central purposes.

First, computational representation is necessary for the ongoing discovery and development of general principles of operation and understanding within the general theoretical mechanics of neural networks and neuroelectric signalling. This is illustrated clearly by the computational applications accompanying the development of broad theoretical views in each of the next three parts of this book on neuronal firing levels, and dynamic firing patterns in local and composite neural networks.

Second, computational representation is necessary to help guide the interpretation of data from particular experimental studies. There is a central need in current neural and brain sciences for the development of an understanding of the network context within which recorded neuroelectric signals appear. The need here is for a means to make effective the process wherein "theory guides, and experiment decides." Computational representation is a necessary ingredient to link the broader theories of neural network operation to the partial experimental data from some of their component parts. This need for computational representation here is directly reflective of the sheer complexity of these systems. It is discussed more broadly in Chapter 18 on the nature of composite neural networks.

Third, realistic computational representation of the mechanics of neuroelectric signalling can be used to link the broad fields of computational neural networks and neurobiology. The foundations for this link can be accomplished rather easily and directly simply by constructing the computational neural network programs on basis of the central governing neurobiological equations and procedures outlined here.

The Approach to a Neurobiologically Founded Computational Representation of Neural Networks

The outline presented here is based directly on the discussions of the previous chapters of this book.

1. Identify the membrane equation (4.1f) as the central governing equation for neuroelectric signals in all integrative neuronal compartments.
2. Represent entire neurons either by (a) multicompartmental models whose dynamics and spatial interrelations are defined by the coupled equation system shown in Eq. (5.4), or (b) a single patch for the neuronal soma, represented mathematically by Eq. (4.1f), wherein dendritic interactions are not treated explicitly and dendritic input is represented by the *SC* term for the soma compartment.
3. Represent input synaptic activations by modulations of the active membrane conductance terms, g_i, in the receiving compartments of the targeted neurons; represent the fundamental nature of the synapse

by its equilibrium potential (excitatory or inhibitory); represent the individual chemistry and properties of the synapse in the conductance changes to the extent desired (amplitude, decay rate, possible short-term plasticity).

4. If necessary, represent the production of calcium-mediated action potentials in dendritic regions by Eq. (6.11) or by more elaborate equation systems of the type of Eq. (6.2).
5. Represent the production of output action potentials by either (a) the state-variable model described by Eq. (6.3), or (b) more elaborate descriptions of their underlying conductances, as for example, in equations of the type of Eq. (6.2).
6. Link together cells into networks by linking output spike trains in constituent neurons to the occurrences of postsynaptic conductance modulations in target receiving cells, according to the connectivity prescribed for the network. The connectivities may incorporate spatial relationships simply by relating cell identity numbers to spatial locations.
7. Link networks into composite networks or systems by a similar link of output spike trains to input synaptic conductances.
8. Identify clearly the arrangement of all significant relevant neurobiological factors within your computational scheme. Be sure these are clearly represented by a minimum number of parameters and that each has a clear and direct physical interpretation.
9. Add or suppress various neurobiological detail as desired or significant for your case and add, as necessary, any various ad hoc processes of effects relevant to your case or interests (interactions by field currents, processes of learning by synaptic plasticity, global influences of drugs or chemical regulators, and so on). Again, be sure these are clearly represented by a minimum number of parameters and that each has a clear and direct physical interpretation.
10. Collect the essential systemic and input parameters clearly at the head of your program.

A Useful Integration Technique for Neuronal Governing Equations

One may use any of a number of standard techniques for numerically integrating the neuroelectric governing equations. However, since all of them are of first-order ordinary differential equations, they lend themselves to a particular method of integration that is especially well-adapted to the needs of modeling neuroelectric signalling. On the one hand, the approach can be made as accurate as desired by simply decreasing the integration step size; on the other hand, rougher but acceptably representative approximations can be obtained with high computer efficiency by increasing integration step sizes up to as much as, say, a millisecond. The approach gives reasonable accuracy with larger step sizes because the response patterns are typically sluggish exponential curves characteristic of first-order systems, wherein the typical durations of the active processes are on the order of 0.1 to 0.2 of the membrane time constant.

The Target Value

This integration approach is based on a simple physical understanding of the behavior of the first-order governing equations. Consider the physical drive to which the variable E is subjected in the generic first-order neuronal governing equation shown in Eq. (7.1):

$$\frac{\mathrm{d}E}{\mathrm{d}t} = -A(t)E + B(t) \tag{7.1}$$

The instantaneous trajectory of E at any given instant in time, t, is determined by three factors: its current value, its target value, and how it is going to get to the target value. The target value can be recognized as the steady-state value E would attain if the instantaneous values of $A(t^*)$ and $B(t^*)$ were held constant over a sufficiently long time interval. This target value can be obtained from Eq. (7.1) by setting the time derivative of E equal to zero and solving the remaining equation for E. This target value can thus be readily seen to be equal to $B(t^*)/A(t^*)$.

Because of the first-order governing equation, the approach to the target value, at any instant, t^*, is exponential with a time constant of $1/A(t^*)$.

One can therefore say that the instantaneous trajectory of E at t^* is given by Eq. (7.2):

$$E(t) = E(t^*)\exp\left[-A(t)(t-t^*)\right] + \frac{B(t^*)}{A(t^*)}\{1 - \exp\left[-A(t^*)(t-t^*)\right]\} \tag{7.2}$$

This is simply a conceptual generalization of the elementary solution of the linear first-order equation with constant coefficients.

Numerical Integration

We can use this physical interpretation of the instantaneous trajectory of E

to produce accurate and computer efficient simulation of the general case with time-varying parameters as follows. First, represent the general continuously time-varying situation with a stepwise constant approximation, wherein all variations in $A(t)$ and $B(t)$ are taken as constant over time intervals of some size, Δ. Then Eq. (7.2) can be used to form the iterative equation, Eq. (7.3), for the value of E_{i+1}, given E_i and the values of the driving functions, A_i and B_i, which are constant over that time interval.

$$E_{i+\Delta} = E_i \exp[-A_i\Delta] + \frac{B_i}{A_i}\{1 - \exp[-A_i\Delta]\} \tag{7.3}$$

Similar expressions can be written for any and all of the first-order governing equations. These techniques and many example programs are given in NBM.

Many useful simulations can be performed with high computer efficiency by using large step sizes (about 1 msec) if all relevant time constants are, say, 5 msec or larger. These simulations can produce rough but acceptable accuracy for a wide range of applications to neuroelectric signalling. They produce the correct trends of influence of the various constituent parameters and therefore allow meaningful comparisons across cell types and interpretations of the influence of the various parameters. Such results can speak directly to the principles of engineering design and the engineering organization of the systems being studied. Further, one may test significant conceptual conclusions or try to match closely particular experimental data by taking the step size smaller for a small number of runs. Moreover, this approximation can be made as accurate desired simply by taking the step size, Δ, increasingly small.

Representative Computational Programs for Neuroelectric Signalling

The programs presented here are representative of the general neurobiologically oriented approach to computational neural networks presented in this chapter. Clearly, many other types of computational representation of the ideas advocated here are possible.

The four program listings given here represent (a) a point neuron generating output action potentials according to the state-variable model when driven by applied currents (PTNRN10), (b) a recurrently connected local neural network with excitatory and inhibitory pools of neurons (NTWRK11), (c) an arbitrary system of interconnected neural networks (SYSTM11), and (d) a neuron with a arbitrarily constructed dendritic tree (DENDR21).

These are useful programs. They, and many variations on them, are presented and discussed in NBM. The important variables and parameters in these programs are defined in a glossary in an appendix of that book.

The last program included here (SYSTM5G) is a recent extension of SYSTM11 that allows one to define three levels of complexity for the synaptic junctional matrices between the participating populations. The most elaborate level includes the embedding of multiple sequential configurations, which are discussed in Chapter 14 of this book. The variables and parameters of this program can be easily interpreted in terms of those of SYSTM11 and the discussion of Chapter 14.

```
/JOB
A224.
FAKE.
PDFV.
MNF(Y).
LGO.
/EOR
      PROGRAM PTNRN10 (INPUT,OUTPUT,TAPE6=OUTPUT)
*
****     THIS PROGRAM SIMULATES THE RESPONSES OF A THREE-STATE-
*        VARIABLE MODEL OF A SINGLE POINT NEURON TO APPLIED STEP
*        AND RAMP INPUT CURRENTS.
****
*     INTEGER SCSTRT,SCSTP
      PARAMETER STEP=1.,EK=-10.
      DIMENSION P(500)
*
****     THIS SECTION READS AND WRITES THE INPUT PARAMETERS
*
      READ 5010, C,TTH,B,TGK,TH0,TMEM
      READ 5020, SC,SL;SCSTRT,SCSTP
      READ 5030, LTSTOP
*
      WRITE(6,6000)
*     WRITE(6,6010) C,TTH,B,TGK,TH0,TMEM
      WRITE(6,6020) SC,SL,SCSTRT,SCSTP
*     WRITE(6,6030) LTSTOP
*
****     THIS SECTION INITIALIZES VARIABLES
*
      E=0. $TH=TH0 $S=0. $GK=0.
      DCTH=EXP(-STEP/TTH) $DGK=EXP(-STEP/TGK)
*
**** THIS SECTION UPDATES STATE VARIABLES AT EACH TIME STEP
```

```
*
      DO 1000 L=1,LTSTOP
*
      SCN=0. $IF(L.GE.SCSTRT.AND.L.LT.SCSTP) SCN=SC+SL*FLOAT(L-SCSTRT)
*
      GK=GK*DGK+B*S*(1.-DGK) $GTOT=1.+GK $DCE=EXP(-GTOT*STEP/
        TMEM)
      E=E*DCE+(SCN+GK*EK)*(1.-DCE)/GTOT
      TH=TH0+(TH-TH0)*DCTH+C*E*(1.-DCTH)
      S=0. $IF(E.GE.TH) S=1. $P(L)=E+S*(50.-E)
1000  CONTINUE
*
****    THIS SECTION WRITES OUT ACTIVITY VARIABLES
*
      WRITE(6,6050) (P(L),L=1,LTSTOP)
*
****    THESE ARE THE FORMATS
*
 5010 FORMAT(12F6.2)
 5020 FORMAT(2F6.2,2I6)
 5030 FORMAT(I6)
 6000 FORMAT(1H1)
 6010 FORMAT(5X,*C,TTH,B,TGK,TH0,TMEM=*,6F7.2)
 6020 FORMAT(5X,*SC,SL,SCSTRT,SCSTP=*,2F7.2,2I7)
 6030 FORMAT(5X,*LSTSTOP=*,I7)
 6050 FORMAT(/,(2X,20F6.2))
      STOP
      END
/EOR
     .5    30.    5.    3.    10.    5.
  20.    0.    1    100
     100
/EOR
/EOI


/JOB
A224, CM=60000, TL=700.
FAKE.
PDFV.
MNF (Y)
LGO, SST081.
SAVE, SST081.
EXIT.
SAVE, SST081.
\EOR
      PROGRAM    NTWRK11    (NTOUT,INPUT,OUTPUT,TAPE7=NTOUT,TAPE6
```

```
      =OUTPUT)
*
****    THIS PROGRAM SIMULATES THE ACTIVITY OF A RECURRENTLY-
        CONNECTED
****    POOL OF EXCITATORY AND INHIBITORY NEURONS.
*
      PARAMETER NCLS=600,NEX=300,NIN=300,MCTP1=6,STEP=1.
      DIMENSION NEF(140),NIF(140),T(140),TTH(2),TGK(2)
      COMMON SCIT(NCLS,MCTP1),GI(NCLS,MCTP1)
      COMMON/SV/ E(NCLS),TH(NCLS),S(NCLS),GK(NCLS),B(2),C(2),
      *              DCG(2),DCTH(2),THO(2),TMEM(2)
      COMMON/TR/ NT(2,2),NCT(2,2),STR(2,2),NCL(2),INSED(2,2)
*
****    THIS SECTION READS AND WRITES THE INPUT PARAMETERS
*
      READ 2010, C(1),TTH(1),B(1),TGK(1),THO(1),TMEM(1)
      READ 2010, C(2),TTH(2),B(2),TGK(2),THO(2),TMEM(2)
      READ 2020, NT(1,1),NT(1,2),NT(2,1)
      READ 2020, NCT(1,1),NCT(1,2),NCT(2,1)
      READ 2010, STR(1,1),STR(1,2),STR(2,1)
      READ 2020, INSED(1,1),INSED(1,2),INSED(2,1),INSED(2,2)
      READ 2030, P,STRIP,LIPSTP
      READ 2020, LTSTOP
*
      WRITE(6,6000)
      WRITE(6,6010) C(1),TTH(1),B(1),TGK(1),THO(1),TMEM(1)
      WRITE(6,6015) C(2),TTH(2),B(2),TGK(2),THO(2),TMEM(2)
      WRITE(6,6020) NT(1,1),NT(1,2),NT(2,1)
      WRITE(6,6022) NCT(1,1),NCT(1,2),NCT(2,1)
      WRITE(6,6024) STR(1,1),STR(1,2),STR(2,1)
      WRITE(6,6030) P,STRIP,LIPSTP
      WRITE(6,6035) INSED(1,1),INSED(1,2),INSED(2,1),INSED(2,2)
      WRITE(6,6040) LTSTOP
*
****    THIS SECTION INITIALIZES VARIABLES
*
      NCL(1)=NEX $NCL(2)=NIN
      DO 102 I=1,2
      DCTH(I)=EXP(-STEP/TTH(I)) $DCG(I)=EXP(-STEP/TGK(I))
102   CONTINUE
      DO 110 I=1,NCLS
      E(I)=0. $S(I)=0. $GK(I)=0.
      IF(I.LE.NEX) THEN $TH(I)=THO(1)
      ELSE $TH(I)=THO(2) $ENDIF
      DO 115 J=1,MCTP1
      SCIT(I,J)=0. $GI(I,J)=0.
115   CONTINUE
```

```
110   CONTINUE
*
****    THIS SECTION UPDATES STATE VARIABLES AT EACH TIME STEP
*
      DO 1000 L=1,LTSTOP
      NEF(L)=0 $NIF(L)=0
*
****                 UPDATING THE EXCITATORY CELLS
*
      CALL RANSET(INSED(2,2))
      DO 1500 I=1,NEX
      IF(L.LE.LIPSTP.AND.RANF(0.).LE.P) SCIT(I,1)=SCIT(I,1)+STRIP
      CALL SVUPDT(1,I)
      IF(E(I).LT.TH(I)) GO TO 1490
      S(I)=1. $NEF(L)=NEF(L)+1 $CALL RANGET(INSED(2,2))
      CALL TRNSMT(1,1,I)
      CALL TRNSMT(1,2,I) $CALL RANSET(INSED(2,2))
1490  CONTINUE
1500  CONTINUE $CALL RANGET(INSED(2,2))
*
****                 UPDATING THE INHIBITORY CELLS
*
      DO 1700 I=NEX+1,NCLS
      CALL SVUPDT(2,I)
      S(I)=0. $IF(E(I).LT.TH(I)) GO TO 1690
      S(I)=1. $NIF(L)=NIF(L)+1
      CALL TRNSMT(2,1,I-NEX)
1690  CONTINUE
1700  CONTINUE
*
      DO 1910 I=1,NCLS $DO 1920 J=1,MCTP1-1
      SCIT(I,J)=SCIT(I,J+1) $GI(I,J)=GI(I,J+1)
1920  CONTINUE $SCIT(I,MCTP1)=0. $GI(I,MCTP1)=0.
1910  CONTINUE
      T(L)=L
*
1000  CONTINUE
*
****                 THIS SECTION WRITES OUT ACTIVITY VARIABLES
*
      WRITE(6,6105) $WRITE(6,6110) (NEF(L),L=1,LTSTOP)
      WRITE(6,6115) $WRITE(6,6110) (NIF(L),L=1,LTSTOP)
      WRITE(7,7110) (NEF(L),L=1,LTSTOP)
      WRITE(7,7105) $WRITE(7,7110) (NIF(L),L=1,LTSTOP)
*
      STOP
****                 THESE ARE THE FORMATS
```

```
*
2010  FORMAT(12F6.2)
2020  FORMAT(12I6)
2030  FORMAT(2F6.2,I6)
6000  FORMAT(1H1)
6010  FORMAT(5X,*EXC:C,TTH,B,TGK,THO,TMEM = *,6F7.2)
6015  FORMAT(5X,*INH:C,TTH,B,TGK,THO,TMEM = *,6F7.2)
6020  FORMAT(5X,*NT11,NT12,NT21 = *,3I6)
6022  FORMAT(5X,*NCT11,NCT12,NCT21 = *,3I6)
6024  FORMAT(5X,*STR11,STR12,STR21 = *,3F7.2)
6030  FORMAT(5X,*P,STRIP,LIPSTP = *,2F7.2,I7)
6035  FORMAT(5X,*INSEDS = *,4I7)
6040  FORMAT(5X,*LTSTOP = *,I7)
6105  FORMAT(/,10X,*HERE ARE THE OUTPUTS:*)
6110  FORMAT(5X,20I6)
6115  FORMAT(/)
7105  FORMAT(/)
7110  FORMAT(20I4)
      END
*
      SUBROUTINE SVUPDT(IS,K)
*
****  THIS SUBROUTINE UPDATES THE STATE VARIABLES OF AN INDIVIDUAL
        NEURON
*
      PARAMETER NCLS = 600,NEX = 300,MCTP1 = 6,STEP = 1.,EI = -10.,EK = -10.
      COMMON SCIT(NCLS,MCTP1),GI(NCLS,MCTP1)
      COMMON/SV/ E(NCLS),TH(NCLS),S(NCLS),GK(NCLS),B(2),C(2),
      *                DCG(2),DCTH(2),THO(2),TMEM(2)
*
      GK(K) = GK(K)*DCG(IS) + B(IS)*(1.-DCG(IS))*S(K)$GTOT = 1. + GK(K) + GI(K,1)
      DCE = EXP(-GTOT*STEP/TMEM(IS))
      E(K) = E(K)*DCE + (SCIT(K,1) + GI(K,1)*EI + GK(K)*EK)*(1.-DCE)/GTOT
      TH(K) = THO(IS) + (TH(K)-THO(IS))*DCTH(IS) + C(IS)*E(K)*(1.-DCTH(IS))
      S(K) = 0.
      RETURN
      END
*
      SUBROUTINE TRNSMT(IS,IR,K)
****     THIS SUBROUTINE PROJECTS SYNAPTIC ACTIVATION FROM A SEND-
         ING
****     FIBER TO ITS TARGETED RECEIVING CELLS.
*
      PARAMETER NCLS = 600,NEX = 300,MCTP1 = 6
      COMMON SCIT(NCLS,MCTP1),GI(NCLS,MCTP1)
      COMMON/TR/NT(2,2),NCT(2,2),STR(2,2),NCL(2),INSED(2,2)
*
```

```
      CALL RANSET(K*INSED(IS,IR))
      DO 100 J=1,NT(IS,IR)
      NREC=INT(RANF(0.)*FLOAT(NCL(IR)))+1
      IF(IR.EQ.2) NREC=NREC+NEX
      KCT=INT(RANF(0.)*FLOAT(NCT(IS,IR)))+2
      IF(IS.EQ.1) SCIT(NREC,KCT)=SCIT(NREC,KCT)+STR(IS,IR)
      IF(IS.EQ.2) GI(NREC,KCT)=GI(NREC,KCT)+STR(IS,IR)
100   CONTINUE
      RETURN
      END
*
/EOR
    0.    25.    10.    2.    10.    5.
    0.    25.    10.    2.    10.    5.
      12      26     12
       4       2      2
    2.     1.       .35
    5629   7143   3517   2243
     .06  10.            4
     100
/EOR


      PROGRAM SYSTM11 (SYSTM,INPUT,OUTPUT,TAPE6=OUTPUT,
      TAPE7=SYSTM)
*
***   THIS PROGRAM SIMULATES THE ACTIVITY OF AN ARBITRARY
***   NUMBER OF NEURON POOLS ACTIVATED BY AN ARBITRARY NUMBER
***   OF INPUT FIBER SYSTEMS.
*
      INTEGER TYPE,SNTP
      PARAMETER NTPOPS=5,NFPOPS=2,NCPOPS=3,NCLS=50,MCTP1=2,
      *             SNTP=2,STEP=1.,NTGMX=2
      DIMENSION P(NFPOPS),INSTR(NFPOPS),INSTP(NFPOPS),INFSED(NFPOPS),
      *             IDF(NCLS),TTH(NCPOPS),TGK(NCPOPS),T(SNTP),DCS(SNTP),
      *             NF(NTPOPS),NTGR(NTPOPS)
      COMMON G(SNTP,NCPOPS,NCLS,MCTP1)
      COMMON/SV/E(NCPOPS,NCLS),TH(NCPOPS,NCLS),S(NCPOPS,NCLS),
      *             GK(NCPOPS,NCLS),B(NCPOPS),C(NCPOPS),DCG(NCPOPS),
      *             DCTH(NCPOPS),TMEM(NCPOPS),THO(NCPOPS),EQ(SNTP)
      COMMON/TR/NCT(NTPOPS,NTGMX),NT(NTPOPS,NTGMX),STR(NTPOPS,
      *             NTGMX),INSED(NTPOPS,NTGMX),N(NTPOPS),TYPE(NTPOPS,
      *             NTGMX),IRCP(NTPOPS,NTGMX)
*
****          THIS SECTION READS AND WRITES THE INPUT PARAMETERS
*
      READ 5110, IFLG1,IFLG2
```

```
      READ 5120, ((EQ(I),T(I)),I=1,SNTP)
      DO 110 I=1,NTPOPS $IF(I.LE.NCPOPS) THEN
      READ 5130, N(I),C(I),TTH(I),B(I),TGK(I),TMEM(I),THO(I),
      *                 NTGR(I) $ELSE $II=I-NCPOPS
      READ 5140, N(I),P(II),INSTR(II),INSTP(II),INFSED(II),
      *                 NTGR(I) $ENDIF
      DO 120 J=1,NTGR(I)
120   READ 5150, IRCP(I,J),TYPE(I,J),NCT(I,J),NT(I,J),
      *                 STR(I,J),INSED(I,J)
110   CONTINUE
      READ 5110, LTSTOP
*
      WRITE(6,6000) $WRITE(6,6005) NTPOPS,NFPOPS,NCPOPS,NCLS,MCTP1,
      * NTGMX,IFLG1,IFLG2
      WRITE(6,6110) ((EQ(I),T(I)),I=1,SNTP)
      WRITE(6,6112) $WRITE(6,6116) $WRITE(6,6118)
      DO 210 I=1,NTPOPS $IF(I.LE.NCPOPS) THEN
      WRITE(6,6130) N(I),C(I),TTH(I),B(I),TGK(I),TMEM(I),THO(I),
      *                 NTGR(I) $ELSE $II=I-NCPOPS
      WRITE(6,6140) N(I),P(II),INSTR(II),INSTP(II),INFSED(II),
      *                 NTGR(I) $ENDIF
      DO 220 J=1,NTGR(I)
220   WRITE(6,6150) IRCP(I,J),TYPE(I,J),NCT(I,J),NT(I,J),
      *                 STR(I,J),INSED(I,J)
210   CONTINUE
      WRITE(6,6160) LTSTOP
*
****  THIS SECTION INITIALIZES VARIABLES
*
      DO 310 I=1,NCPOPS
      DCTH(I)=EXP(-STEP/TTH(I)) $DCG(I)=EXP(-STEP/TGK(I))
      DO 310 J=1,NCLS
      E(I,J)=0. $TH(I,J)=THO(I) $S(I,J)=0. $GK(I,J)=0.
      DO 310 K=1,MCTP1 $DO 310 L=1,SNTP
      G(L,I,J,K)=0.
310   CONTINUE
      DO 320 I=1,SNTP
320   DCS(I)=EXP(-STEP/T(I))
*
****  THIS SECTION UPDATES STATE VARIABLES AT EACH TIME STEP
*
      DO 1000 L=1,LTSTOP
*
*
****                 UPDATING THE CELLS
*
      DO 2000 I=1,NCPOPS
```

```
      NF(I)=0 $IF(IFLG1.GT.0) WRITE(7,7060) L,I
      DO 2900 J=1,N(I)
      CALL SVUPDT(I,J)
      IF(E(I,J).LT.TH(I,J)) GO TO 2850
      S(I,J)=1. $NF(I)=NF(I)+1 $IDF(NF(I))=J
      DO 2800 K=1,NTGR(I)
      IF(NT(I,K).GT.0) CALL TRNSMT(I,K,J)
2800  CONTINUE
2850  CONTINUE
2900  CONTINUE $IF(IFLG1.GT.0) WRITE(7,7060) (IDF(J),J=1,NF(I))
2000  CONTINUE
*
****                 UPDATING THE INPUT FIBERS
*
      DO 3000 I=1,NFPOPS
      II=I+NCPOPS $NF(II)=0 $IF(IFLG1.GT.0) WRITE(7,7060) L,II
      IF(L.LT.INSTR(I).OR.L.GE.INSTP(I)) GO TO 3000
      CALL RANSET(INFSED(I))
      DO 3900 J=1,N(II)
      IF(RANF(0.).GT.P(I)) GO TO 3850 $CALL RANGET(INFSED(I))
      NF(II)=NF(II)+1 $IDF(NF(II))=J
      DO 3800 K=1,NTGR(II)
      IF(NT(II,K).GT.0) CALL TRNSMT(II,K,J)
3800  CONTINUE $CALL RANSET(INFSED(I))
3850  CONTINUE
3900  CONTINUE $IF(IFLG1.GT.0) WRITE(7,7060) (IDF(J),J=1,NF(II))
      CALL RANGET(INFSED(I))
3000  CONTINUE
*
      WRITE(6,6510) L,(NF(I),I=1,NTPOPS)
      WRITE(7,7510) L,(NF(I),I=1,NTPOPS)
*
      DO 1910 I=1,NCPOPS $DO 1910 J=1,NCLS $DO 1910 M=1,SNTP
      G(M,I,J,1)=G(M,I,J,1)*DCS(M)+G(M,I,J,2) $DO 1920 K=2,MCTP1-1
      G(M,I,J,K)=G(M,I,J,K+1)
1920  CONTINUE $G(M,I,J,MCTP1)=0.
1910  CONTINUE
*
1000  CONTINUE
*
      STOP
*
****  THESE ARE THE FORMATS
*
5110  FORMAT(12I6)
5120  FORMAT(12F6.2)
5130  FORMAT(I6,6F6.2,I6)
```

```
5140  FORMAT(I6,F6.4,4I6)
5150  FORMAT(4I6,F6.2,I6)
6000  FORMAT(1H1)
6005  FORMAT(5X,*NTPOPS,NFPOPS,NCPOPS,NCLS,MCTP1,NTGMX,IFLG1,*
      *,*IFLG2=:*,8I7)
6110  FORMAT(5X,*EQ,T=*,12F7.2)
6112  FORMAT(//,5X,*HERE ARE, FOR EACH POPULATION:,*,/,10X,*N,C,TTH,B,
      TG*K,TMEM,THO,NTGR*,/,20X,*OR*)
6116  FORMAT(10X,*N,P,INSTR,INSTP,INFSED,NTGR*,/,20X,*AND*)
6118  FORMAT(10X,*IRCP,TYPE,NCT,NT,STR,INSED..FOR EACH TG*)
6130  FORMAT(/,5X,I7,6F7.2,I7)
6140  FORMAT(/,5X,I7,F7.4,4I7)
6150  FORMAT(4I7,F7.2,I7)
6160  FORMAT(5X,*LTSTOP=*,I7)
6510  FORMAT(5X,15I7)
7510  FORMAT(15I5)
7060  FORMAT(20I4)
      END
*
      SUBROUTINE SVUPDT(I,K)
*
****  THIS SUBROUTINE UPDATES THE STATE VARIABLES OF AN INDIVIDUAL
      NEURON
*
      INTEGER SNTP
      PARAMETER NCPOPS=3,NCLS=50,MCTP1=2,SNTP=2,EK=-10,STEP=1.
      COMMON G(SNTP,NCPOPS,NCLS,MCTP1)
      COMMON/SV/E(NCPOPS,NCLS),TH(NCPOPS,NCLS),S(NCPOPS,NCLS),
      *              GK(NCPOPS,NCLS),B(NCPOPS),C(NCPOPS),DCG(NCPOPS),
      *              DCTH(NCPOPS),TMEM(NCPOPS),THO(NCPOPS),EQ(SNTP)
*
      GS=0. $SYN=0. $DO 10 J=1,SNTP
      GS=GS+G(J,I,K,1) $SYN=SYN+G(J,I,K,1)*EQ(J)
10    CONTINUE
      GK(I,K)=GK(I,K)*DCG(I)+B(I)*(1.-DCG(I))*S(I,K)
      GTOT=1.+GK(I,K)+GS $DCE=EXP(-GTOT*STEP/TMEM(I))
      E(I,K)=E(I,K)*DCE+(SYN+GK(I,K)*EK)*(1.-DCE)/GTOT
      TH(I,K)=THO(I)+(TH(I,K)-THO(I))*DCTH(I)+C(I)*E(I,K)*(1.-DCTH(I))
      S(I,K)=0.
*
      RETURN
      END
*
      SUBROUTINE TRNSMT(IS,L,K)
*
****     THIS SUBROUTINE PROJECTS SYNAPTIC ACTIVATION FROM A
****     SENDING FIBER TO ITS TARGETED RECEIVING CELLS.
```

```
*
      INTEGER TYPE,SNTP
      PARAMETER NTPOPS=5,NCPOPS=3,NCLS=50,MCTP1=2,SNTP=2,
      NTGMX=2
      COMMON G(SNTP,NCPOPS,NCLS,MCTP1)
      COMMON/TR/NCT(NTPOPS,NTGMX),NT(NTPOPS,NTGMX),STR(NTPOPS,
      *          NTGMX),INSED(NTPOPS,NTGMX),N(NTPOPS),TYPE(NTPOPS,
      *          NTGMX),IRCP(NTPOPS,NTGMX)
*
      IR=IRCP(IS,L)
      CALL RANSET(K*INSED(IS,L))
      DO 100 J=1,NT(IS,L)
      NREC=INT(RANF(0.)*FLOAT(N(IR)))+1
      KCT=INT(RANF(0.)*FLOAT(NCT(IS,L)))+2
      G(TYPE(IS,L),IR,NREC,KCT)=G(TYPE(IS,L),IR,NREC,KCT)+STR(IS,L)
100   CONTINUE
*
      RETURN
      END
/EOR
      0     0
 70.     1.    -10.    5.
     50  0.     25.   20.    5.    5.    10.    2
      3     1     1     8  .5   1227
      2     1     1     8  .5   1936
     50   .5    25.   20.    5.    5.    10.    0
     50  0.     25.   20.    5.    5.    10.    1
      1     2     1    10 1.    3421
     50   .1          1   999 4165    1
      1     1     1    10  .5   8917
     50   .1         15    50 3315    1
      3   1       1    10  .5  6373
     20
/EOR
        PROGRAM DENDR21

*
****    THIS PROGRAM SIMULATES THE ONGOING INPUT-OUTPUT ACTIVITY OF
*       A SINGLE NEURON WITH A PASSIVE DENDRITIC TREE OF ARBITRARY
*       MORPHOLOGY ACTIVATED BY AN ARBITRARY NUMBER OF SYNAP-
        TIC INPUT
*       SYSTEMS WHICH ARE INDIVIDUALLY ADJUSTABLE. SUBROUTINE
        TREE
****    COMPUTES THE TOTAL RESISTANCE OF THE DENDRITIC TREE.
*
      INTEGER RG,TPCN,BTCN,DDN,CDN,SNTP,TYPE,SCSTRT,SCSTP
      REAL LBRNCH,LSEG
```

```
      PARAMETER NDN=30,NREG=4,NCMPT=10,NSTEPS=500,STEP=1.,
      *              NDSPY=2,SNTP=1,NFPOPS=1,NTGS=1,EK=-10.,FD=1.,FS=1.
      DIMENSION  DDN(NDSPY),CDN(NDSPY),GDD(NREG),GTRT(NREG),
      *              EQ(SNTP),TG(SNTP),DG(SNTP),
      *              TYPE(NFPOPS),N(NFPOPS),NT(NFPOPS),NBC(NFPOPS),
      *              P(NFPOPS),STR(NFPOPS),LSTRT(NFPOPS),LSTP(NFPOPS),
      *              INSED1(NFPOPS),INSED2(NFPOPS),NTG(NFPOPS),
      *              IBC(NFPOPS,NDN),ITB(NFPOPS,NTGS),U(NDN,NCMPT),
      *              V(NDN,NCMPT),G(SNTP,NDN,NCMPT),GT(NDN,NCMPT),
      *              Z(NDN,NCMPT),GS(SNTP),SYN(NDN,NCMPT),PS(5,1000),
      *              NF(NFPOPS),IDF(100),G0(NDN,NCMPT),PT(NFPOPS,NTGS)
      COMMON/TREE/ RST,LBRNCH,LSEG,CON,RES,RG(NDN),TPCN(2,NDN),
      *              BTCN(2,NDN),RS,SC
*
****          THIS SECTION READS AND WRITES THE INPUT PARAMETERS
*
      READ 5000, IFLG1,IFLG2
      READ 5010, C,TTH,B,TGK,TMEM,TH0
      READ 5020, RS,NST,RST,LBRNCH
      DO 130 I=1,NDN
130   READ 5000, J,RG(I),TPCN(1,I),TPCN(2,I),BTCN(1,I),BTCN(2,I)
      READ 5000, ((DDN(I),CDN(I)),I=1,NDSPY)
*
      RES=700000. $CON=1./(TMEM*10.**11) $LSEG=LBRNCH/FLOAT(NCMPT)
      AS=4.*3.14159*(RS**2) $R=RST $AM=2.*3.14159*R*LSEG
      RSD=RES*(LSEG/2./(R**2)+RS/((RS/SQRT(2.))**2))/3.14159*FS
      GDS=1./(RSD*CON*AS) $GSD=1./(RSD*CON*AM)
      RBB=RES*(LSEG/(R**2)+RS/((RS/SQRT(2.))**2))/3.14159*FS
      GBB=1./(RBB*CON*AM)
      DO 160 I=1,NREG
      IF(I.GT.1) R=R/(2.**(2./3.))
      AM=2.*3.14159*R*LSEG $RDD=RES*LSEG/3.14159/R**2*FD
      GDD(I)=1./(RDD*CON*AM)
      IF(I.LT.NREG) RTRB=RES*LSEG*(1./R**2+1./((R/(2.**(2./3.)))**2))
      * /2./3.14159*FD
      RTRT=RES*LSEG*(1./R**2+1./((R*(2.**(2./3.)))**2))/2./3.14159*FD
      IF(I.LT.NREG) GTRB(I)=1./(RTRB*CON*AM)
      GTRT(I)=1./(RTRT*CON*AM)
      IF(I.EQ.NREG) GLK=3.14159*(R**2)/AM
160   CONTINUE
*
      READ 5010, ((EQ(I),TG(I)),I=1,SNTP)
      DO 210 I=1,NFPOPS
      READ 5030, TYPE(I),N(I),NT(I),NBC(I),P(I),STR(I),LSTRT(I),LSTP(I),
      * INSED1(I),INSED2(I),NTG(I)
      READ 5000, (IBC(I,J),J=1,NBC(I))
      DO 220 J=1,NTG(I)
```

```
220     READ 5240, ITB(I,J),PT(I,J)
210     CONTINUE
        READ 5250, SC,SCSTRT,SCSTP
        READ 5000, LTSTOP
*
        WRITE(6,6000)
        WRITE(6,6010) NDN,NREG,NCMPT,NSTEPS,STEP,RES,NDSPY,SNTP,
        *             NFPOPS,NTGS,IFLG1,IFLG2,FD,FS
        WRITE(6,6020) C,TTH,B,TGK,TMEM,TH0
        WRITE(6,6030) RS,NST,RST,LBRNCH
        WRITE(6,6040) $DO 310 I=1,NDN
310     WRITE(6,6050) I,RG(I),TPCN(1,I),TPCN(2,I),BTCN(1,I),BTCN(2,I)
        WRITE(6,6060) ((DDN(I),CDN(I)),I=1,NDSPY)
        WRITE(6,6070)
        WRITE(6,6075) GDS,GSD,GBB,((GTRT(I),GDD(I),GTRB(I)),I=1,NREG),GLK
        WRITE(6,6110) ((EQ(I),TG(I)),I=1,SNTP)
        WRITE(6,6120) $DO 320 I=1,NFPOPS
        WRITE(6,6130) TYPE(I),N(I),NT(I),NBC(I),P(I),STR(I),LSTRT(I),
        *             LSTP(I),INSED1(I),INSED2(I),NTG(I)
        WRITE(6,6075) (IBC(I,J),J=1,NBC(I))
        WRITE(6,6140)
320     WRITE(6,6150) ((ITB(I,J),PT(I,J)),J=1,NTG(I))
        WRITE(6,6160) SC,SCSTRT,SCSTP $WRITE(6,6170) LTSTOP
        IF(IFLG2.GT.0) CALL TREE $WRITE(6,6000)
*
****            THIS SECTION INITIALIZES VARIABLES
*
        G0S=1.+FLOAT(NST)*GDS
        DO 400 I=1,NDN $K=RG(I)
        IF(TPCN(1,I).LE.0) THEN
        G0(I,1)=1.+GSD+FLOAT(NST-1)*GBB+GDD(K) $ELSE
        IF(TPCN(2,I).EQ.0) G0(I,1)=1.+2.*GDD(K)
        IF(TPCN(2,I).GT.0) G0(I,1)=1.+GTRT(K)+2.*GDD(K) $ENDIF
        DO 410 J=2,NCMPT-1
410     G0(I,J)=1.+2.*GDD(K)
        IF(BTCN(1,I).LE.0) THEN $G0(I,NCMPT)=1.+GLK+GDD(K) $ELSE
        IF(BTCN(2,I).EQ.0) G0(I,NCMPT)=1.+2.*GDD(K)
        IF(BTCN(2,I).GT.0) G0(I,NCMPT)=1.+2.*GTRB(K)+GDD(K) $ENDIF
400     CONTINUE
*
        DO 500 I=1,NDN $DO 500 J=1,NCMPT
        U(I,J)=0. $V(I,J)=0. $DO 500 K=1,SNTP $G(K,I,J)=0.
500     CONTINUE $US=0. $VS=0. $GK=0. $TH=TH0
        DGK=EXP(-1./TGK) $DTH=EXP(-1./TTH)
        DO 510 K=1,SNTP $DG(K)=EXP(-STEP/TG(K))
510     GS(K)=0.
*
```

```
****            THIS SECTION UPDATES STATE VARIABLES AT EACH TIME STEP
*
      DO 1000 L=1,LTSTOP
      SCN=0. $IF(L.GE.SCSTRT.AND.L.LT.SCSTP) SCN=SC
*
****            UPDATING THE INPUT FIBERS
*
      DO 2000 I=1,NFPOPS
      NF(I)=0 $IF(L.LT.LSTRT(I).OR.L.GE.LSTP(I)) GO TO 2000
      DO 2900 J=1,N(I)
      CALL RANSET(J*L*INSED1(I))
      IF(RANF(0.).GT.P(I)) GO TO 2900
      NF(I)=NF(I)+1 $IDF(NF(I))=J
      DO 2800 K=1,NT(I)
      IDN=IBC(I,INT(RANF(0.)*FLOAT(NBC(I)))+1)
      IF(IDN.GT.0) THEN $ICN=INT(RANF(0.)*FLOAT(NCMPT))+1
      G(TYPE(I),IDN,ICN)=G(TYPE(I),IDN,ICN)+STR(I)
      ELSE $GS(TYPE(I))=GS(TYPE(I))+STR(I) $ENDIF
2800  CONTINUE
2900  CONTINUE
      DO 2600 J=1,NTG(I)
      CALL RANSET(J*L*INSED2(I))
      IF(RANF(0.).GT.PT(I,J)) GO TO 2600
      NF(I)=NF(I)+1 $IDF(NF(I))=J+N(I)
      IF(ITB(I,J).GT.0) THEN $ICN=INT(RANF(0.)*FLOAT(NCMPT))+1
      G(TYPE(I),ITB(I,J),ICN)=G(TYPE(I),ITB(I,J),ICN)+STR(I)
      ELSE $GS(TYPE(I))=GS(TYPE(I))+STR(I) $ENDIF
2600  CONTINUE
2000  CONTINUE
*
*
****            UPDATING THE CELLS
*
      GSM=GK $SYNS=GK*EK $DO 3000 K=1,SNTP
      SYNS=SYNS+GS(K)*EQ(K)
3000  GSM=GSM+GS(K) $GTS=G0S+GSM $ZS=EXP(-GTS*STEP/TMEM/
      FLOAT(NSTEPS))
      DO 3200 I=1,NDN $DO 3200 J=1,NCMPT $SYNG=0. $SYN(I,J)=0.
      DO 3210 K=1,SNTP
      SYN(I,J)=SYN(I,J)+G(K,I,J)*EQ(K)
3210  SYNG=SYNG+G(K,I,J) $GT(I,J)=G0(I,J)+SYNG
3200  Z(I,J)=EXP(-GT(I,J)*STEP/TMEM/FLOAT(NSTEPS))
*
*
****            THIS SUB-SECTION PERFORMS FINE-GRAINED INTEGRATION
****            OF COMPARTMENT POTENTIALS.
*
```

```
      DO 4000 IS=1,NSTEPS
      DSC=0. $DO 4010 I=1,NST
4010  DSC=DSC+U(I,1)*GDS
      VS=US*ZS+(DSC+SYNS+SCN)*(1.-ZS)/GTS
*
      SDP=0. $DO 4020 J=1,NST
4020  SDP=SDP+U(J,1)
      DO 4100 I=1,NDN $K=RG(I)
*
      IF(TPCN(1,I).LE.0) THEN $SDT=GBB*(SDP-U(I,1))
      V(I,1)=U(I,1)*Z(I,1)+(GSD*US+SDT+GDD(K)*U(I,2)+SYN(I,1))*
      * (1.-Z(I,1))/GT(I,1) $ELSE
      IF(TPCN(2,I).EQ.0) V(I,1)=U(I,1)*Z(I,1)+(GDD(K)*(U(TPCN(1,I)
      *          ,NCMPT)+U(I,2))+SYN(I,1))*(1.-Z(I,1))/GT(I,1)
      IF(TPCN(2,I).GT.0) V(I,1)=U(I,1)*Z(I,1)+(GTRT(K)*U(TPCN(1,I),
      *          NCMPT)+GDD(K)*(U(TPCN(2,I),1)+U(I,2))+SYN(I,1))*(1.-Z(I,1))/
      *          GT(I,1)
      ENDIF
*
      DO 4120 J=2,NCMPT-1
4120  V(I,J)=U(I,J)*Z(I,J)+(GDD(K)*(U(I,J-1)+U(I,J+1))+SYN(I,J))*
      * (1.-Z(I,J))/GT(I,J)
*
      IF(BTCN(1,I).LE.0) THEN
      V(I,NCMPT)=U(I,NCMPT)*Z(I,NCMPT)+(GDD(K)*U(I,NCMPT-1)+SYN(I,
      *          NCMPT))*(1.-Z(I,NCMPT))/GT(I,NCMPT) $ELSE
      IF(BTCN(2,I).EQ.0) V(I,NCMPT)=U(I,NCMPT)*Z(I,NCMPT)+(GDD(K)*
      *          (U(BTCN(1,I),1)+U(I,NCMPT-1))+SYN(I,NCMPT))*(1.-Z(I,NCMPT))/
      *          GT(I,NCMPT)
      IF(BTCN(2,I).GT.0) V(I,NCMPT)=U(I,NCMPT)*Z(I,NCMPT)+(GTRB(K)*
      *          (U(BTCN(1,I),1)+U(BTCN(2,I),1))+GDD(K)*U(I,NCMPT-1)+SYN(I,
      *          NCMPT))*(1.-Z(I,NCMPT))/GT(I,NCMPT) $ENDIF
*
4100  CONTINUE
      US=VS $DO 4200 I=1,NDN $DO 4200 J=1,NCMPT
4200  U(I,J)=V(I,J)
*
4000  CONTINUE
*
      TH=TH*DTH+(TH0+C*VS)*(1.-DTH) $GK=GK*DGK
      S=0. $IF(VS.LT.TH) GO TO 4300
      S=1. $GK=GK+B*(1.-DGK)
4300  CONTINUE $PS(1,L)=VS+S*(50.-VS)
*
      IF(IFLG1.GT.0) THEN $WRITE(6,6280) VS $DO 4400 I=1,NDN
4400  WRITE(6,6280) (U(I,J),J=1,NCMPT) $ENDIF
      DO 4500 I=1,NDSPY
```

```
4500  PS(I+1,L)=V(DDN(I),CDN(I))
*
      DO 4610 I=1,NDN $DO 4610 J=1,NCMPT $DO 4610 K=1,SNTP
4610  G(K,I,J)=G(K,I,J)*DG(K)
1000  CONTINUE
*
****          THIS SECTION WRITES OUT ACTIVITY VARIABLES
*
      DO 1600 I=1,NDSPY+1 $WRITE(6,6002)
1600  WRITE(6,6280) (PS(I,L),L=1,LTSTOP)
      DO 1700 I=1,NDSPY+1 $WRITE(7,6002)
1700  WRITE(7,7280) (PS(I,L),L=1,LTSTOP)
*
      STOP
****  THESE ARE THE FORMATS
*
5000  FORMAT(12I6)
5010  FORMAT(12F6.2)
5020  FORMAT(F6.2,I6,2F6.2)
5030  FORMAT(4I6,2F6.2,5I6)
5240  FORMAT(I6,F6.2)
5250  FORMAT(F6.2,2I6)
6000  FORMAT(1H1)
6002  FORMAT(/)
6010  FORMAT(5X,*NDN,NREG,NCMPT,NSTEPS,STEP,RES,NDSPY,SNTP,NFPOPS,*
     *          ,*NTGS,IFLG1,ILFG2,FD,FS=*,/,10X,4I7,2F10.2,6I7,2F7.2)
6020  FORMAT(5X,*C,TTH,B,TGK,TMEM,TH0,=*,6F7.2)
6030  FORMAT(5X,*RS,NST,R,LBRNCH=*,F7.2,I7,2F7.2)
6040  FORMAT(/,5X,*BR,RG,TPCNS,BTCNS=*)
6050  FORMAT(5X,6I7)
6060  FORMAT(5X,*DSP=*,2I7)
6070  FORMAT(/,5X,*GDS,GSD,GBB,(GTRT-GDD-GTRB),GLK=*)
6075  FORMAT(3X,16F8.2)
6110  FORMAT(5X,*EQS,TGS=*,12F7.2)
6120  FORMAT(5X,*SNT,N,NT,NBC,P,STR,LSTRT,LSTP,INSED1,INSED2,NTG,IBCS=*)
6130  FORMAT(10X,4I7,2F7.2,6I7)
6140  FORMAT(5X,*TARGETS:BRNCH,RATE=*)
6150  FORMAT(5X,I7,F7.2)
6160  FORMAT(5X,*SC,SCSTRT,SCSTP=*,F7.2,2I7)
6170  FORMAT(5X,*LTSTOP=*,I7)
6280  FORMAT(5X,20F6.2)
7280  FORMAT(13F6.2)
      END
*
      SUBROUTINE TREE
*
****      THIS SUBROUTINE COMPUTES THE TOTAL RESISTANCE AND OTHER
```

```
****    QUANTITIES OF INTEREST FOR A BIFURCATING DENDRITIC TREE.
*
        INTEGER RG,TPCN,BTCN,USED
        REAL LBRNCH,LSEG,LC
        PARAMETER NDN=30,NREG=4
        COMMON/TREE/ RST,LBRNCH,LSEG,CON,RES,RG(NDN),TPCN(2,NDN),
        *           BTCN(2,NDN),RS,SC
        DIMENSION LC(NREG),RL(NREG),AD(NREG),USED(NDN),YN(NDN),NM(NDN)
*
        AS=4.*3.14159*RS**2
        WRITE(6,6102) RS,AS $WRITE(7,6102) RS,AS
        WRITE(6,6105) $WRITE(7,6105)
        DO 160 I=1,NREG
        IF(I.EQ.1) R=RST
        IF(I.GT.1) R=R/(2.**(2./3.))
        AM=2.*3.14159*R*LSEG
        LC(I)=SQRT(R/(2.*RES*CON))
        RL(I)=RES/(3.14159*R**2)
        AD(I)=(TANH(LBRNCH/LC(I))+R/2./LC(I))/
        *           ((1.+TANH(LBRNCH/LC(I))*R/2./LC(I))*LC(I)*RL(I))
        DCS=EXP(-LSEG/LC(I)) $DCB=EXP(-LBRNCH/LC(I))
        WRITE(6,6110) I,R,AM,RL(I),AD(I),LC(I),DCS,DCB
        WRITE(7,6110) I,R,AM,RL(I),AD(I),LC(I),DCS,DCB
160     CONTINUE
*
        N=0 $DO 10 J=1,NDN
10      USED(J)=0
        DO 100 I=1,NDN
        IF(BTCN(1,I).GT.0.OR.USED(I).GT.0) GO TO 100
        N=N+1 $NM(N)=I $YN(N)=AD(RG(I)) $USED(I)=1
        IF(TPCN(2,I).EQ.0) GO TO 90
        J=TPCN(2,I) $USED(J)=1 $YN(N)=YN(N)+AD(RG(J))
90      CONTINUE $IF(TPCN(1,I).EQ.0) GO TO 100
        J=TPCN(1,I) $K=RG(J) $NM(N)=J
        YN(N)=(YN(N)+AD(K))/(1.+YN(N)*AD(K)*(LC(K)*RL(K))**2)
100     CONTINUE $NN=N
*
        DO 300 L=1,NREG-2 $NN=0
        DO 200 I=1,N $IF(USED(NM(I)).GT.0) GO TO 200 $NN=NN+1
        IF(TPCN(2,NM(I)).EQ.0) GO TO 190
        YN(I)=YN(I)+YN(I+1) $USED(TPCN(2,NM(I)))=1
190     J=TPCN(1,NM(I)) $K=RG(J) $NM(NN)=J
        YN(NN)=(YN(I)+AD(K))/(1.+YN(I)*AD(K)*(LC(K)*RL(K))**2)
200     CONTINUE $N=NN
300     CONTINUE
*
        YS=CON*4.*3.14159*RS**2 $YD=0.
```

```
      DO 400 I=1,NN
400   YD=YD+YN(I)
      SIG=YD/YS $RTOT=1./(YS+YD) $RSMA=1./YS $RDN=1./YD $ESS=SC/
      (1.+SIG)
      WRITE(6,6120) SIG,YS,YD,RTOT,RSMA,RDN,ESS
      WRITE(7,6120) SIG,YS,YD,RTOT,RSMA,RDN,ESS
*
6102  FORMAT(5X,*RS,AS=*,2F7.2)
6105  FORMAT(/,5X,*RG,R,AM,RL,AD,LC,DC,DCB=:*)
6110  FORMAT(5X,I7,2F7.2,2E12.2,F7.1,2F7.4)
6120  FORMAT(5X,*SIG,YS,YD,RTOT,RSMA,RDN,ESS=*,/,5X,F7.2,5E10.2,F7.2)
*
      RETURN
      END
/EOR
       0      1
      0.    25.    400.     10.    11.    24.
      5.5      3   1.25    158.
       1      1      0       0      4      5
       2      1      0       0      6      7
       3      1      0       0      8      9
       4      2      1       5     10     11
       5      2      1       4     12      0
       6      2      2       7     13     14
       7      2      2       6     15      0
       8      2      3       9     16     17
       9      2      3       8     18      0
      10      3      4      11     19     20
      11      3      4      10     21      0
      12      2      5       0     22      0
      13      3      6      14     23     24
      14      3      6      13     25      0
      15      2      7       0     26      0
      16      3      8      17     27      0
      17      3      8      16     28      0
      18      2      9       0     29     30
      19      4     10      20      0      0
      20      4     10      19      0      0
      21      3     11       0      0      0
      22      2     12       0      0      0
      23      4     13      24      0      0
      24      4     13      23      0      0
      25      3     14       0      0      0
      26      2     15       0      0      0
      27      3     16       0      0      0
      28      3     17       0      0      0
      29      3     18      30      0      0
```

```
      30        3       18       29         0         0
       2        1       25        5
   70.      3.
       1        0        0        1    0.      0.        0         0   1227   1936 1
       0
       0     0.
3500.  1     9999
   15
/EOR
/EOI


      PROGRAM SYSTM5G
*
****  THIS PROGRAM SIMULATES THE ACTIVITY OF AN ARBITRARY NUMBER OF
****  NEURON POOLS ACTIVATED BY AN ARBITRARY NUMBER OF INPUT
      FIBER SYSTEMS.
****  MODIFIED, AUGUST, 1992 TO INCLUDE SYNAPTIC JUNCTIONAL MATRICES.
*
*     NOTE: When specifying sq's for a 4 x 4 recurrent junction (ee, ei, ie, ii),
*        specify JTYPE as 3 for the ee junction and as 44 for the
*        ee, ie, ii junctions; also specify NTGRFL for the ee junction as 44.
*
***************
*
      INTEGER TYPE,SNTP
      PARAMETER(NTPOPS=6,NFPOPS=1,NCPOPS=5,NCLS=100,MCTP1=2,
     *          SNTP=2,STEP=1.,NTGMX=5,NSTMX=100,NSTPMX=10)
      DIMENSION P(NFPOPS),INSTR(NFPOPS),INSTP(NFPOPS),INFSED(NFPOPS),
     *          IDF(NCLS),TTH(NCPOPS),TGK(NCPOPS),T(SNTP),DCS(SNTP),
     *          NF(NTPOPS),ISTM(NCPOPS),LSTIM
     *          (NSTPMX),INTERV(NSTPMX),LSTP(NSTPMX),NSTIM(NSTPMX),
     *          NSTIM1(NSTPMX),SI(NSTPMX),MARKER(NSTPMX)
      COMMON/A1/G(SNTP,NCPOPS, NCLS,MCTP1)
      COMMON/A2/EQ(SNTP)
      COMMON/BB/E(NCPOPS,NCLS),TH(NCPOPS,NCLS)
      COMMON/CC/LOCATN(NSTPMX, NSTMX)
      COMMON/SV/S(NCPOPS,NCLS),
     *          GK(NCPOPS,NCLS),B(NCPOPS),C(NCPOPS),F(NCPOPS),
     *          DCG(NCPOPS),DCTH(NCPOPS),TMEM(NCPOPS),THO(NCPOPS)
      COMMON/J1/NCT(NTPOPS,NTGMX),STR(NTPOPS,
     *          NTGMX,NCLS,NCLS),NTGR(NTPOPS,N(NTPOPS),
     *          TYPE(NTPOPS,NTGMX),IRCP(NTPOPS,NTGMX)
      COMMON/J2/NT(NTPOPS,NTGMX),INSED(NTPOPS,NTGMX)
     *          ,JTYPE(NTPOPS,NTGMX)
      COMMON/J3/SF(NTPOPS,NTGMX)
```

```
*
      OPEN(UNIT=3,FILE='s5g.in',STATUS='OLD')
      OPEN(UNIT=4,FILE='s5g.sq',STATUS='OLD')
      OPEN(UNIT=6,FILE='s5g.o1',STATUS='UNKNOWN')
      OPEN(UNIT=7,FILE='s5g.o2',STATUS='UNKNOWN')
*
****        THIS SECTION READS AND WRITES THE INPUT PARAMETERS
*
      READ(3,*)IFLG1,IFLG2
      DO 105 I=1, SNTP
105   READ(3,*)EQ(I),T(I)
      DO 110 I=1, NTPOPS
      IF (I.LE.NCPOPS) THEN
      READ(3,*)N(I),C(I),F(I),TTH(I),TGK(I),TMEM(I),
      *         THO(I),NTGR(I)
      ELSE
      II=I-NCPOPS
      READ(3,*)N(I),P(II),INSTR(II),INSTP(II),INFSED(II),
      *         NTGR(I)
      ENDIF
      DO 120 J=1,NTGR(I)
120   READ(3,*)IRCP(I,J),TYPE(I,J),NCT(I,J),NT(I,J),
      *         JTYPE(I,J),INSED(I,J)
110   CONTINUE
      READ(3,*) LTSTOP
*
      WRITE(6,6005) NTPOPS,NFPOPS,NCPOPS,NCLS,MCTP1,
      *         NTGMX,IFLG1,IFLG2
      WRITE(6,6110) ((EQ(I), T(I)),I=1,SNTP)
      WRITE(6,6112)
      WRITE(6,6116)
      WRITE(6,6118)
      DO 210 I=1,NTPOPS
      IF(I.LE.NCPOPS) THEN
      WRITE(6,6130) N(I),C(I)F(I),TTH(I),B(I),TGK(I),TMEM(I),
      *         THO(I),NTGR(I)
      ELSE II=I-NCPOPS
      WRITE(6,6140) N(I),P(II),INSTR(II),INSTP(II),INFSED(II),
      *         NTGR(I)
      ENDIF
      DO 220 J=1, NTGR(I)
220   WRITE(6,6150) IRCP(I,J),TYPE(I,J),NCT(I,J),NT(I,J),
                JTYPE(I,J),INSED(I,J)
210   CONTINUE
      WRITE(6,6160) LTSTOP
*
****      THIS SECTION GENERATES THE SYNAPTIC JUNCTIONAL MATRICES
```

```
*
      DO 400 I=1,NTPOPS
      DO 400 J=1,NTGR(I)
      NSQS=0
      IF(JTYPE(I,J).EQ.1.OR.JTYPE(I,J).EQ.44) GO TO 400
      IF(JTYPE(I,J).EQ.2) THEN
      IR=IRCP(I,J)
      DO 410 K=1,N(I)
      DO 411 L=1,N(IR)
411   STR(I,J,K,L)=0.
      DO 410 L=1,NT(I,J)
      RNOW=random(INSED(I,J))
      NREC=INT(RNOW*FLOAT(N(IR)))+1
410   STR(I,J,K,NREC)=1.
ENDIF
*
      IF(JTYPE(I,J).EQ.3.AND.NSQS.EQ.0) THEN
      NSQS=1
      CALL SYNJNCT(I)
      ENDIF
*
400   CONTINUE
*
9000  CONTINUE
****            THIS SECTION INITIALIZES VARIABLES
*
      DO 302 I=1,NTPOPS
      IF(NTGR(I).GT.0) THEN
      READ(4,*) (SF(I,J),J=1,NTGR(I))
      IF(SF(I,1).LT.0.) GO TO 9999
      WRITE(6,7020) (SF(I,J),J=1,NTGR(I))
      ENDIF
302   CONTINUE
*
****            READING THE STIMULATION PATTERNS
*
      READ(4,*) NSTMP
      IF(NSTMP.LT.0) GO TO 9999
      DO 300 I=1, NSTMP
      READ(4,*) ISTM(I),LSTIM(I),INTERV(I),LSTP(I),NSTIM(I),
      *         NSTIM1(I),SI(I)
      WRITE(6,7030) ISTM(I),LSTIM(I),INTERV(I),LSTP(I),NSTIM(I),
      *         NSTIM1(I),SI(I)
      IF(NSTIM(I).GT.0) READ(4,*) (LOCATN(I,K),K=1,NSTIM(I))
      WRITE(6,7060) (LOCATN(I,K),K=1,NSTIM(I))
      MARKER(I)=0
      LSTM(I)=1
```

```
300   CONTINUE
*
      DO 310 I = 1,NCPOPS
      DCTH(I) = EXP(-STEP/TTH(I))
      DCG(I) = EXP(-STEP/TGH(I))
      DO 310 J = 1,NCLS
      E(I,J) = 0.
      TH(I,J) = THO(I)
      S(I,J) = 0.
      GK(I,J) = 0.
      DO 310 K = 1,MCTP1
      DO 310 L = 1, SNTP
      G(L,I,J,K) = 0.
310   CONTINUE
      DO 320 I = 1, SNTP
320   DCS(I) = EXP(-STEP/T(I))
*
****          THIS SECTION UPDATES STATE VARIABLES AT EACH TIME STEP
*
      DO 1000 LT = 1,LTSTOP
*
****          APPLYING THE FIXED STIMULATION
*
      DO 910 I = 1, NSTMP
      IF (LT.EQ.LSTIM(I).AND.LT.LE.LSTP(I)) THEN
      IF(IFLG2.GT.0)
      *       WRITE(6,7060) ISTM(I),(LOCATN(I,J + MARKER(I)),J = 1,NSTIM1(I))
              CALL STIMULS(I,ISTM(I),LSTIM(I),INTERV(I),NSTIM(I),NSTIM1(I),
      *       SI(I),MARKER(I))
      ENDIF
910   CONTINUE
*
****          UPDATING THE CELLS
*
      DO 2000 I = 1,NCPOPS
      NF(I) = 0
      DO 2900 J = 1,N(I)
      CALL SVUPDT(I,J)
      IF(E(I,J).LT.TH(I,J)) GO TO 2850
      S(I,J) = 1.
      NF(I) = NF(I) + 1
      IDF(NF(I)) = J
      DO 2800 K = 1,NTGR(I)
      IR = IRCP(I,K)
      IF(JTYPE(I,K).EQ.1) THEN
      NSEED = INSED(I,K) + J**2
      PC = FLOAT(NT(I,K))/FLOAT(N(IR))
```

```
      DO 2701 L=1,N(IR)
      RNOW=random(NSEED)
      IF(RNOW.GE.PC) GO TO 2701
      G(TYPE(I,K),IR,L,NCT(I,K)+1)=G(TYPE(I,K),IR,L,NCT(I,K)+1)
      *     +SF(I,K)
2701  CONTINUE
      ELSE
      DO 2700 L=1,N(IR)
      G(TYPE(I,K),IR,L,NCT(I,K)+1)=G(TYPE(I,K),IR,L,NCT(I,K)+1)
      *     +SF(I,K)*STR(I,K,J,L)
2700  CONTINUE
      ENDIF
2800  CONTINUE
2850  CONTINUE
2900  CONTINUE
      IF(IFLG1.GT.0) WRITE(6,7060) LT,I,NF(I),((IDF(J)),J=1,NF(I))
2000  CONTINUE
*
****            UPDATING THE INPUT FIBERS
*
      DO 3000 I=1,NFPOPS
      II=I+NCPOPS
      NF(II)=0
      IF(LT.LT.INSTR(I).OR.LT.GE.INSTP(I)) GO TO 3000
      DO 3900 J=1,N(II)
      RNOW=random(INFSED(I))
      IF(RNOW.GT.P(I)) GO TO 3850
      NF(II)=NF(II)+1
      IDF(NF(II))=J
      DO 3800 K=1,NTGR(II)
      IR=IRCP(II,K)
      IF(JTYPE(II,K).EQ.1) THEN
      NSEED=INSED(II,K)+J**2
      PCI=FLOAT(NT(II,K))/FLOAT(N(IR))
      DO 3701 L=1,N(IR)
      RNOW=random(NSEED)
      IF(RNOW.GE.PCI) GO TO 3701
      G(TYPE(II,K),IR,L,NCT(II,K)+1)=G(TYPE(II,K),IR,L,NCT(II,K)+1)
      *     +SF(II,K)
3701  CONTINUE
      ELSE
      DO 3700 L=1,N(IR)
      G(TYPE(II,K),IR,L,NCT(II,K)+1)=G(TYPE(II,K),IR,L,NCT(II,K)+1)
      *     +SF(II,K)*STR(II,K,J,L)
3700  CONTINUE
      ENDIF
3800  CONTINUE
```

```
3850  CONTINUE
3900  CONTINUE
      IF(IFLG1.GT.0) WRITE(6,7060) LT,II,NF(II),((IDF(J)),J = 1,NF(II))
3000  CONTINUE
*
      WRITE(6,6510) LT,(NF(I),I = 1,NTPOPS)
      WRITE(7,7510) LT,(NF(I),I = 1,NTPOPS)
*
      DO 1910 I = 1,NCPOPS
      DO 1910 J = 1,NCLS
      DO 1910 M = 1,SNTP
      G(M,I,J,1) = G(M,I,J,1)*DCS(M) + G(M,I,J,2)
      DO 1920 K = 2,MCTP1-1
      G(M,I,J,K) = G(M,I,J,K + 1)
1920  CONTINUE
      G(M,I,J,MCTP1) = 0.
1910  CONTINUE
*
1000  CONTINUE
*
      GO TO 9000
9999  CONTINUE
*
      ENDFILE(6)
      ENDFILE(7)
      CLOSE(3)
      CLOSE(4)
      CLOSE(6)
      CLOSE(7)
*
      STOP
*
****           THESE ARE THE FORMATS
*
6005  FORMAT('NTPOPS,NFPOPS,NCPOPS,NCLS,MCTP1,NTGMX,IFLG1',
      *',IFLG2 = :',/,5x,817)
6110  FORMAT('EQ,T = ',8F7.2)
6112  FORMAT('HERE ARE, FOR EACH POPULATION:,',/,10X,'N,C,F,TTH,B,
      *TGK,TMEM,THO,NTGR',/,20X,'OR')
6116  FORMAT(10X,'N,P,INSTR,INSTP,INFSED,NTGR',/,20X,'AND')
6118  FORMAT(10X,'IRCP,TYPE,NCT,NT,JTYPE,INSED..FOR EACH TG')
6130  FORMAT(5X,I7,7F7.2,I7)
6140  FORMAT(5X,I7,F7.4,4I7)
6150  FORMAT(6I7)
6160  FORMAT(5X,'LTSTOP = ',I7)
6510  FORMAT(5X,10I7)
7020  FORMAT('SFs:',10F7.3)
```

```
7030 FORMAT('ISTM,LSTM,INTERV,LSTP,NSTIM,NSTIM1:',6I6)
7510 FORMAT(10I5)
7060 FORMAT(20I4)
     END
*
     SUBROUTINE SVUPDT(I,K)
*
****         THIS SUBROUTINE UPDATES THE STATE VARIABLES OF AN
             INDIVIDUAL NEURON
*
     INTEGER SNTP
     PARAMETER(NCPOPS=5,NCLS=100,MCTP1=2,SNTP=2,EK=-10,STEP=1.,
     *    NSTMX=100)
     COMMON/A1/G(SNTP,NCPOPS,NCLS,MCTP1)
     COMMON/A2/EQ(SNTP)
     COMMON/BB/E(NCPOPS,NCLS),TH(NCPOPS,NCLS)
     COMMON/SV/                     S(NCPOPS,NCLS),
     *       GK(NCPOPS,NCLS),B(NCPOPS),C(NCPOPS),F(NCPOPS),
     *       DCG(NCPOPS),DCTH(NCPOPS),TMEM(NCPOPS),THO(NCPOPS)
*
     GS=0.
     SYN=0.
     DO 10 J=1, SNTP
     GS=GS+G(J,I,K,1)
     SYN=SYN+G(J,I,K,1)*EQ(J)
10   CONTINUE
     GK(I,K)=GK(I,K)*DCG(I)+B(I)*(1.-DCG(I))*S(I,K)
     GTOT=1.+GK(I,K)+GS
     DCE=EXP(-GTOT*STEP/TMEM(I))
     E(I,K)=E(I,K)*DCE+(SYN+GK(I,K)*EK)*(1.-DCE)/GTOT
     TH(I,K)=THO(I)+(TH(I,K)-THO(I))*DCTH(I)+
     *       (C(I)*E(I,K)+F(I)*S(I,K)*THO(I))*(1.-DCTH(I))
     S(I,K)=0.
*
     RETURN
     END
*********
     SUBROUTINE SYNJNCT(I)
*
**** THIS SUBROUTINE COMPUTES THE SYNAPTIC JUNCTIONAL MATRICES
****   BETWEEN POPULATION I AND ITS JTH TARGET POPULATION.
*
     INTEGER TYPE,SNTP
     PARAMETER(NTPOPS=6,NCPOPS=5,NCLS=100,MCTP1=2,SNTP=2,
     NTGMX=5)
     DIMENSION NTGRFL(NTGMX),NS(30,50),NR(30,50)
     COMMON/A1/G(SNTP,NCPOPS,NCLS,MCTP1)
```

```
      COMMON/A2/EQ(SNTP)
      COMMON/J1/NCT(NTPOPS,NTGMX),STR(NTPOPS,
     *          NTGMX,NCLS,NCLS),NTGR(NTPOPS),N(NTPOPS),
     *          TYPE(NTPOPS,NTGMX),IRCP(NTPOPS,NTGMX)
      COMMON/J2/NT(NTPOPS,NTGMX),INSED(NTPOPS,NTGMX)
     *        ,JTYPE(NTPOPS,NTGMX)
*
      READ(4,*) (NTRGFL(J),J=1,NTGR(I))
      DO 580 IT=1,NTGR(I)
      IF(NTRGFL(IT).EQ.0) GO TO 580
      IF(NTRGFL(IT).EQ.44) THEN
      DO 410 K1=1,2
      DO 410 I1=1,N(I+1)
      DO 410 J1=1,N(I+K1-1)
  410 STR(I+1,K1,I1,J1)=1.
      ELSE
      IF(EQ(TYPE(I,IT)).LE.0.) THEN
      DO 420 I1=1,N(I)
      DO 420 J1=1,N(IRCP(I,IT))
  420 STR(I,IT,I1,J1)=1.
      ENDIF
      ENDIF
*
      READ(4,*) NPTRN
      DO 560 NP=1,NPTRN
      READ(4,*) NFS,LPER,LNKD,DC
      READ(4,*) ((NS(J,K),K=1,NFS),J=1,LPER)
      WRITE(6,6110) NFS,LPER,LNKD,DC
      WRITE(6,6120) ((NS(J,K),K=1,NFS),J=1,LPER)
*
      READ(4,*) NFR
      READ(4,*) ((NR(J,K),K=1,NFR),J=1,LPER)
      READ(4,*) IFLSW,KS,KR,JLO,JUP,KLO,KUP
      WRITE(6,6130) NFR
      WRITE(6,6120) ((NR(J,K),K=1,NFR),J=1,LPER)
*
      INUP=1
      IF(NTGRFL(IT).EQ.44) INUP=4
      DO 555 IN=1,INUP
      IF(IN.LE.2) NF1=NFS
      IF(IN.GT.2) NF1=NFR
      IF(IN.EQ.1.OR.IN.EQ.3) NF2=NFR
      IF(IN.EQ.2.OR.IN.EQ.4) NF2=NFS
*
      DO 550 LS=1,LPER
      DO 540 IS=1,NF1
      DO 530 K=1,LNKD
```

```
      LR=MOD(LS+K-1,LPER)+1
      ET=EXP(-DC*FLOAT(K-1)/FLOAT(LNKD))
      DO 520 L=1,NF2
      IF(IN.LE.2) MS=NS(LS,IS)
      IF(IN.GT.2)MS=NR(LS,IS)
      IF(IN.EQ.1.OR.IN.EQ.3) MR=NR(LR,L)
      IF(IN.EQ.2.OR.IN.EQ.4) MR=NS(LR,L)
      IF(NTGRFL(IT).EQ.44) THEN
      IF(IN.LE.2) STR(I,IN,MS,MR)=MAX(STR(I,IN,MS,MR),ET)
      IF(IN.GT.2) STR(I+1,IN-2,MS,MR)=MIN(STR(I+1,IN-2,MS,MR),1.-ET)
      ELSE
      TI=STR(I,IT,MS,MR)
      IF(EQ(TYPE(I,IT)).GT.0.) STR(I,IT,MS,MR)=MAX(TI,ET)
      IF(EQ(TYPE(I,IT)).LE.0.) STR(I,IT,MS,MR)=MIN(TI,1.-ET)
      ENDIF
520   CONTINUE
530   CONTINUE
540   CONTINUE
550   CONTINUE
*
555   CONTINUE
*
      IF(IFLSW.GT.0) THEN
      WRITE(6,6140) KS,KR,JLO,JUP,KLO,KUP
      DO 610 J=JLO,JUP
      WRITE(6,6142) ((STR(KS,KR,J,K)),K=KLO,KUP)
610   CONTINUE
      ENDIF
*
560   CONTINUE
580   CONTINUE
      RETURN
      STOP
6110  FORMAT('NFS,LPER,LNKD,DC=',3I6,F6.3,5X,'here are the firers:')
6120  FORMAT(20I4)
6130  FORMAT('NFR=',I6,5X,'here are the firers')
6140  FORMAT('KS,KR,JLO,JUP,KLO,KUP=',6I4,4X,
      *'its strength matrix in jlo,jup–klo,kup')
6142  FORMAT (12F6.3)
      END
*********
*
      REAL FUNCTION random(lseed)
      INTEGER hi,lo,test
      PARAMETER(la=16807,m=2147483647,lq=127773,
      *lr=2836)
*
```

```
      hi = lseed/lq
      lo = mod(lseed,lq)
      test = la*lo-lr*hi
      if (test.GE.0.0) then
      lseed = test
      else
      lseed = test + m
      endif
      random = FLOAT(lseed)/FLOAT(m)
*
      RETURN
      END
*
      SUBROUTINE STIMULS(I,IR,LSTIM,INTERV,NSTIM,NSTIM1,SI,MARKER)
*
      INTEGER TYPE,SNTP
      PARAMETER(NTPOPS = 6,NCPOPS = 5,NCLS = 100,MCTP1 = 2,
      *         SNTP = 2,NTGMX = 5,NSTMX = 100,NSTPMX = 10)
      COMMON/A1/G(SNTP,NCPOPS,NCLS,MCTP1)
      COMMON/BB/E(NCPOPS,NCLS),TH(NCPOP,NCLS)
      COMMON/CC/LOCATN(NSTPMX,NSTMX)
      COMMON/J1/NCT(NTPOPS,NTGMX),STR(NTPOPS,
      *         NTGMX,NCLS,NCLS),NTGR(NTPOPS),N(NTPOPS),
      *         TYPE(NTPOPS,NTGMX),IRCP(NTPOPS,NTGMX)
      COMMON/J3/SF(NTPOPS,NTGMX)
*
      DO 10 I = 1,NSTIM1
      J = 1 + MARKER
      IF(IR.LE.NCPOPS) THEN
      E(IR,LOCATN(I,J)) = TH(IR,LOCATN(I,J)) + SI
      ELSE
      DO 8 K = 1,NTGR(IR)
      IRR = IRCP(IR,K)
      DO 7 L = 1,N(IRR)
      G(TYPE(IR,K),IRR,L,NCT(IR,K) + 1) = G(TYPE(IR,K),IRR,L,
      *         NCT(IR,K) + 1) + SF(IR,K)*STR(IR,K,LOCATN(I,J)L)
7     CONTINUE
8     CONTINUE
      ENDIF
10    CONTINUE
      MARKER = MARKER + NSTIM1
      IF(MARKER.GE.NSTIM) MARKER = 0
      LSTIM = LSTIM + INTERV
      RETURN
      END
******************
```

II

Dynamic Similarity Theory of Neuronal Firing

Chapter 8 An Overview of the Dynamic Similarity Approach to Neuronal Firing Levels

The purpose of this second part of the book is to provide a theoretical mechanics for neuronal firing, formulated in the dynamic similarity language of characteristic nondimensional parameters. Conversion from input synaptic bombardment to output firing is the fundamental neuroelectric operation performed by single neurons. A succinct universal theoretical characterization of the mechanics and overall properties of this operation is central to a theoretical mechanics of neural networks.

It is instructive to consider the relationship of the theory developed in this second part to that developed in the first part. The first part has developed a theoretical mechanics for the foundations of neuroelectric signaling in neurons. Equations describing the spatial and temporal development of these signals and their relationships to underlying neurobiological processes have been carefully derived and hierarchically organized. We have shown how to represent these equations (and the processes they represent) succinctly in digital computer simulation. We have indicated where the equations can be elaborated to include further neurobiological detail as desired.

These efforts provide the foundations from which development can be made in two main directions. On the one hand, one may formulate computer simulations. With the approach of the first part, one is able to make such computations as elaborate as desired and produce quantitative predictions and analysis of individual or collective neuronal activity, limited only by current shortcomings in neurobiological knowledge or in computational power.

However, such computational efforts are most instructive when used in conjunction with a guiding theory. This is true for applications both to specific experimental situations and within broad theoretical development. Computation should be seen as a tool to enhance the interface of theory and experiment in the classic scientific mode of "theory guides, experiment decides." Indeed, the goal of the entire process is a theory that correctly explains the phenomena in terms of its central causal agents, with the various complicating factors seen in their proper perspective. Within these constraints, the simpler the theory is, the better.

This part of the book, then, attempts to provide just such a central guiding theory for special case of the mechanics of neuronal firing. This theory is based on the membrane equation, the compartmental approach to dendritic geometry, and the threshold rule for firing, all of which are discussed in Part I. It is intended to provide broad theoretical guidance for experimental and computational studies of neuronal firing. We are particularly interested in showing how the central characteristics of the mechanics of firing depend on a neuron's physiological and anatomical parameters, so as to begin the understanding of the comparative operational structural design of neurons.

The first chapter of this part provides a broad, conceptual overview of this theory. The value and significance of the dynamic similarity approach are discussed, the overall conceptual structure of the theory and its central results are surveyed, and its general scope is considered. The remaining four chapters of the section develop the mathematical substance of the theory.

Value and Significance of the Dynamic Similarity Theory for Neurons

There are four powerful values of the dynamic similarity characterization of physical systems, all of which can contribute significantly to our deeper understanding of the operations and functional design of neurons.

Parameter compression is the first value of the dynamic similarity approach. Physical systems typically exhibit a number of parameters, but these parameters usually influence the behavior of the system only in certain groupings. Such groupings, called the *characteristic numbers* of the system, can be found by manipulating the central equations that govern the physical operations of the system. They are advantageously defined to be nondimensional. The number of characteristic nondimensional numbers is less than the number of individual parameters. Moreover, the characteristic numbers can be identified as measuring a centrally significant physical mechanism within the operations of the system.

For example, the Reynolds number can be found in the equation for momentum transfer in fluids, to group together the fluid density (ρ), the fluid viscosity (μ) , the free stream fluid velocity (V), and a characteristic length of the geometry (L), in the ratio $\epsilon V^* L/\mu$. This parameter alone governs the transfer of momentum in the system, and it effectively reduces the number of free parameters from four to one. The ratio can physically be interpreted as the ratio of the inertial forces (which tend to keep the fluid particles in constant motion) to the viscous forces (which tend to alter relative motions).

The direct result of this parameter compression is twofold. First, one has reduced the complexity of parameter influence in the problem, rendering mathematical exploration considerably easier. Second, and considerably more significant, one has gained controlling insight into those minimal essential processes and measures that determine the behavior of the system. One has identified its deepest operative causal agents: one knows, in the most succinct way possible, how the system works and how its processes depend on its properties. This understanding is the essential key to its engineering design.

We will show in this part, from the membrane equation, that a neuron's firing sensitivities are governed by a few characteristic non-dimensional numbers involving its resting and synaptic conductances, threshold, refractory properties, and geometry.

Prediction of behavioral levels and domains is the second significant contribution of the dynamic similarity approach. One may theoretically map out the behavioral repertoire of a system in terms of its characteristic nondimensional numbers, by using its governing equations. The domain or level of behavior of a particular system for a given level of stimulation may then be identified according to the particular values of the nondimensional numbers that apply for that system. Once the general repertoire is mapped out theoretically, the behavior of individual systems for specified stimulation can be predicted on the basis of its characteristic numbers alone. For example, fluid mechanics predicts that the transition from smooth to turbulent flow occurs at particular values of the Reynold's number. One may predict, therefore, whether a fluid will exhibit smooth or turbulent flow for a given free-stream velocity simply by calculating its Reynold's number in this situation.

Comparative operational character is this third contribution of the dynamic similarity approach. The characteristic nondimensional numbers of systems define different operational propensities. Systems with different parameters will respond differently to the same stimulation. The nondimensional numbers define these differences. One may compare the behavioral propensities of different systems, by comparing the numerical values of their characteristic numbers.

This feature is particularly significant with regard to the theoretical mechanics of neuroelectric signaling. One of the most striking features of the brain and nervous system is the high degree of often idiosyncratic order, design, and regularity in its component neural networks. Neural networks in the hippocampus, the cerebral cortex, the reticular formation, the cerebellum, the basal ganglia, the thalamus, the spinal cord, and many others exhibit highly ordered, regular design in both their special classes of often elaborately structured neurons and their systematic patterns of interconnection. This order is highly suggestive of intelligent engineering design, wherein structure and physiology are arranged to serve operation and function. So far, however, we have only the first most general ideas for interpreting some of the interconnection patterns (as discussed further in Chapter 11) and very little foundation for interpreting the design of individual constituent neurons.

The dynamic similarity approach for neurons presented here is directed precisely toward this level of comparative operational interpretation of various cell classes across various neural networks.

Construction of scaled or functionally equivalent systems is the fourth value of the dynamic similarity approach. In fluid mechanics, for example, it is very useful to construct small, scaled-down models of airplanes, so as to perform extensive experimentation with minimum danger, expense, and inconvenience. The scaled-down laboratory models are constructed to exhibit the same characteristic nondimensional numbers as the full-size airplane. The resulting physical results are therefore predicted to be the same in both the scaled down and full size cases.

This principle applies also to the theoretical formulation of neuronal activity. Model neurons can be designed to represent particular biological neurons on the basis of equating their characteristic nondimensional numbers. Alternatively, different biological neurons that have similar characteristic numbers can be expected to exhibit similar dynamic propensities.

Overall Approach and Structure of the Theory

What is wanted is a conceptual characterization of neuronal firing that captures the essential features of its central governing anatomical and physiological processes, but at the same time is not overly obscured by the wealth of possible neurobiological specificity and idiosyncrasy. Similarly, one wants mathematical usage that is faithful to the essential physical processes but not made moribund by needless and misplaced complexity and illusions of precision. The approach followed here seeks representations of the anatomy, physiology, and stimulation conditions, and mathematical manipulations that incorporate this philosophy.

The theory is developed in the following four chapters. Chapter 9 develops the general mathematical foundation of the approach. Chapter 10 determines the parametric characterization for spatially selective and spatially uniform distributions of input. Chapters 11 and 12 develop the universal firing rate transfer curve for various temporal patterns of spatially uniform input.

General Mathematical Foundation

The first section of Chapter 9 applies the techniques developed in Chapter 5 to derive an extended membrane equation for driving the soma potential by synaptic input distributed arbitrarily over an arbitrarily structured dendritic tree. This fundamental result is represented by Eq. (9.5). For convenience of subsequent generalized characterization, it is cast in terms of levels of branching in the dendritic tree.

The second section of Chapter 9 develops the foundations for general parametric characterization of neuronal firing levels on the basis of the extended membrane equation. First, this equation is effectively integrated over time by evaluating it at instants of firing and taking the intervals between firings as the reciprocal of the representative output firing rate. This mathematical device also allows us to separate the equation into two equal parts, one representing the representative effective input levels and one containing the representative output firing rate through the medium of the effective accumulative relative refractoriness it produces.

This separation of the equation is mathematically formalized by the definition of a universal nondimensional characteristic function, ρ^*. The representative output firing rate is uniquely determined by the value of ρ^*. This allows the determination of a universal firing rate transfer curve wherein output firing rate is given in terms of ρ^*. This formulation is given in Eq. (9.7).

On the other hand, the input activation of the neuron can be succinctly interpreted in terms of the value produced for ρ^*, as modulated by the essential anatomical and physiological parameters of the neuron. The final parts of Chapter 9 compress this parametric dependence of ρ^* to its simplest and most direct expression. This is shown in Eq. (9.8), particularly in its final canonical form, Eq. (9.8e).

This chapter, then, produces a succinct mathematical relationship between representative input and output firing rates in arbitrarily structured neurons in a universal language that applies all neurons and includes the means to show clearly how the individual anatomy and physiology of a neuron influence its firing rate sensitivities. Note that the approach is not designed to deal with the fine-grained temporal structure of either output or input trains; this can be done by computer simulation. The approach rather strives to offer us a broad, comparative view of the overall engineering design of input–output firing sensitivities in neurons.

Application to Spatially Uniform and Spatially Selective Input

The first section of Chapter 10 applies the general governing Eq. (9.8), to the case of input distributed uniformly over the compartments of neurons. This case is shown to be equivalent mathematically to that of a point neuron. Its input–output curve is determined by the succinct form shown in Eq. (10.1c). The curve is shown to depend on three characteristic parameters: the effective weighting factors of excitatory and inhibitory synaptic inputs, and the effective refractory leakage parameter. The mathematics shows that the threshold control input firing rate is the reciprocal of the normalized excitatory synaptic weighting factor.

The second section of Chapter 10 applies the general governing Eq. (9.8) to the case of spatially selective input. These cases are comprehensively described by Eq. (10.3). The representation of effective current flow and leakage throughout the dendritic tree by the parameters of this equation system are discussed physically and mathematically. Figures 10-1 and 10-2 and Eq. (10.4) exemplify this discussion.

The chapter concludes with a prototypical parametric representation of a universal canonical neuron with symmetric dendritic tree. It is suggested

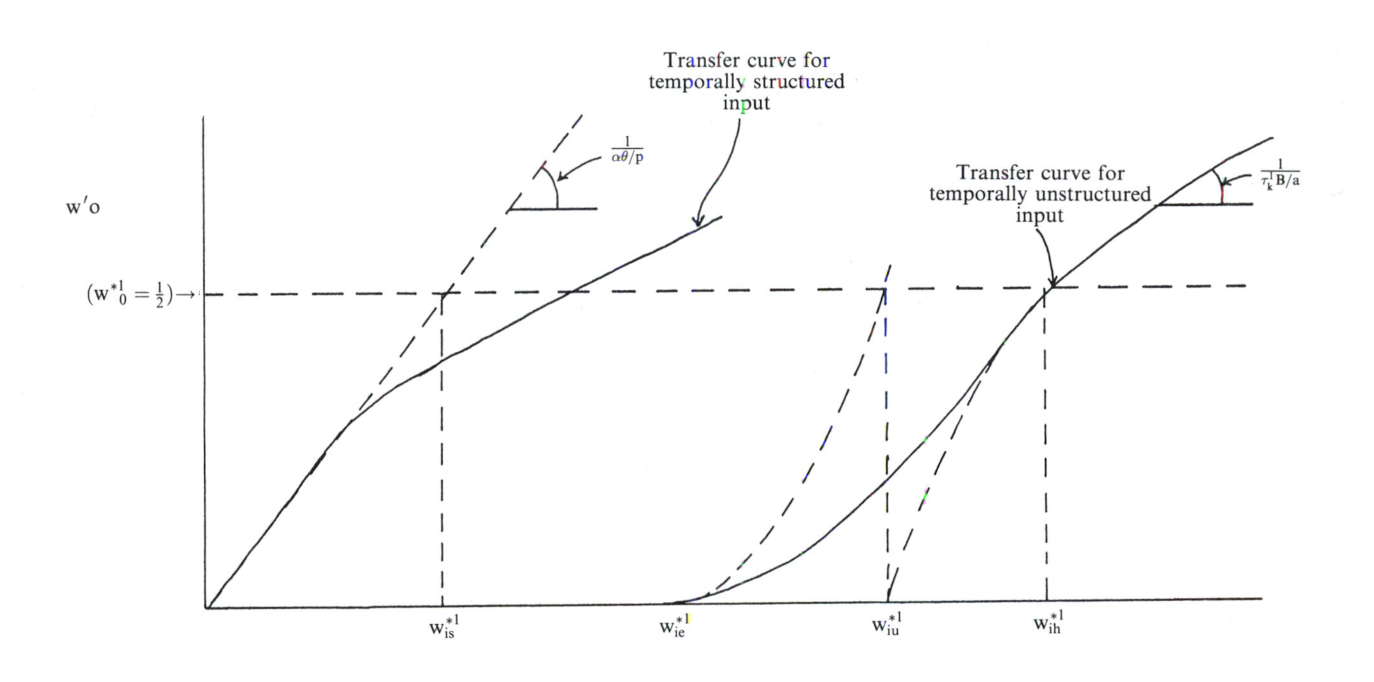

FIGURE 8.1. Universal firing rate curve from dynamic similarity analysis (curves are developed in Chapters 11 and 12).

that this idealization can serve as a template to characterize, in at least approximate terms, the absolute and relative firing rate sensitivities of most, if not all, neurons to their various individual input activation systems. A typical neuron should be representable in terms of about a dozen characteristic effective synaptic weighting factors.

Universal Firing Rate Transfer Curves

Chapters 11 and 12 apply Eq. (10.1c) for spatially uniform input or point neurons to determine explicit parameter-dependence in input–output firing rate transfer for particular temporal patterns of input. The overall firing rate curve obtained from these analyses is shown here as Figure 8.1.

This curve is pieced together on the basis of a neuron's response to three types of input temporal pattern. The first is an optimally temporally structured pattern wherein every input pulse contributes maximally to output firing. This response is shown in the left side of Figure 8.2. It brings about output firing at lower ranges of input rate. This pattern is discussed in the last part of Chapter 11.

The second and third input patterns are temporally unstructured. Deterministic representations in terms of average conductance and summed regular PSP trains produce the right-most part of the transfer curve for unstructured input. These are the least effective inputs for generating output firing. They are discussed in the first section of Chapter 11. The lower portion of the transfer curve for temporally-unstructured input is analyzed in terms of a probabilistic representation of input firing in terms of a Poisson distribution of input intervals. This representation and the resulting transfer curve are shown to merge at higher firing rates. The probabilistic analysis is presented in Chapter 12.

The purpose of the universal transfer curve shown in Figure 8.2 is to relate, in broad comparative terms, the representative input and output firing rates of neurons in terms of their constituent anatomical and physiological parameters. The curve gives a broad mapping of firing domains in terms of a neuron's characteristic output firing rate, $w_o^{*\prime}$, and several characteristic input rates, $w_{is}^{*\prime}$, $w_{il}^{*\prime}$, $w_{iu}^{*\prime}$, and $w_{ih}^{*\prime}$.

For example, one may say on the basis of Figure 8.2 that a given neuron will approach firing at a significant fraction of its characteristic output rate, $w_o^{*\prime}$, when temporally structured input is presented at about w_{is}^{*} or temporally unstructured input is presented at about $w_{iu}^{*\prime}$. All these characteristic rates, in turn, are given in terms of the neuron's parameters in Chapters 11 and 12. The characteristic output rate, for example, is defined

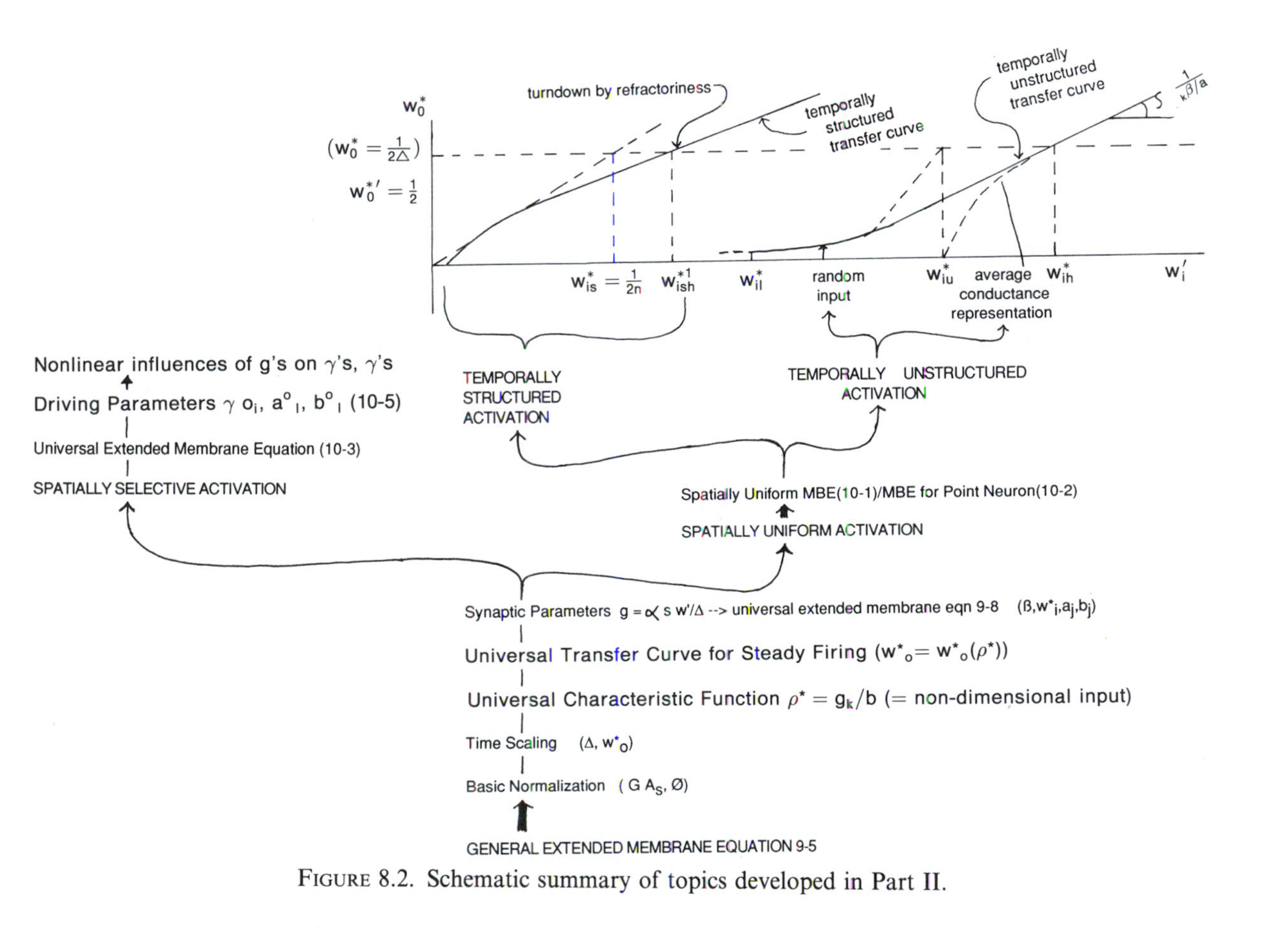

FIGURE 8.2. Schematic summary of topics developed in Part II.

as one-half the reciprocal of the neuron's effective integration interval, which is approximately $1/(2\tau)$. This value can be taken as representative of a neuron's being "strongly on" on both theoretical and empirical grounds.

Comments on the Scope of the Theory

Figure 8.2 summarizes the topics just outlined above and taken up in detail in the next four chapters. The theory developed here can be applied universally to most types of neurons when properly interpreted. Further, the theory can be intelligently modified to apply to situations involving biological factors not explicitly considered here. These statements are considered again at the end of Chapter 12.

Chapter 9 Parametric Characterization of Neuronal Firing: General Mathematical Formulation

This chapter spells out the mathematical formulation of the general parametric characterization of the processes governing neuronal firing. The first section of the chapter derives a general extended membrane equation to describe the potential at the neuronal soma for synaptic activation distributed arbitrarily over an arbitrarily configured dendritic tree. The second section of the chapter rearranges this equation in dynamic similarity terms and identifies its characteristic nondimensional parameters.

THE EXTENDED MEMBRANE EQUATION FOR A NEURON WITH ARBITRARY DENDRITIC TREE

The extended membrane equation for the potential at the neuronal soma in terms of input distributed over the entire dendritic tree is derived in this part according to the compartmental approach discussed in Chapter 5. For all of this chapter, no restrictions are made on the geometry of the dendritic tree nor on the distribution of its synaptic input or degree of synaptic variability.

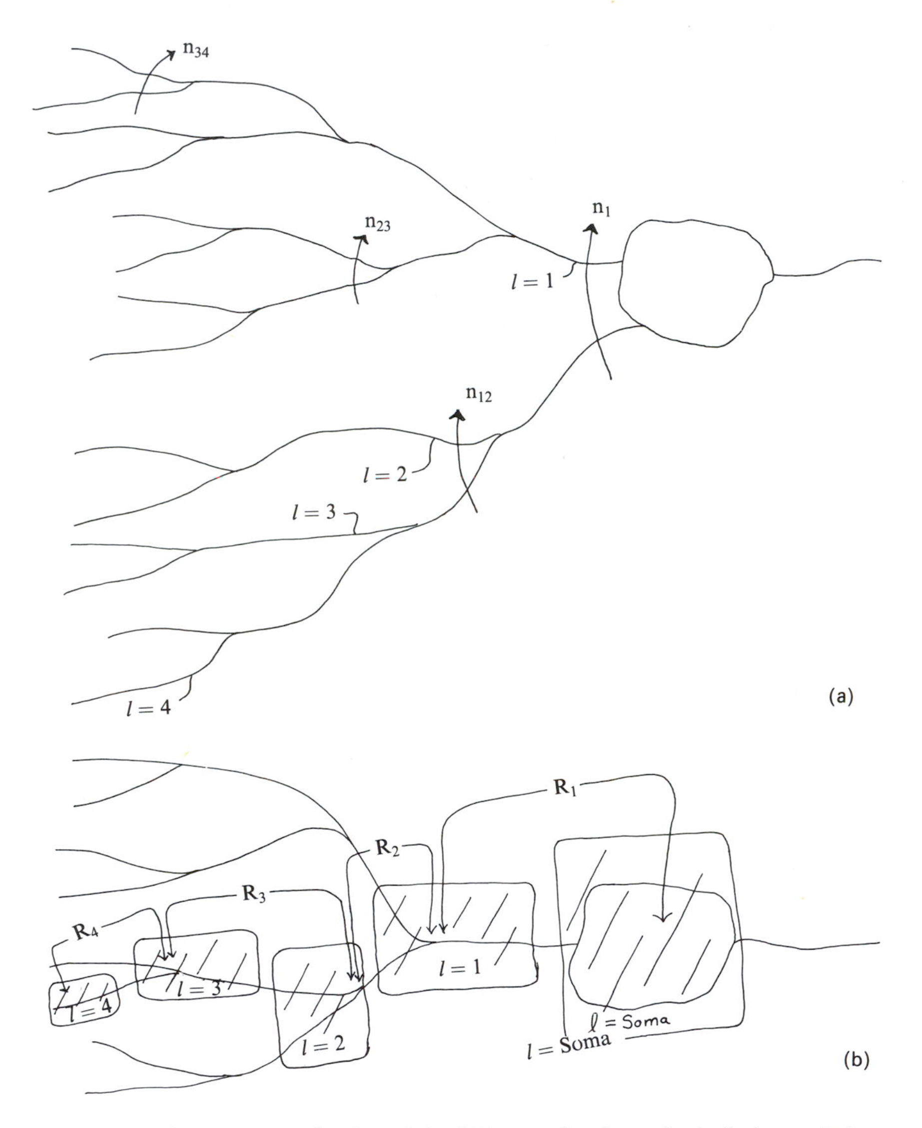

FIGURE 9.1. Compartmentalization of dendritic tree for dynamic similarity analysis: a. arbitrary dendritic tree (identification of levels of branching); b. identification of compartments and intercompartmental conductances; c. equivalent circuit representation.

Such an arbitrarily structured dendritic tree can be represented as shown in Figure 9.1.

This tree is here represented by compartments for each unbranching dendritic segment and the soma, such that all branching points are at

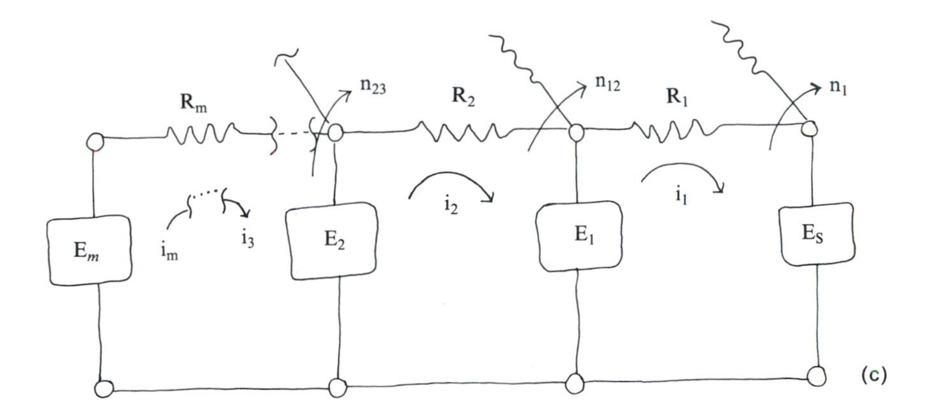

FIGURE 9.1. Continued.

effective compartment centers. The resulting compartmental equivalent circuit representation is illustrated in Figure 9.1b. The unbranching dendritic segments can be identified as occurring within each of m stages, each stage being defined by the number of branching points between it and the soma. Suppose there are m levels of branching with n_l branches at each level and a total of n_b branches. The microstructure of the branching can be defined in terms of numbers of branches at each branching point, n_{lk}, where l identifies the branch level and k identifies the individual branch at the lth level. The total number of compartments, n, is $nb + 1$.The relations among n_l, n_b, n_{lk}, and n are shown in Eq. (9.1):

$$\begin{aligned} n_l &= \sum n_{l-1,l} \\ n_b &= n_1 + n_2 + \ldots + n_m \\ n &= 1 + nb \end{aligned} \tag{9.1}$$

Individual branches can be identified either by the two subscripts, l and k, or by a single subscript j. The double subscripting (l and k), referring to branch levels, is necessary for the mathematical derivation, but the single subscripting (j) is considerably clearer for most purposes.

Following the general methods of Chapter 5, the governing equations for individual compartment at each of the lth levels can be written as shown in Eq. (9.2):

$$a_j \frac{\mathrm{d}E_j}{\mathrm{d}t} = -\left\{ a_j + \sum \frac{1}{R'_{ji}} \right\} E_j + \sum g_j[E^* - E_j] + \sum \frac{E_i}{R'_{ji}} \tag{9.2a}$$

$$a_j = A_j/A_s, \qquad g_j = \frac{g_j \text{ total}}{GA_s}, \qquad SC' = \frac{SC}{GA_s\theta}$$
$$R'_{ji} = R_{ji}\, GA_s, \qquad R_{ji} = \frac{\rho L_{lj}}{\prod r_{lj}^2} \tag{9.2b}$$

$$g_k[E_s - E_k^*] = -E_s + g_s[E^* - E_s] + \sum \frac{E_1 - E_s}{R'_1} - \xi_s + SC'$$

$$E_1 = \frac{g_1 E_1^* + E_s/R'_1 + \sum E_2/R'_2 - \xi_1}{a_1 + g_1 + 1/R'_1 + \sum 1/R'_2}$$
$$\vdots \tag{9.2c}$$
$$E_l = \frac{g_l E_l^* + E_{l-1}/R'_l + \sum E_{l+1}/R'_{l+1} - \xi_l}{a_l + g_l + 1/R'_l + \sum 1/R'_{l+1}}$$

$$E_m = \frac{g_m E_m^* + E_{m-1}/R'_m - \xi_m}{a_m + g_m + 1/R'_m}$$

Equation 9.2a are the membrane equation for each compartment. Note that each E_l is understood to represent the potentials for all the n_l branches at that level; to emphasize conceptual structure rather than mathematical detail and to avoid excessively cumbersome notation, the additional k subscripts are not written explicitly.

Recall that the terms in this equation have been normalized to the total conductance of the soma, G^*A_s, and to the neuronal firing threshold, θ, as shown in Eq. (9.2b). The a_l terms represent the membrane surface areas of the individual compartments normalized to the surface area of the soma. This interpretation assumes that the resting membrane conductance is uniform throughout the neuron. The g_l terms represent the active conductance modulations in each compartment, normalized to the total conductance of the soma. These may include any number of active synaptic types in any compartment, although for ease of representation will be represent these by the one symbol, g_l. The potentials are normalized to firing threshold.

Equations (9.2c) expand the generic form, (9.2a), to the specific branching levels of this representation. Note that each equation in (9.2c), except the first, represents a number of different equations, one for each of the unbranched dendritic segments at that stage. The sums in these equations are over the multiple branchings at the distal ends of each dendritic segment.

The equation of central interest here is the first of Eq. (9.2c), the equation for the soma. The active conductances for the soma have been expanded to include the postfiring refractory conductance, g_k, as well as various synaptic activations of the soma, g_s. The capacitive terms in the equations generally will be neglected and the results interpreted in terms of steady or target levels of activity. The corrections required by this simplification are indicated by the terms, ξ_l, in Eq. (9.2c). These terms and the term for applied activating current, SC', in the soma equation, will be carried along to show where such terms can be inserted when desired. These terms for the capacitive effect, ξ_l, and the applied current will not be carried through the analysis of this chapter but will be reintroduced briefly in Chapter 11.

Since the potentials at any branch level, l, can be computed in terms of direct interactions involving only potentials in the $l-1$ and $l+1$ stages, Eq. (9.2) can be conveniently rearranged into the form shown in Eq. (9.3):

$$g_k[E_s - E_k^*] = -E_s + g_s[E_s^* - E_s] + \sum \frac{E_1 - E_s}{R_1'}$$

$$E_1 = \frac{E_s/R_1' + g_1 E_1^* + (g_2^* E_2^*)}{(a_1 + g_1 + 1/R_1') + G_2^*}$$

$$\vdots \tag{9.3a}$$

$$E_l = \frac{g_l E_l^* + (g_{l+1}^* E_{l+1}^*) + \sum E_{l-1}/R_l'}{(a_l + g_l + 1/R_l') + G_{l+1}}$$

$$E_m = \frac{g_m E_m^* + E_{m-1}/R_m'}{(a_m + g_m + 1/R_m')}$$

$$g_l^* E_l^* = \frac{\sum g_l E_l^* + (g_{l+1}^* E_{l+1}^*)}{R_l'(a_l + g_l + G_{l+1}^*) + 1}$$

$$G_l^* = \sum \frac{a_l + g_l}{R_l'(a_l + g_l + G_{l+1}^* + 1} \tag{9.3b}$$

$$g_{m+1}^* E_{m+1}^* = G_{m+1}^* = 0$$

This form lends itself to an explicit iterative solution of the system and begins to show the characteristic groupings of some of the main parameters. The terms, $g_l^* E_l^*$, and G_l^* are defined for iterative convenience in Eq. (9.3b).

Equations (9.3) allow one to write an explicit solution for the extended membrane equation for the soma, by iteratively substituting for the E_l, and making use of the definitions in (9.3b). The solution for four stages of branching is shown explicitly in Eq. (9.4a)

$$
\begin{aligned}
g_k'(E_s - E_k^*) &= g_s(E_s^* - E_s) - E_s - \sum_{\textcircled{1}}^{n_1} \frac{E_s}{R_1'} \\
&+ \sum_{\textcircled{1}} \left\{ \frac{\dfrac{E_s}{R_1'} + g_1' E_1^* + \sum\limits_{\textcircled{2}} \dfrac{g_2' E_2^* + \sum\limits_{\textcircled{3}} \dfrac{g_3' E_3^* + \sum\limits_{\textcircled{4}} \dfrac{g'4E_4^*}{1 + R_4'(a_4 + g_4')}}{1 + R_3'\left\{ (a_3 + g_3') + \sum\limits_{\textcircled{4}} \dfrac{(a_4 + g_4')}{1 + R_4'(a_4 + g_4')} \right\}}}{1 + R_2'\left[(a_2 + g_2') + \sum\limits_{\textcircled{3}} \dfrac{(a_3 + g_3')}{1 + R_3'\left\{ (a_3 + g_3') + \sum\limits_{\textcircled{4}} \dfrac{(a_4 + g_4')}{1 + R_4'(a_4 + g_4')} \right\}} \right]}}{1 + R_1'\left\{ (a_1 + g_1') + \sum\limits_{\textcircled{2}} \dfrac{(a_2 + g_2')}{1 + R_2'\left[(a_2 + g_2') + \sum\limits_{\textcircled{3}} \dfrac{(a_3 + g_3')}{1 + R_3'\left\{ (a_3 + g_3') + \sum\limits_{\textcircled{4}} \dfrac{(a_4 + g_4')}{1 + R_4'(a_4 + g_4')} \right\}} \right]} \right\}} \right\}
\end{aligned}
\tag{9.4a}
$$

$$g_k[E_s - E_k^*] = g_s[E_S^* - E_s] - E_s - \sum E_s/R_1' + \sum(\alpha_1/R_1')E_s + \sum \alpha_1 g_1 E_1^*$$
$$+ \sum \alpha_1 \sum \alpha_2 g_2 E_2^* + \sum \alpha_1 \sum \alpha_2 \sum \alpha_3 g_3 E_3^* \qquad (9.4b)$$
$$+ \sum \alpha_1 \sum \alpha_2 \sum \alpha_3 \sum \alpha_4 g_4 E_4^*$$

$$\alpha_4 = \frac{1}{1 + R_4'(a_4 + g_4)}$$
$$\alpha_3 = \frac{1}{1 + R_3'[(a_3 + g_3) + \sum \alpha_4(a_4 + g_4)]}$$
$$\alpha_2 = \frac{1}{1 + R_2'[(a_2 + g_2) + \sum \alpha_3(a_3 + g_3)]} \qquad (9.4c)$$
$$\alpha_1 = \frac{1}{1 + R_1'[(a_1 + g_1) + \sum \alpha_2(a_2 + g_2)]}$$

The general solution for higher numbers of branch levels can be written immediately by inference from Eq. (9.4a). Notice that this solution includes four levels of nested summation, as it must to represent the four stages of branching. Summation over the *m*th branching points is indicated by the symbol Σ.

This solution can be conveniently recast as shown in Eq. (9.4b), by defining the characteristic structural parameters, α_l, as shown in (9.4c). A fundamental fact of neuronal activation by conductance modulations is that the resistive structure of neurons is dependent on the magnitudes and distribution of its active conductance modulations at any given time. This fact is shown clearly by the presence of the g_l terms in the definitions for the values of α in Eq. (9.4c). These Equations also clearly show that the appropriate way to deal with this feature of conductance activation, is to consider that the $g_l E_l^*$ terms represent the central driving influences of synaptic activation, and they treat the additional effects of the g_l as secondary influences on the neuron's resistive structure.

Equation (9.4b) is a relatively direct representation of the influence on the soma potential of input applied over the entire dendritic tree expressed in terms of the primary geometrical and physiological parameters of the neuron. We can now begin to identify the physical characteristics implied by this equation and to make it more compact. First, note that in addition to the leakage of current directly out through the soma membrane (indicated by the lone E_s term) there is now additional leakage into the dendritic tree represented by the combined term, $\Sigma\{[1 - \alpha_1]/R_1'\}$.

The synaptic drive components from individual dendritic compartments are represented directly by the $g_l^* E_l$ terms. These terms are weighted with individual weighting factors for each individual compartment in terms of the values of α. The terms, α, in turn, can be given a very direct and useful physical interpretation: any given α_l represents the proportion of current input to its compartment that is channeled into its $l-1$th parent branch. The remainder of this current flows out both through the compartment membrane and into the peripheral branches of the tree to which it is connected. Therefore, α_2 is the percentage of input current to a second-level branch that gets to its parent first-level branch, and α_1 is the percentage of current to a first-level branch that gets to the soma.

It follows immediately that the percentage of current that gets to the soma from any given compartment in the tree is equal to the product of the α terms of all the branches that interconnect the compartment with the soma. These products are exactly the sums shown as the weighting factors in Eq. (9.4b). We can therefore write Eq. (9.4b) more succinctly as Eq. (9.5a).

Equations (9.5) have been the goal of this section. These equations

$$g_k[E_s - E_k^*] = -E_s[1 + D] + g_s[E_s^* - E_s] + \sum \gamma_{1k} g_1 E_1^* + \sum \gamma_{2k} g_2 E_2^* + \ldots + \sum \gamma_{2m} g_m E_m^* \tag{9.5a}$$

$$D = \sum [1 - \gamma_{1k}]/R_1', \qquad \gamma_j = \gamma_{lk} = \alpha_{1k}\alpha_{2k}\ldots\alpha_{lk} \tag{9.5b}$$

give us a compact representation of the influences on the soma potential of input distributed arbitrarily over the neuron, in terms of its constituent anatomical and physiological parameters. The influences of these parameters are represented succinctly in the soma–dendritic leakage term, D, and the individual compartmental weighting factors for synaptic input, (γ_{lk} or γ_j).

The fact that the synaptic weighting factors are influenced by the synaptic conductances does not detract from the usefulness of this characterization. One may draw most all the meaningful interpretations and comparisons on the basis of idealized control cases, where these nonlinear effects are ignored. One may understand conceptually that the effect of the nonlinearities is to diminish the effectiveness of input by allowing more current leakage. These effects can be approximated if desired. One may compute the influence of non-linear effects precisely from the equations whenever it is necessary.

In the next section, we will further manipulate the extended membrane equation (9.5a) to apply to input-output firing relationships in neurons. From here on we will generally refer to γ and other individual compartmental variables with the single subscript, j.

General Parametric Characterization of the Extended Membrane Equation for Neuronal Firing

This part adapts the extended membrane equation just developed to input–output firing relationships in neurons, and applies the dynamic similarity formulation of parameter characterization to this relationship. This chapter formulates this parameter characterization in most general terms. Subsequent chapters work out the specifics for main classes of spatial and temporal distributions of input. Here we consider the adaptation of the equation to firing rates, the general form of the universal firing rate transfer curve, and universal normalization of the equation and its parameters.

Adaptation to Firing Rates

Equation (9.5a) gives us the somatic potential, E_s, in terms of its driving input activation. We can manipulate this equation to give us the broader output variable, the neuron's output firing rate. The first step is to imagine the case of steady output firing and evaluate Eq. (9.5a) at the instant, t^*, at which an output action potential is being initiated. The soma potential at this instant is equal to threshold, so $E_s = 1$. Applying Eq (9.5a) to this case produces Eq (9.6):

$$g_k/b = \rho^* = \frac{-[1+D] + g_s[E_s^* - 1] + \sum \gamma_j g_j E_j^*}{b[1 - E_k^*]} \tag{9.6}$$

Notice that we have also separated the equation into two components by defining the universal nondimensional characteristic function, ρ^*. We will now show that the lefthand side of this equation, represented by the post-firing refractoriness, g_K, depends on and defines the output firing rate, whereas the righthand side, containing the input terms, depends on and defines the input firing rates. We have divided the equation by the unit relative refractory magnitude $b(1 - E_k^*)$, to include this term in the definition of the universal characteristic function, ρ^*. Since E_k is negative, it helps avoid possible confusion by writing it in terms of its absolute value.

Universal Refractory Curve

The relation of ρ^* to output firing rate can be considered on the basis of the representation of steady output firing at rate, w_o, indicated in Figure 9.2. This figure shows the accumulated relative refractory conductance due to potassium during steady output firing. Suppose that the cell has fired z times

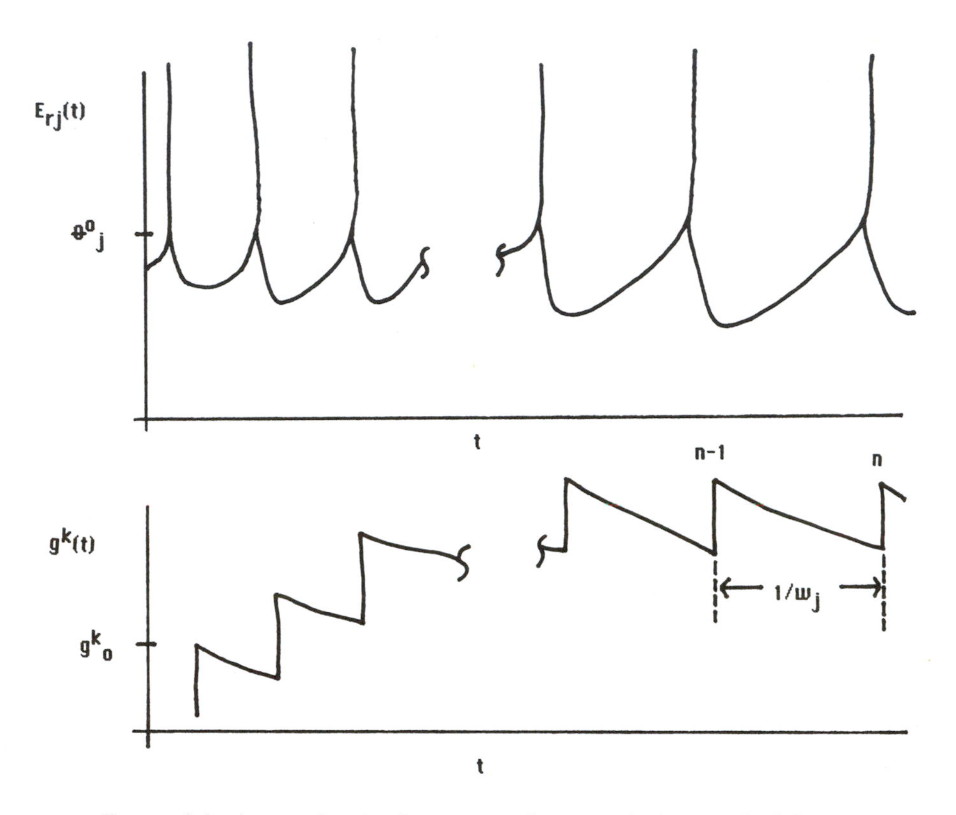

FIGURE 9.2. Accumulated refractory conductance during steady firing.

at regular intervals of $1/w_o$. Consider in particular, its value at a time, t^*, an instant before the cell is to fire again. We can write the accumulative value for g_k at t^* in terms of exponential decays at multiple intervals of $(1/w_o)$ as shown in Eq. (9.7a).

$$g_k(t^*) = b\exp\,[-1/(w_0\tau_k)] + b\exp[-2/(w_o\tau_k)] + \ldots b\exp[-z/w_0\tau_k)] \tag{9.7a}$$

$$g_k(t^*) = b\exp\,[-1/(w_0\tau_k)]\sum\exp[-i/(w_o\tau_k)] \tag{9.7b}$$

$$g_k(t^*) = b\exp\,[-1/(w_0\tau_k)]\{1-\exp\,[-z/(w_o\tau_k)]\}/\{1-\exp\,[-1/(w_0\tau_k)]\} \tag{9.7c}$$

$$g_k/b = \frac{1}{\exp\,(1/(w_0\tau_k)] - 1} = \rho^* \tag{9.7d}$$

$$w_0 = \frac{1}{\tau_k \ln\,[1+1/\rho^*]}, \quad \text{for } \rho^* > 0; \qquad = 0 \quad \text{for } \rho^* \le 0 \tag{9.7e}$$

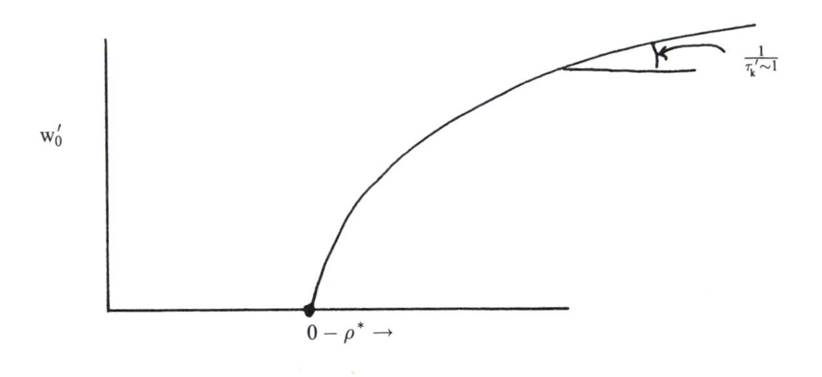

FIGURE 9.3. Universal output firing rate as a function of ρ^*.

$$w_0' = \frac{1}{\tau_k' \ln\,[1 + 1/\rho^*]}, \qquad \rho^* > 0 \tag{9.7f}$$

$$\frac{\partial w_0'}{\partial \rho^*} = 1/\tau_k' \approx 1 \tag{9.7g}$$

This expression can be compressed notationally as shown in (9.7b). The form, however, is a geometric series in exp $(-1/(w_{ok}))$. Therefore g_k can be expressed very succinctly as shown in (9.7c).

Since the neuron is to fire at time t^*, the membrane potential is at threshold, so we can apply Equation 9.6. This means that g_k/b can be set equal to ρ^*, as shown in Equation (9.7d). We have taken the case where the neuron has been firing steadily for some time, $z \to \infty$. Eq. (9.7d) can be solved explicitly for w_o as shown in (9.7e). This equation can be cast in the normalized, universal form shown in Eq. (9.7f), where both w_o and τ_k have been normalized with respect to a characteristic time interval, Δ. This interval will be taken as the representative integration interval of the neuron. One usually takes the membrane time constant as its value. Typically, τ_k' will be not much different from 1.

The universal characteristic steady firing rate curve as a function of ρ^* given by Eq. (9.7f) is shown in Figure 9.3. Physiologically, this curve depends only on the assumption that relative refractoriness is mediated by accumulative exponentially decaying excursions in potassium conductance. It is reasonable to take the firing rate $w_o = 1/\Delta$ as a nominal upper limit on the normal firing of a neuron, and $w_o = 1/(2\Delta)$ as an estimate on the medium-high side of its normal operative rate. These correspond to values of w_o' of 1 and 1/2 and, by the universal equation (9.7f), to values of ρ^* of

0.582 and 0.157, respectively. If τ_k can be approximated by the membrane time constant the slope of the universal steady firing curve approaches 1 at high ρ^*.

Universal Normalization

Equation (9.6) has allowed us to link the characteristic function, ρ^*, to the universal steady output firing rate curve shown in Figure 9.3. We now will manipulate the input side of ρ^* to attain the best characterization of the forces driving towards and resisting spike generation.

Equations (9.8a) introduce the significant synaptic parameters into the equation governing neuronal firing.

$$g_j = \delta_j s_j w_j = \frac{\delta_j s_j w_j}{\Delta(1/\Delta)} = \frac{\delta_j s_j w_j'}{\Delta} \tag{9.8a}$$

$$\begin{aligned}\rho^* = \Big\{ &- [1 + D] + [(\delta_s s_s/\Delta)[E_s^{*\prime} - 1]w_s' + \sum \gamma_j(\delta_j s_j/\Delta)E_j^{*\prime}w_j'] \\ &+ [(\delta_s^i s_s^i/\Delta)[1 + |E_s^{*\prime}|]w_s^{i\prime} + \sum \gamma_j(\delta_j^i s_j^i/\Delta)|E_j^{*i}|w_j^{i\prime}]\Big\}\Big/(b[1 + |E_k^{*\prime}|])\end{aligned} \tag{9.8b}$$

$$\rho^* = \frac{[a_s w_s' + \sum a_j w_j'] - [b_s w_s^{i\prime} + \sum b_j w_j^{i\prime}] - (1 + X)}{\beta} \tag{9.8c}$$

$$a_j = \frac{\gamma_j \delta_j s_j E_j^{*\prime}}{(1 + D)\Delta}, \qquad b_j = \frac{\gamma_j \delta_j s_j |E_j^{*i\prime}|}{(1 + D)\Delta},$$

$$a_s = \frac{\delta_s s_s [E_s^{*\prime} - 1]}{(1 + D)\Delta}, \qquad b_s = \frac{\delta_s s_s^i [1 + |E_s^{*i\prime}|]}{(1 + D)\Delta} \tag{9.8d}$$

$$\beta = b[1 + |E_k^{*\prime}|]/(1 + D)$$

$$\rho^* = \frac{[a_s w_s' + \sum a_j w_j'] - [b_s w_s^{i\prime} + \sum b_j w_j^{i\prime}] - a w_i^*}{\beta} \tag{9.8e}$$

$$w_i^* = [1 + X]/a, \qquad a = \text{representative value of } (a_s, a_j) \tag{9.8f}$$

First, we represent the synaptic conductance terms by the expressions shown in Eq. 9.8a. Here, δ_j and s_j, are, respectively, the duration and amplitude of the individual synaptic conductance modulations at the synapses of the jth compartment. The w_j can be taken as the jth input firing rate, so that these equations represent average synaptic conductance values over time. Alternatively, one may think of the w_j simply as flags to identify the individual synapses.

Substituting Eq. (9.8a) into (9.6) results in (9.8b). In this expression we have explicitly expressed two classes of synapses (excitatory and inhibitory) in each compartment. This can be generalized to include any number of distinct synaptic types in any compartment in obvious ways whenever desired. The subscripts s refer to the soma; the superscripts i refer to inhibition; unsuperscripted synaptic quantities refer to excitatory synapses.

The first steps in the characteristic normalization process are indicated in Eqs. (9.8c) and (9.8d). First, we define the effective nondimensional synaptic weighting factors, a_j and b_j, for excitatory and inhibitory synapses, respectively, as shown in Eq. (9.8d). These measure the effective unit input current seen at the soma from synapses in the jth compartment, normalized for current leakage and expressed in terms of the governing parameters.

These equations also define the universal nondimensional refractory parameter, β, which measures the leakage of current from the soma due to refractoriness relative to resting leakage from the soma both through the soma membrane and into the dendritic tree.

Note that Eq. (9.8c) has been generalized by the inclusion of the term X, which is a catch-all generalization term that allows one to place in the theory additional driving terms may sometimes show up in the membrane equation, such as terms for applied current, SC', or capacitive terms, as represented by ξ in Eq. 9.2, and other similar effects. We will find it useful in Chapter 11 to subsume the synaptic inhibitory terms within X. However, X normally can be taken as zero.

Also note that it is very useful to define the characteristic input firing rate, $w_i^{*\prime}$, in terms of a representative synaptic weighting factor, a, as shown in Eq. (9.8f). Mathematically, the term $a^* w_i^{*\prime}$ is simply $1 + X$. Physically, however, $w_i^{*\prime}$ has the very useful quality of directly measuring the threshold value of the representative input firing rate (i.e., the value of input firing rate at which ρ^* becomes greater than 0, and therefore produces positive output firing).

Thus, Eq. (9.8e) expresses the universal input activation function, ρ^*, for arbitrary activation of any arbitrarily structured dendritic tree by excitatory and inhibitory synapses in terms of universal non-dimensional weighting factors for all individual synaptic types in each of the dendritic branches of the neuron and the universal refractory leakage parameter, β. The equation also defines the characteristic input firing rate, $w_i^{*\prime}$, as the reciprocal of the effective excitatory synaptic weighting factor.

Eqs (9.8e), (9.6), and (9.7f), comprise a universal description applicable in principle to any neuron with any activation. The next chapter will consider their application to the cases of spatially selective and spatially uniform synaptic activation in order to further illuminate these characteristic numbers and their usefulness in describing neuronal firing behavior. The final two chapters of the section will specifically apply the approach to explicit universal firing rate transfer curves for temporally regular, and temporally irregular input uniformly distributed.

Chapter 10 Parametric Characterization of Neuronal Firing with Spatially Distributed Input

This chapter extends the general equations developed in the preceding chapter for parametric characterization of neuronal firing to include explicit consideration of the spatial distribution of input firing patterns. The first section of the chapter applies to the case of input firing distributed uniformly over the compartments of the neuron. This case will be shown to be essentially equivalent to that of a point neuron and characterizable in terms of three nondimensional parameters: the effective synaptic weighting factors for the excitatory and inhibitory inputs, and the effective leakage through the refractory potassium channels; or equivalently, the control threshold input firing rate (seen as the reciprocal of the excitatory synaptic weighting factor), the ratio of inhibitory to excitatory synaptic weighting factors, and the ratio of the refractory leakage to the excitatory synaptic weighting factor. The next two, and last, chapters of this part apply these equations to developing explicit equations governing a universal input–output transfer curve for neuronal firing rates.

The second part of this chapter looks more closely at the mechanics for describing the influence of dendritic geometry in cases of selectively distributed inputs. The equations and parameters governing the responses to inputs applied at specific individual compartments are presented explicitly and discussed. In addition, the family of characteristic parameters that form the dynamic similarity description of prototypical symmetric neuronal dendritic trees are derived.

Spatially Uniform Activation

Input activation distributed uniformly across the unbranching dendritic segments of a neuron can be expressed for our compartmental representation as shown in Eq. (10.1a):

$$w'_s = w'_j = w'/n, \qquad w^{i\prime}_s = w^{i\prime}_j = w^{i\prime}/n \tag{10.1a}$$

$$g_k/b = \rho^* = [aw' - bw^{i\prime} - aw^*_{io}]/\beta$$
$$a = \left(a_s + \sum a_j\right)/n, \qquad b = (b_s + \sum b_j)/n, \qquad w^{*\prime}_{io} = 1/a \tag{10.1b}$$

$$g_k/b = \rho^* = [w' - (b/a)w^{i\prime} - w^*_{io}]/(\beta/a) \tag{10.1c}$$

The salient feature of the dynamic similarity characterization of this situation is its extreme simplicity. Substitution of the input distribution of Eq. (10.1a) into the general governing equation, (9-8e), produces the succinct form shown in Eq. (10.1b). The effective synaptic strengths—excitatory, a, and inhibitory, b—are clearly defined as the corresponding average strengths over the neuronal compartments.

This equation is governed by the three parameters a, b, and β (remember that w^*_i is simply $1/a$). These three parameters are relative current leakage parameters: they measure, respectively, the relative current leakage in through excitatory synapses, out through inhibitory synapses, and out through postfiring refractory leakages—all relative to the combined leakage from the soma out through its membrane and into its dendritic tree. The physiological controlling factors for the membrane potential and the resulting neuronal firing are the membrane and longitudinal current channels; it is not surprising that these define the characteristic parameters of the governing equation.

Equation (10.1b) can be expressed even more succinctly by Eq. (10.1c). This equation characterizes the behavior in terms of the three parameters: w^*_i, b/a, and β/a. These parameters are, respectively, the control threshold excitatory input firing rate, the relative inhibitory synaptic strength (as compared to excitatory synaptic strength), and the relative refractory leakage (as compared to excitatory synaptic strength).

Similarity to a Point Neuron

This governing equation for spatially uniform activation is functionally equivalent to that for a point neuron as may be readily justified by Eq. (10.2):

$$\frac{dE}{dt} = \left[-E + \sum g^i[E^{i*} - E] + SC'\right]/\tau \tag{10.2a}$$

$$g_k/b = \rho^* = \frac{g^e[E^* - 1] - g^i[1 - E^{i*}] - 1 + SC' - \xi}{b[1 + |E_k^*|]} \tag{10.2b}$$

$$g_k/b = \rho^* = \frac{(\delta_s/\Delta)w'[E^* - 1] - (\delta^i s^i/\Delta)w^{i\prime}[1 + |E^{i*}|] - 1 + SC' - \xi}{b[1 + |E_k^*|]} \tag{10.2c}$$

$$\rho^* = [a_p w' - b_p w^{i\prime} - a w_i^*]/(b[1 + |E_k^*|]) \tag{10.2d}$$

$$\begin{aligned} a_p &= (\delta s/\Delta)[E^* - 1], \qquad b_p = (\delta^i s^i/\Delta)[1 + |E^{i*}|], \\ X &= SC' + \xi, \qquad w_i^{*\prime} = (1 + X)/a \end{aligned} \tag{10.2e}$$

Equation (10.2a) is the membrane equation for a point neuron consisting of a soma compartment only; it can be readily put into the form shown in (10.2b), by breaking the active conductances into excitatory and inhibitory synaptic components and a potassium refractory component and representing the capacitive term by ξ. Since there are no dendrites SC' is either 0 or reflects current supplied by a stimulating electrode. If conductances are represented by the products of duration, amplitude, and input rates as in Eq. (9-8a), then Eq. (10.2c) results. This equation can be compressed into the form shown in Eq. (10.2d).

By comparing Eq. (10.1c) and (10.2d), we can see that the more general case of activation uniformly distributed over a dendritic tree generalizes the effective synaptic weighting factors to include the somatically seen effectiveness of all compartments, and it generalizes the somatic leakage to include leakage into the dendritic tree (represented by D). With these simple modifications, the governing equations for the two cases are identical.

The only physical difference in behavior is that the effective synaptic strengths vary in favor of inhibition at higher input firing rates for the neuron with dendritic tree. This is because of the increased effective membrane leakage due to active synapses throughout the tree. This nonlinear effect is expressed mathematically in the governing equations,

(10.1), by the g_j' terms in the denominators of the α terms. With this small caveat, the basic forms of the governing equations are essential identical, and the overall characteristics of their behavior are fundamentally similar. Most significant theoretical characterizations of neurons and comparisons across neurons can be made neglecting these effects. The effects can be estimated or computed when desired.

Chapters 11 and 12 will apply Eq. (10.1) to the development of universal firing rate transfer curves for neurons subjected to temporally regular and temporally irregular input, respectively.

Selectively Distributed Input

Selective distribution of synaptic input over dendritic trees may include great variety, intricacy, and idiosyncrasy, and it may involve much complexity in analysis. The present purpose is not to spell out and elaborate these kinds of intricate interactions. They can be better dealt with by digital computer simulation as developed in Part I. Rather, our purpose here is to determine the salient groupings of anatomical and physiological parameters that govern the firing rate responsiveness in neurons. These will give us insight into the mechanics of interaction of variously placed excitatory and inhibitory inputs, and their effectiveness, seen against the receptive morphology and physiology of the receiving neuron. This in turn will allow us to make clearer interpretations of the neuron's operations and to make interpretative and predictive comparisons of operational characteristics across types of neurons.

We now show how the mechanics developed in Chapter 9 should be applied to various situations involving dendritic trees and how it can be used to succinctly characterize the relative dynamic effectiveness of input systems placed in particular regions of dendritic trees. We consider and discuss the applications of Eq. (9.8) to particular dendritic regions generally, analyze in detail the response of a simple dendritic tree to a single input channel and, lastly, identify the family of canonical parameters that comprehensively define the comparative distributed responsiveness of prototypical dendritic trees.

General Interpretation of the Governing Equations for Dendritic Input

Figure 10.1 shows a prototypical dendritic tree with four levels of branching and simple bifurcations at every branch point. The general mechanics

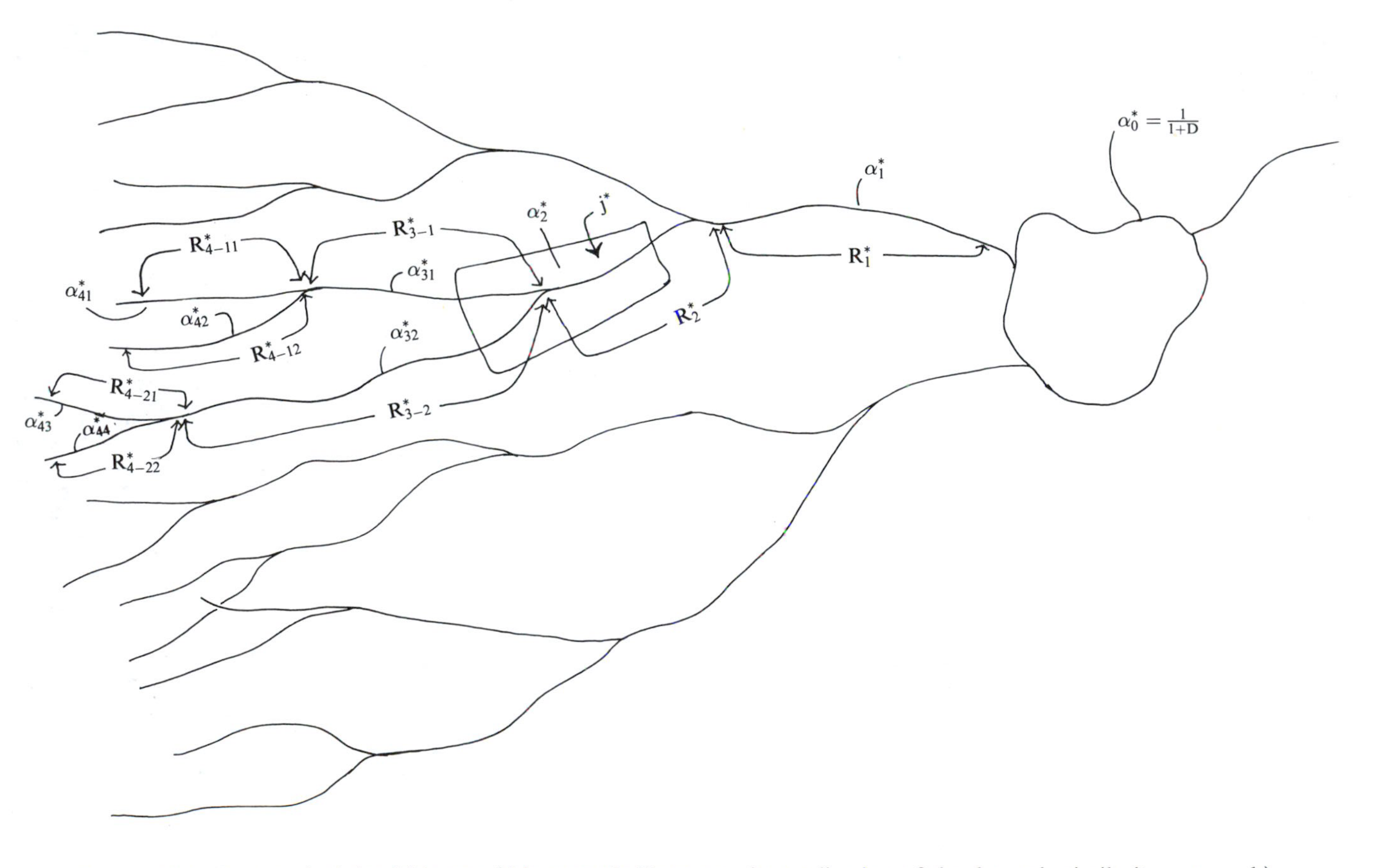

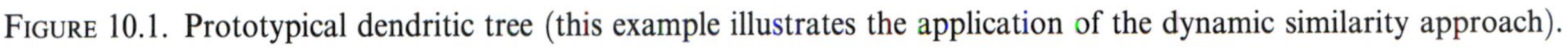
FIGURE 10.1. Prototypical dendritic tree (this example illustrates the application of the dynamic similarity approach).

governing the dynamic responses in such a tree are described by Equations 10.3, which are reproduced from Chapter 9 where they were developed:

$$\rho^* = \frac{[a_s w_s' + \sum a_j w_j'] - [b_s w_s^{i\,\prime} + \sum b_j w_j^{i\prime}] - a w_i^*}{\beta} \tag{10.3a}$$

$$a_j = \frac{\gamma_j \delta_j s_j E_j^{*\,\prime}}{(1+D)\Delta}, \qquad b_j = \frac{\gamma_j \delta_j^i s_j^i |E_j^{*\,i\prime}|}{(1+D)\Delta},$$

$$a_s = \frac{\delta_s s_s [E_s^{*\,\prime} - 1]}{(1+D)\Delta}, \qquad b_s = \frac{\delta_s s_s^i [1 + |E_s^{*\,i\prime}|]}{(1+D)\Delta} \tag{10.3b}$$

$$\beta = b[1 + |E_k^{*\,\prime}|]/(1+D), \qquad w_i^* = [1+X]/a, \qquad X = SC' + \xi + \ldots$$

$$a = \text{representative value of } (a_s, a_j)$$

$$\gamma_j = \gamma_{lk} = \alpha_1 \alpha_2 \ldots \alpha_l$$

$$\alpha_l = \frac{1}{1 + R_l'[(a_l + g_l) + \sum \alpha_{l+1}(a_{l+1} + g_{l+1}')]}$$

$$g_l' = g_l/(GA_s), \qquad a_l = A_l/A_s, \tag{10.3c}$$

$$R_l' = R_l G A_s$$

$$= \text{longitudinal resistance between } l\text{th and } (l-1)\text{th compartment}$$

To clarify the notation and interpretation of this general mechanics, consider the application of Eq. (10.3) to a particular representative compartment, j^*, taken in the second level of branching.

The interest here is on the physical interpretation of the characteristic synaptic weighting factors, a_j and b_j, which are defined in Eq. (10.3b). This equation says that a_j is equal to the product of the active synaptic conductance change (normalized to the resting conductance of the soma, GA_s), s_j; the equilibrium potential of the synapse (normalized to resting threshold, Θ_o), $E_j^{*\,\prime}$; the ratio of the effective synaptic duration, δ_j, to the effective integration interval of the neuron, Δ; and the effective compartment-to-soma transmission factor, $\gamma_j/(1+D)$. The first three terms should be clear; our interest here is in clarifying the transmission factor.

Equation (10.3c) reiterates our dual notation for referring to compartments: a given compartment may be referred to by the single subscript j, which is convenient most of the time; the same compartment may be referred to by two subscripts, l and k, where l denotes the branch level of the compartment and k denotes the particular branch at that level. When the branch level description is used, the k terms are usually not written explicitly to avoid unnecessary, cumbersome notation. In this case, compartment terms labeled with l apply to the lth branch; those with $(l+1)$ apply to the $(l+1)$th branch, and so on. One simply understands that any particular branch generally has different numerical values for these common symbols.

Consider, for example, the transmission factor for the j^*th compartment, γ_j^*. Recall that this factor physically represents the proportion of current from compartment j^*th transmitted to the soma. The γ_j^* factor is further broken down into a product of several α terms, in this case the two terms: α_1^* and α_2^*. Figure 10.1 shows in bold lines those compartments that directly determine the values of the particular α_1 and α_2 implied here and therefore the transmission factor, γ_j^*. Physically, this α_2^* is the percentage of current from compartment j^* (in branch level 2) transmitted successfully to its parent compartment at branch level 1. Note that physically this must depend on the longitudinal resistance between these compartments, R_2'; on the local membrane conductance in the parent, j^* branch, a_l and g_l'; and on the backward current leakage into all distal compartments that connect to compartment j^* through its sibling branches in the third level of branching. All these factors are explicitly and succinctly represented in Eq. (10.1c). Note the branch selection logic implied when applying this equation to any given compartment.

The factor α_2^* gives the input current from compartment j^* transmitted to the distal end of its parent branch in level 1. Therefore, the leakage of current from that parent branch into the other second-level compartments to which it is connected do not directly influence α_2^*. Leakage into these compartments does, however, influence the percentage of current coming into this parent branch that gets to its output end, which is the input end of the soma. More exactly, the net current transmitted to the soma from j^* is the amount transmitted to the input end of its parent first-level compartment, α_2^*, multiplied by α_1^* for this compartment. One can see, by applying Eq. (10.3c) to the term α_1^* for this case, that this term depends on longitudinal resistance from compartment 1 to the soma, on leakage out through the membrane of compartment, and leakage back into all the second-level compartments to which this parent first-level branch connects.

The overall conductance factor to the soma, γ_j^*, is the product of α_1^* and α_2^*, and therefore embodies all these leakages. Finally, the synaptic weighting factor, a_j^*, depends on the product of γ_j^* and the term $(1/(1+D))$. This latter term reflects the percentage of input current to the soma that drives the soma potential above threshold. The term $(1+\mathrm{D})$ represents the current that flows through the resting soma membrane to drive it up to threshold (1) and the current that flows out of the soma to its contiguous dendritic branches.

These general interpretations will be spelled out in explicit mathematical detail for a simple case.

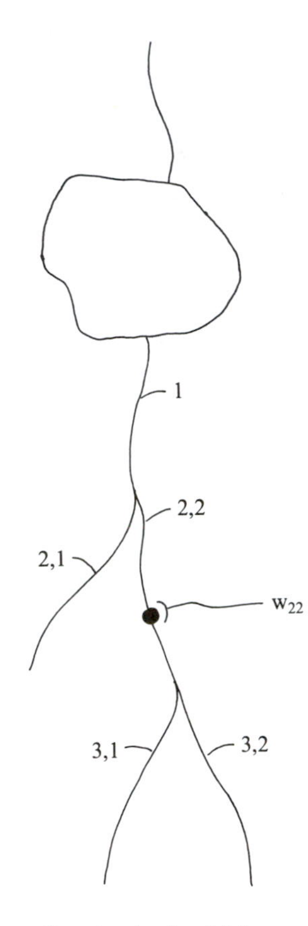

FIGURE 10.2. Activation of a simple dendritic tree by a single input channel.

Activation of a Simple Dendritic Tree by a Single Input Channel

Consider the simple third-level dendritic tree with five branches shown in Figure 10.2. Suppose this tree is presently activated by a single active input channel in its (2, 2) branch as shown in the figure. The general equation system, Eq. (10.3), reduces for this case to Eq. (10.4):

$$g_k/b = \rho^* = [a_{22}w'_{22} - aw^*_i]/\beta = [a_{22}w'_{22} - a_{22}w^*_i]/\beta \tag{10.4a}$$

$$\rho^* = (a_{22}/\beta)[w'_{22} - w^*_i] = \frac{\gamma_{22}\delta_{22}s_{22}E^*_{22}[w'_{22} - w^*_i]}{\Delta b[1 + |E^*_k|]}$$

$$w^*_i = [1 + D]/(\gamma_{22}\delta_{22}s_{22}E^{*\prime}_{22}), \qquad \beta = b[1 + |E^*_k|]/(1 + D) \tag{10.4b}$$

$$\rho^* = \frac{[\gamma_{22}\delta_{22}s_{22}E^{*\prime}_{22}/\{(1 + D)\Delta\}w'_{22} - 1]}{b[1 + |E^*_k|]/(1 + D)} \tag{10.4c}$$

$$\gamma_{22} = \alpha_1\alpha_{22}, \qquad D = \sum(1-\alpha_1)/R'_1 = (1-\alpha_1)/R'_1$$

$$\alpha_1 = \frac{1}{1 + R'_1[a_1 + \sum \alpha_2 a_2]} = \frac{1}{1 + R'_{22}[a_1 + (\alpha_{21}a_{21} + \alpha_{22}a_{22})]}$$

$$\alpha_{21} = \frac{1}{1 + R'_{21}(a_{21})} \tag{10.4d}$$

$$\alpha_{22} = \frac{1}{1 + R'_{22}[a_{22} + \sum \alpha_3 a_3]} = \frac{1}{1 + R'_{22}[a_{22} + (\alpha_{31}a_{31} + \alpha_{32}a_{32})]}$$

$$a_j = A_j/A_s, \qquad A_j = 2\prod r_j l_j, \qquad A_s = 4\prod r_s^2$$

$$R'_{ij} = R_{ij}GA_s, \qquad R_{ij} = \frac{\rho l_i/2}{\prod r_i^2}\left[1 + \frac{1}{(r_o/r_i)^2 - 1}\right] \tag{10.4e}$$

$$\text{if } i = \text{soma}, \quad l_i = r_i = r_s/\sqrt{2}$$

The effective synaptic weighting factor here, a_{22}, is $\gamma_{22}\delta_{22}s_{22}E_{22}^{*\prime}/[(1+D)\Delta]$. The control threshold input firing rate, $w_i^{*\prime}$, is simply the reciprocal of this term.

The various theoretical geometric factors, γ_{22}, α_1, α_{22}, and so on are given explicitly in terms of the immediate geometry of the neuron in Eqs. (10.4d) and (10.4e). The notation here is the same as in Chapter 5.

Comprehensive Comparative Characterization of Prototypical Dendritic Trees

The methods and interpretations surrounding Figures 10.1 and 10.2 allow one to develop the characteristic synaptic weighting factors for any dendritic tree. The salient characteristics of most dendritic trees, however, can be approximated more easily by the set of canonical characteristic weighting factors presented in this for prototypical idealized symmetric trees. Any naturally occurring real dendritic tree can be approximated by mapping its salient geometric features and distribution of synaptic inputs onto such a prototypical symmetric tree. The resulting values for the characteristic synaptic weighting factors allows one to compare the various input systems on the neuron and to compare various types of neurons on this basis.

The prototypical idealized dendritic tree is indicated in Figure 10.3. Four levels of branching are shown, but this may be easily extended to any number of branchings. The only restrictions are that the branches at any one level are taken as equal in length and radius and that the number of branches originating from parent branches at any given branch level are the same. The figure shows bifurcations at every branch point; there could, in fact, be more branches at any levels. (Note also that there might be only one

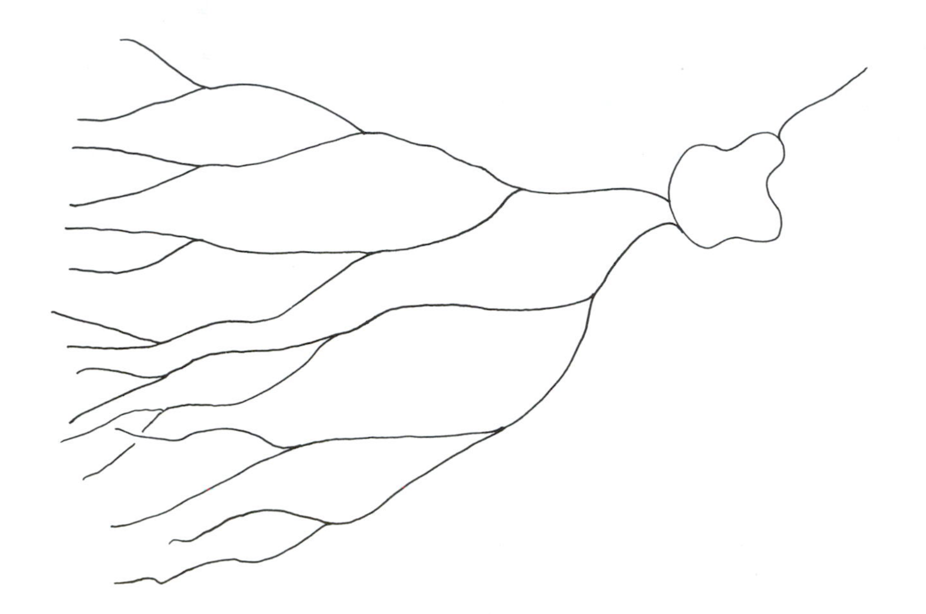

FIGURE 10.3. Canonical dendritic tree for characterization by dynamic similarity parameters.

branch from given parent branches, which means essentially that a single dendritic branch could be partitioned into as many compartments as one likes. This in turn means that the theory developed here can be applied to as fine a geometrical partitioning as desired.)

These assumptions mean that the geometrical parameters of the theory, (the values of γ, α, R, etc.) will be the same for all compartments at each branch level. Let us further consider the idealized control case, where no synapses are active and, hence, all nonlinear leakage factors can be ignored. In this prototypical idealized case, the values of α for the four levels are given in Eq (10.5a).

$$
\begin{aligned}
\alpha_4^{\text{o}} &= \frac{1}{1 + R_4' a_4} \\
\alpha_3^{\text{o}} &= \frac{1}{1 + R_3'[a_3 + n_{34}\alpha_4^{\text{o}} a_4]} \\
\alpha_2^{\text{o}} &= \frac{1}{1 + R_2'[a_2 + n_{23}\alpha_3^{\text{o}} a_3]} \\
\alpha_1^{\text{o}} &= \frac{1}{1 + R_1'[a_1 + n_{12}\alpha_2^{\text{o}} a_2]}
\end{aligned}
\tag{10.5a}
$$

$$R'_1 = R\,\text{soma} \rightarrow 1GA_s, \qquad R'_l = R_{l-1,l}GA_s, \qquad a_i = A_i/A_s$$

$$\begin{aligned} \gamma_1^{\text{o}} &= \alpha_1^{\text{o}} \\ \gamma_2^{\text{o}} &= \alpha_1^{\text{o}}\alpha_2^{\text{o}} \\ \gamma_3^{\text{o}} &= \alpha_1^{\text{o}}\alpha_2^{\text{o}}\alpha_3^{\text{o}} \\ \gamma_4^{\text{o}} &= \alpha_1^{\text{o}}\alpha_2^{\text{o}}\alpha_3^{\text{o}}\alpha_4^{\text{o}} \end{aligned} \tag{10.5b}$$

$$\begin{aligned} a_l^{\text{o}} &= \gamma_l^{\text{o}}\delta_l s_l E_l^{*\prime}/([1+D]\Delta), \qquad a_s = \delta_s s_s[E_s^* - 1]/([1+D]\Delta) \\ b_l^{\text{o}} &= \gamma_l^{\text{o}}\delta_l^i s_l^i|E_l^{i*\prime}|/([1+D]\Delta), \qquad b_s = \delta_s^i s_s^i[1 + |E_s^{i*\prime}|]/([1+D]\Delta) \end{aligned} \tag{10.5c}$$

$$w_i^* = 1/a, \qquad \beta = b[1 + |E_k^{*\prime}|]/(1+D) \tag{10.5d}$$

Here the $n_{l,l+1}$ terms represent the number of branches issuing from lth-level parent fibers. The superscript 0 refers to this canonical idealized control representation.

The overall transmission factors for each branching level for this case can then be given as in Eq. (10.5b).

Finally, the canonical synaptic weighting factors are given in Eq. (10.5c). Equations (10.d) complete the description with expressions for the threshold input firing rate and the relative refractory leakage parameter.

The canonical neuron is therefore comprehensively described by at most $[m(1+n)+2]$ characteristic dynamic similarity parameters. Here, n is the number of levels of branching in the dendritic tree, and m is the number of distinct physiological types of synapses active on the neuron. (Equations (10.c) are written for two types of synapses, excitatory and inhibitory, but this can be immediately generalized.) These parameters are the threshold input firing rate, w_i^*; the refractory leakage parameter, β; and $[m(1+n)]$ synaptic weighting factors, a_j (b_j).

To apply this approach to naturally occuring neurons, one would take average numbers for the sizes of branches in each compartment and for the numbers of branchings at branch points at each level. The characteristic numbers should be indicative and useful even where deviations occur. Indeed, the deviations may be better interpretable in some cases against the background of the canonical form.

One should anticipate describing moderately elaborate neurons with a dozen or so characteristic numbers, depending primarily on the physiological richness and anatomical distribution of their synaptic input. If, for example, there are two synaptic types, each distributed to all of four levels of branching, there will be twelve characteristic parameters.

Chapter 11 Universal Firing Rate Transfer Curve for Temporally Regular Input

This chapter and the next chapter develop the universal firing rate transfer curve illustrated in Figure 8.2. This curve and its nondimensional characterization are very useful in interpreting the overall input–output dynamic sensitivities of neurons, their dependencies on constituent anatomical and physiological parameters, and their comparisons across types of neurons.

The analysis will be performed for the case of spatially uniform distribution of input and therefore governed by the dynamic similarity equation in the form of Eq. 10.1. Recall that this form is equivalent to that of a point neuron. The transfer curve will be pieced together from analyses for temporally unstructured and temporally structured input.

Temporally unstructured input is relatively inefficient in activating receiving neurons. Maximum temporal summation of inputs occurs when pulses occur close together in time. Higher output rates are obtained when input pulses are grouped together in time, rather than spread out with more uniform intervals.

We will consider three representations of temporally unstructured input.

The first two are the average conductance representation, and a representation of the summation of uniformly spaced PSPs by an equivalent envelope. These can be easily represented by the approaches already developed and are considered now. Both are particularly inefficient means of generating output firings when measured in terms of the required numbers of input pulses. Therefore, these results apply to the same general highish input side of the overall firing rate transfer curve.

The third type of unstructured input considered consists of random trains of PSPs distributed with Poisson distribution. This provides some chance groupings of pulses and thereby is a somewhat more effective activating system than the average conductance or regularly spaced input systems. It is still, however, considerably less effective in activating receiving cells than highly temporally structured input. At higher input rates the response to the Poisson representation merges with the those from the other two unstructured input systems, because at higher rates, there is much temporal summation and refractoriness; the individual differences in timing become insignificant.

The development of the Poisson representation requires a probabilistic mathematical approach and is therefore dealt with separately in Chapter 12. All the other cases are considered in this chapter.

Firing Rate Transfer Curves with Temporally Unstructured Input

The three types of activation considered here are directly formulated in terms of the universal nondimensional characteristic function, ρ^*, and the corresponding universal transfer curve introduced in Eq. (9.7) and Figure 9.3. Equation 10.1 will be applied in this context in all these cases.

Firing Rate With Average Conductance Activation

The average conductance representation is mathematically defined by the general representation of synaptic conductances introduced in Eq. (9.8a), and therefore embodied in all the subsequent mathematical development in Chapters 9 and 10. Here, we explicitly interpret the w' terms as the mean input firing rates and take the conductance values as constant corresponding to constant levels of w'.

The characteristic firing rate transfer curves will be considered here as functions of the excitatory input firing rates at given fixed levels of inhibitory input rate. One could easily obtain from Eq. (10.1c) other transfer curves where input excitatory and inhibitory rates are coupled in various ways. Firing rate curves for systems with recurrent inhibition can be estimated readily by extending the postfiring potassium parameter, β, to include a component representing postfiring recurrent inhibition, since both act by supplying conductance leakages to a negative equilibrium potential subsequent to output firing.

To represent this case succinctly, consider the restructuring of Equation (10.1c) to include a generalization of $w_i^{*\prime}$ to $w_{iu}^{*\prime}$ as shown in Eq. (11.1):

$$g_k/b = \rho^* = \frac{w_i - [aw_i^{*\prime} + bw^{i\prime}]/a}{\beta/a} \tag{11.1a}$$

$$w_{iu}^* = [aw_i^{*\prime} + bw^{i\prime}]/a \tag{11.1b}$$

$$g_k/b = \rho^* = \frac{w^{i\prime} - w_{iu}^{*\prime}}{\beta/a} \tag{11.1c}$$

$$w_0' = \frac{1}{\tau_k' \ln\left\{1 + \dfrac{\beta/a}{w' - w_{iu}^{*\prime}}\right\}} \tag{11.1d}$$

$$\left.\frac{\partial w_0'}{\partial w'}\right|_\infty = 1/(\tau_k'\beta/a) \tag{11.1e}$$

$$w_0^* = 1/(2\Delta) \tag{11.1f}$$

$$w_{ih}^{*\prime} = w_{iu}^{*\prime} + \frac{\beta/a}{\exp\,[2/\tau_k' - 1]} \tag{11.1g}$$

Equation (11.1a) indicates the generalized grouping for $w_{iu}^{*\prime}$, and (11.1b) defines $w_{iu}^{*\prime}$. The subscript u stands for unstructured.

The resulting governing equation expressed in terms of g_K and ρ^* in Eq. (11.1c), is exceedingly direct and succinct. Since ρ^* is directly proportional to w', we see that the characteristic firing rate transfer curve is equivalent to the universal curve introduced in Eq. (9.7) and Figure 9.3. These are reproduced here in the current application as Eq. (11.1d) and Figure 11.1.

Recall that the value $\rho^* = 0$ corresponds to the output firing threshold. This threshold is seen immediately to correspond to the central characteristic nondimensional parameter, $w_{iu}^{*\prime}$. Indeed, this interpretation gives us our physical interpretation of the term. The parameter $w_{iu}^{*\prime}$, is that value to which input firing rates must be elevated to initiate output responses for totally unstructured inputs. The interpretation of this term will be extended in Chapter 12.

Note second that the slope of the curve at high values of w_i' approaches $1/(\tau_K'^{*}\beta/a)$, as illustrated in Figure 11.1 and shown in Eq. (11.1e).

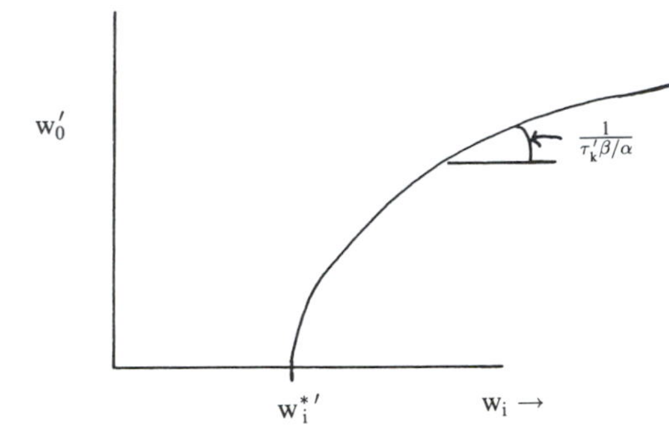

FIGURE 11.1. Universal firing rate transfer curve for representation of activation by average conductance level.

At this point it is very broadly instructive to introduce the characteristic output firing rate of a neuron. This is based on the observation that output firing at the rate of the reciprocal of the neuron integration interval is a good nominal upper limit for normal firing and that half this value is a good nominal value of a strongly "on" state for the neuron. For example, if the neuron's integration interval is taken as its time constant, then for a cell with a membrane time constant of 15 msec (cortical or hippocampal pyramidal cell), the nominal strongly "on" rate would be about 30/sec. For a neuron with a membrane time constant of 5 msec, the nominal strongly "on" rate would be about 100/sec.

We will here define this value of the reciprocal of twice the neuron integration interval (estimated by the membrane time constant) as the characteristic output firing rate of the neuron, and represent it by w_o^*. This definition is given as Eq. (11.1f).

We can now use Eq. (11.1d) together with this definition of the characteristic output rate, w_o^*, to determine that level of input firing rate, w', which is necessary to produce the characteristic (strongly "on") rate of firing in the neuron. This value can be labeled w_{ih}^*, ("characteristic high input rate"), given by Eq. (11.1g).

Another value of these definitions then begins to emerge. We can begin to estimate from the anatomical and physiological parameters of the neuron what various input rates determine its normal operating range and the various regions within that range. This picture will become more complete as we complete the analyses in this and the next chapter.

Firing Rate with Step Current Input

The response to a step current input can be obtained directly by generalizing

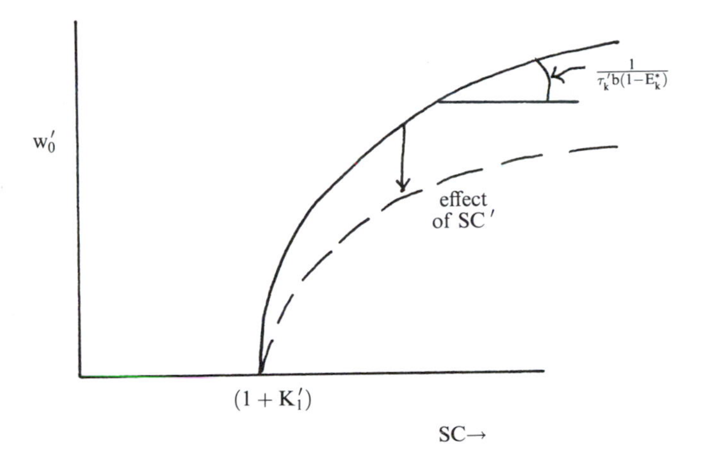

FIGURE 11.2. Universal firing rate curve for step current input.

Eq. (11.1c) to include the SC' term as indicated in the last chapter or, equivalently, by generalizing Eq. (10.2d) to include the dendritic leakage terms as discussed in Chapter 10. In either case the net result is shown in Eq. (11.2):

$$\rho^* = \frac{SC' - (1 + D + \xi)}{b[1 + |E_k^*|]} \tag{11.2a}$$

$$w_o' = \frac{1}{\tau_k' \ln\left\{1 + \dfrac{b[1 + |E_k^*|]}{SC' - (1 + D + \xi)}\right\}} \tag{11.2b}$$

$$SC_\theta' = 1 + D + \xi = 1 \tag{11.2c}$$

$$\left.\frac{\partial w_o'}{\partial SC'}\right|_\infty = \frac{\partial w_o}{\partial \rho^*}\frac{\partial \rho^*}{\partial SC'} = \frac{1}{\tau_k' b[1 + |E_k^*|]} \tag{11.2d}$$

Again, the universal function, ρ^*, is directly proportional to the input, in this case SC'. Here, the threshold current value is the total leakage parameter, $1 + D$, and the asymptotic slope of the firing rate curve is $1/(\tau_k'^* b^*(1 + |E_K^*|))$. This curve is shown in Figure 11.2.

The influence of the capacitive term, ξ, is particularly easy to characterize here. Its influence is as shown in Figure 11.2. At higher levels of transmembrane current a significant fraction of it is going through the capacitive displacement current, as discussed in Chapter 4. This shows up as a temporal change in the transmembrane potential and a lesser amount of current going through the ohmic conductance channels—hence, a lower E. When the current is just barely under threshold level, its slope after a long time will be 0. If we picture just slightly increasing this level of SC' to get the

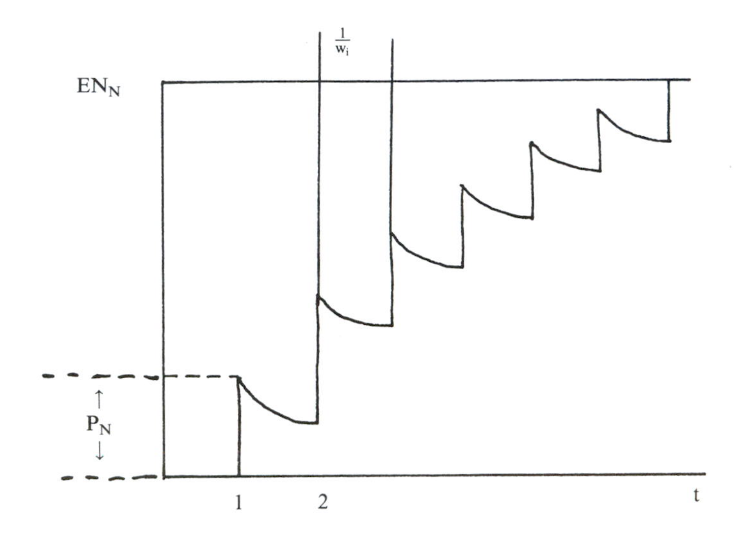

FIGURE 11.3. Linear summation of regularly spaced PSPs.

firing threshold, we can see that the rate of change of E at this firing is still essentially 0, so that the capacitive effect is negligible at threshold. These influences are roughly indicated in Figure 11.2.

Firing Rate With Summation of Regularly Spaced PSPs

Consider now the case of activation by a train of regularly spaced PSPs. If we represent individual PSPs by potential peaks of amplitude P that decay exponentially with the membrane time constant and consider a linear summation of these, we can represent this case as shown in Figure 11.3 and Eq. (11.3).

$$
\begin{aligned}
EN_n &= P\{1 + \exp[-1/(w_i\tau)] + \exp[-2/(w_i\tau)] + \ldots + \exp[-(n-1)/(w_i\tau)] \\
&= P \sum \exp[-i/(w_i\tau)] \qquad (11.3a) \\
&= \frac{P\{1 - \exp[-n/(w_i\tau)]\}}{\{1 - \exp[-1/(w_i\tau)]\}}
\end{aligned}
$$

$$
\begin{aligned}
EN_n' &= \frac{P/\theta}{1 - \exp[-1/(w_i\tau)]}\{1 - \exp[-n/(w_i\tau)\} \\
&= SC_{\text{eff}}'\{1 - \exp[-t/\tau]\}, \qquad w_i = n/t
\end{aligned}
\qquad (11.3b)
$$

$$
SC_{\text{eff}}' = \frac{P/\theta}{1 - \exp[-1/(w_i\tau)]} \qquad (11.3c)
$$

$$\rho^* = \left\{ \frac{P/\theta}{1 - \exp\left[-1/(w_i \tau)\right]} - (1 + \xi) \right\} \Big/ \{b[1 + |E_k^*|] \tag{11.3d}$$

$$w'_{op} = \frac{1}{\tau'_k \ln \left\{ 1 + \dfrac{b[1 + |E_k^*|]}{\dfrac{P/\theta}{1 - \exp\left[-1/(w_i \tau)\right]} - (1 + \xi)} \right\}} \tag{11.3e}$$

$$n = \text{smallest integer} \geq w'_i / w'_{op} \tag{11.3f}$$

$$w'_o = w'_i / n \tag{11.3g}$$

$$w^{*\prime}_{iu} = \frac{1}{\tau'_k \ln \left\{ \dfrac{1}{1 - \dfrac{p/\theta}{1 + \xi}} \right\}} \tag{11.3h}$$

$$\left. \frac{\partial w'_o}{\partial w'_i} \right|_{\infty} = \frac{\partial w_0}{\partial \rho^*} \frac{\partial \rho^*}{\partial w'_i} = \frac{\tau' P/\theta}{\tau'_k b[1 + |E_k^*|]} \tag{11.3i}$$

As in the case of summing exponentially decaying postfiring potassium conductance changes, this summation forms a geometric series and can be represented by the closed form shown in the last of Eq. (11.3a).

The interesting trick here is that the envelope of this PSP summation can be interpreted as the response to an equivalent step current and therefore the response solution for the step current just given can be used to describe this case. To see this, consider that the neuron is receiving inputs at intervals of $1/w_i$. If the first PSP in the summation occurred at the time $1/w_i$ after the time $t = 0$, which would be a reasonable representation for the case of the neuron's last firing being at $t = 0$, then the neuron's firing rate would be n/t, valid at the time of firing of any nth PSP.

On the basis of this interpretation, and making use of the last of Eq. (11.3a), the first portion of (11.3b) may be written. Further, time, t, may be substituted for n/w_i in this equation. This means, however, that the time course of the function shown in Eq. (11.3b) is exactly that of a constant step current applied at $t = 0$. This means we can interpret the first part of the expression on the right as an equivalent effective current whose amplitude is given by Eq. (11.3c).

It follows that we can write the expression for ρ^* given in Eq. (11.3d), and the corresponding solution for w'_{op} given in Equation (11.3e). (The p here stands for preliminary.) Now we must modify the results, however, to recognize that output firing must occur at a time when an input pulse occurs, so that w'_o must be equal to w'_i divided by n, where n is the number of PSPs per output spike. Mathematically, the expressions shown in Eqs. (11.3c) and (11.3d), are continuous representations of the summation envelope, whereas only the values at $t = w_i/n$ are physically valid. The solutions based on this envelope will show a bias toward early firing by some amount less than 1 inter-PSP interval, $1/w_i$. To compensate for this we need simply to adjust the output firing rate, w'_o predicted by Eq. (11.3e), as shown by the rules given in Eqs. (11.3f) and (11.3g).

Equations (11.3h) and (11.3i) show how the threshold firing rate, w^*_{iu}, and limiting slope depend on parameters for this case.

Computer simulations of transfer curves show that the average conductance representation and the PSP summation representation produce very similar results, and both match computer simulations very well.

Firing Rate Transfer for Temporally Structured Input

Temporally structured input is more effective in generating neuronal firing than the unstructured input considered previously and in Chapter 12. As an instructive limiting case, consider that the optimal temporal structuring of input to achieve spike generation (at least at lower rates where refractory effects can be neglected) consists of arranging the input pulses into groups of n, wherein these n simultaneous pulses are sufficient to trigger firing, and then spacing these simultaneous groups equally to minimize the influence of refractoriness between groups.

This representation of an optimally temporally structured input pattern, then, consists of a regularly firing series of individual groups of n spikes wherein the rate of groups is equal to the input rate of individual spikes divided by n. The output rate will be equal to this group input rate, w_i/n, as long as refractoriness or other extraneous factors do not interfere.

The goal here is to characterize this situation mathematically to set an upper limit on the effectiveness of firing rate transfer, to compare this with the transfer for unstructured input, and to show the dependence of this transfer on the characteristic parameters of the neuron.

The mathematical solution for this stimulation is given in Eq (11.4):

$$w'_o = w'_i/n \tag{11.4a}$$

$$\begin{aligned} 1 &= \{\alpha\theta - E_o\}/(P_n) = \{\alpha + g_i/(1+g_i)(|E^*_{in}|/\theta)\}/(P_n/\theta) \\ &= \{\alpha + g_i(1+g_i)(|E^*_{in}|)\}/(P_n) \end{aligned} \tag{11.4b}$$

$$P = \frac{s\{E^* + g_i/(1+g_i)(|E_{in}^*|)\}\{1 - \exp[-\delta/(\tau(1+s+g_i))]\}}{1+s+g_i} \quad (11.4c)$$

$$(P_n) = knP \quad (11.4d)$$

$$k_{\min} = (1+s+g_i)/(1+ns+g_i), \qquad k_{\max} = 1 \quad (11.4e)$$

$$n = \frac{\alpha + g_i/(1+g_i)(|E_{in}^*|)}{ks\dfrac{\{E^* + g_i/(1+g_i)(|E_{in}^*|)\}\{1 - \exp[-\delta/(\tau(1+s+g_i))]\}}{1+s+g_i}} \quad (11.4f)$$

$$w_{is}^{*\prime} = nw_o^{*\prime} = n(1/2)$$

$$= \frac{\alpha + g_i/(1+g_i)(|E_{in}^*|)}{2ks\dfrac{\{E^* + g_i/(1+g_i)(|E_{in}^*|)\}\{1 - \exp[-\delta/(\tau(1+s+g_i))]\}}{1+s+g_i}} \quad (11.4g)$$

Equation (11.4b) is the statement that n simultaneous PSPs must exceed the threshold. The statement is generalized to include an initial level of potential, E_o, which is taken as mediated by a constant average level of inhibitory conductance, g_i. The term P_n is the response of the neuron to n simultaneous inputs. (This is likely to be nonlinear and therefore not directly equal to P times n, where P is the unit PSP response.) The term is a safety factor imagined to be in the range of, say, 1.1 to 1.5. To be compatible with the other formulations in the book, the normalization of these terms to the threshold, ϕ, is shown in Eq. (11.4b).

Equation (11.4c) shows the unit response to an excitatory synaptic conductance change of amplitude s, lasting a duration, δ. (This expression is derived from the membrane equation in Eq. 11.5.) Generally we would estimate that the combined response to n simultaneous synaptic activations of this amplitude will be some factor, k, of n times this response as shown in Eq. (11.4d). Minimum and maximum values of k are given in Eq. (11.4e). The minimal value of k is given for the case where all n synaptic inputs are close enough in time and space to give maximal degrading interaction. This is as if the combined event were a single conductance input of n^*s. The maximum value of k is 1. This might be approximated if all n inputs are sufficiently brief in duration and sufficiently separated in space for no degrading interaction.

Thus, the number of simultaneous synaptic inputs required for firing in this formulation is given by Eq. (11.4f). We recall from Eq. (11.4a) that this is the inverse slope of the firing rate transfer curve. This curve is sketched in Figure 11.4.

We can now define the characteristic number w_{is}^* (read as "characteristic

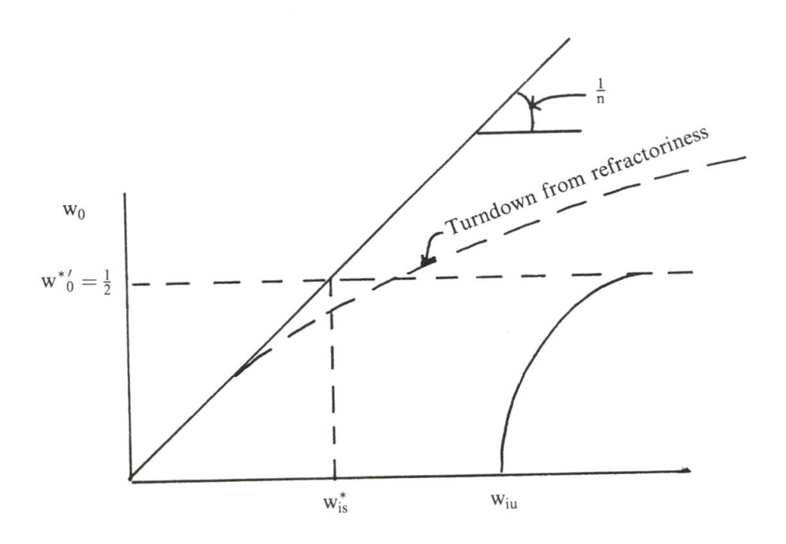

FIGURE 11.4. Universal firing rate transfer curve for highly structured input.

input rate for structured input") by defining the input rate required to produce output firing rates approaching the characteristic output rate of $w_o' = 1/(2\Delta)$. This characteristic rate is shown in Eq. (11.4g).

It is interesting to compare this characteristic parameter for structured input with its counterpart, w_{iu}^*, for unstructured input as defined in Eq. (11.1b). Note in particular that the value for structured input depends less strongly on the level of inhibitory activation. This is because the highly structured input delivers proportionally more excitatory current at active synapses, since they are activated when the membrane potential is considerably farther from threshold and therefore from their equilibrium potentials. This may be the basis of a fundamental design difference between neurons developed to be particularly responsive to highly structured temporal input and those more sensitive to more global input patterns.

Equations (11.5a–d) derive Eq. (11.4c) from the membrane equation.

$$0 = -E + g[E^* - E] - g_i[E - E_{in}^*]$$

(steady-state membrane equation) (11.5a)

$$E_{ss} = (g_e E^* + g_i E_{in}^*)/(1 + g_e + g_i) \quad (11.5b)$$

$$\text{for } g_e = 0, \qquad E_0 = -g_i/(1 + g_i)(-E_{in}^*) \quad (11.5c)$$

for pulse $g_e = s$, of width δ:

$$\begin{aligned}
\text{PSP} &= E_f - E_0 = \{E_{\text{trg}}(1 - \exp[-\delta/\tau_{\text{eq}}]) - E_0 \exp[-\delta/\tau_{\text{eq}}]\} - E_0 \\
&= \{E_{\text{trg}} - E_0\}\{1 - \exp[-\delta/\tau_{\text{eq}}]\} \\
&= \frac{sE^* + g_i E_{in}^*}{1 + s + gi} - \{g_i/(1 + g_i)(-E_{in}^*)\} \qquad (11.5\text{d}) \\
&\quad \times \{1 - \exp[-\delta/(\tau(1 + s + g_i))]\} \\
&= s\frac{\{E^* + g_i/(1 + g_i)(|E_{in}^*|)\}}{1 + s + g_i}\{1 - \exp\{-\delta/(\tau(1 + s + g_i))]\}
\end{aligned}$$

The next chapter turns to a probability model to complete the general composite firing rate transfer curve begun in this chapter.

Chapter 12 Universal Firing Rate Transfer Curve for Randomly Temporally Irregular Input

This chapter presents a probabilistic approach to relating input and output firing levels in neurons when the input train is temporally irregular. This representation will fill out the lower end of the firing rate transfer curve for temporally unstructured input presented in Chapter 11. The analysis is again based on the case of a uniform spatial distribution of input, or point neuron.

The input pattern considered here consists of a Poisson distribution of input pulses. It represents most closely a totally disorganized input pattern. It is likely a more reasonable representation of common neuronal input situations than the temporally regular cases considered in Chapter 11, because most neurons are typically bombarded by activity in thousands of input synapses, many of which are likely to carry pulses that are not tightly locked in time on a small scale.

However, the model is likely not to be a good representation of input–output activities in cases involving representations and operations of specific coded meanings—the signalling of particular dynamic patterns representing particular memory traces and their interactions, for example. Certainly these cases involve the temporal coordination of single spikes across input synapses, thereby violating the assumption of temporal independence of individual inputs made in the mathematical model developed here.

Nonetheless, the model is valid as a reference level against which subsequent studies of the fine structure of input–output coding can be considered. Moreover, it is not unreasonable to anticipate that comparisons of separate types of neurons based on the temporally disorganized model may hold for comparative responses to organized activity as well.

Also within the framework of these overall limitations of the model, it is a strong advantage of the model that it is quite simple both physically and mathematically, but nonetheless houses the essential features of the spike triggering process under this type of bombardment. This means that one may see clearly in simple mathematical expressions the relative significance of the various neuron parameters involved in the process.

The Binomial (Poisson) Representation of Input Temporal Structure

The situation and approach considered here are illustrated in Figure 12.1. The neuron is considered as to be subjected to spatially uniform activation. All excitatory and inhibitory pulses can then be taken as arriving at single equivalent excitatory and inhibitory synapses, as shown in Chapter 10. To characterize the triggering of output pulses by the temporally irregular input train we will partition time into windows of a given duration, Δ, where Δ is a length of time over which there is significant combination (summation) of *e*PSPS. This interval is called the *integration interval* of the neuron and, in fact, is closely related to the membrane time constant, for this latter

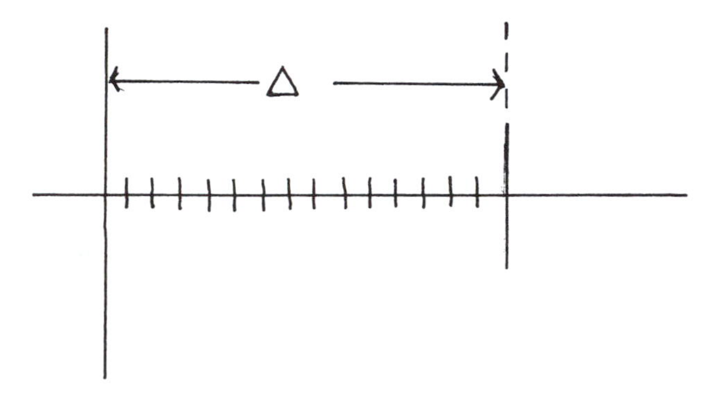

FIGURE 12.1. Time partitioning within a single integration window for the binomial approach to Poisson activation.

measures the lifetime of individual PSPs. For the sake of the analysis, and to leave open the option of generalizing the analysis, we will simply refer to Δ as an individual parameter of a given neuron for now. This Δ is the same integration interval used as a normalizing term in Chapters 9, 10, and 11.

The approach will be to find the number of PSPs that arrive in a given window of length Δ and discover how often this number might be expected to be above threshold. This will relate output firing rate to the mechanics of the input. We will also utilize the membrane equation in the analysis thereby bringing in the basic neurophysiological operations underlying spike production.

As illustrated in Figure 12.1, imagine that the time window, Δ, is partitioned into n steps, each of size z. Moreover, let us suppose that, for a given mean input rate, w_i, the probability of an input event during a given interval z is proportional to the product of w_i and z. These two formulations are given in Equation 12.1.

$$\Delta = nz \tag{12.1a}$$

$$p^* = wz = w\Delta/n = w'/n \tag{12.1b}$$

Throughout this chapter we will continue as appropriate to refer to firing rates normalized to $(1/\Delta)$ and indicate these with primes.

We will further assume that the occurrence of individual input pulses are completely independent and that the step interval z is so small that the probability of two inputs in any one such interval is negligibly small. (This latter assumption allows us to use a binomial approach to this problem that is relatively easy to understand. Moreover, we can in fact let z become arbitrarily small so that this assumption does not limit the theory in any way. This limit of very small values of z corresponds to what is known as the *Poisson limit* of the binomial model, and in fact in this limit we are dealing with what is know as a *Poisson distribution* with no restriction on the number of events that occur in any finite interval.)

This formulation allows us to consider a given Δ window as a particular collection of n cases (z intervals) in each of which there may or may not be an input pulse. The probability of an input pulse (success) in any of these intervals is p, and the probability of the absence of a pulse (failure) is correspondingly $1 - p$. This situation is a classic case in probability theory, called the *Bernoulli trial situation.* Another common example is tossing a coin n times. The key ingredients are that what happens on any given trial is the same for all trials and independent of all other trials.

The solution for the Bernoulli trial situation is well known: it is the binomial distribution. More precisely, one may say that the probability of having exactly x successes (for us, exactly x PSPs) in n trials (n intervals of length z) is given by the probability density function shown in Eq. (12.2a):

$$p(x) = \frac{n!}{x!(n-x)!} p^x (1-p)^{n-x} \tag{12.2a}$$

$$\mu_x = np \tag{12.2b}$$

$$\sigma_x = \sqrt{np(1-p)} \tag{12.2c}$$

$$x = \mu + k\sigma \tag{12.2d}$$

This distribution has a mean, μ_x and standard deviation σ_x given in Eqs. (12.2b) and (12.2c). Note also that we have introduced a fundamental concept here, that of the random variable X, which is the number of PSPs that arrive in any given time window, Δ. We do not know exactly what it will be for any given window, but we do know what its mean or average value will be over many windows. Moreover, we know more specifically what its distribution of values should look like over many such windows. The standard deviation is simply a measure of how many values fall how far away from the mean; that is, how much dispersion is in the process.

For us the general expressions for the mean and standard deviation given in Eq. (12.2) are expressed in terms of the input firing rates and integration interval in Eq. (12.3):

$$\mu = np = nwz = w\Delta = w' \tag{12.3a}$$

$$\sigma = \sqrt{np(1-p)} = \sqrt{w\Delta(1-wz)} = \sqrt{w'(1-wz)} \approx \sqrt{w'} \tag{12.3b}$$

We can see clearly in (12.3b) that if we take z arbitrarily small (no reason we cannot) then the standard deviation becomes equal to the square root of the mean. This is a general result of probability distributions of this type. It has the important physical property that the relative significance of the noise (standard deviation) becomes progressively less significant as the mean gets progressively large. This can be seen by noting that the ratio of the standard deviation over the mean progresses like the reciprocal of the square of the mean, which becomes small as the mean becomes big.

Relation of Physiological Input Activation to Probabilistic Model

From the analysis we have a random variable, X, that represents the number of input pulses arriving during any given integration interval, Δ. Therefore, we can write the expression given in Eq. (12.4a) for the average excitatory synaptic conductance change during such an interval:

$$ge = \frac{X\delta s}{\Delta} \tag{12.4a}$$

$$ge = (\delta s/\Delta)\{\mu + k\sigma\} = (\delta s\Delta)\{w' + k\sqrt{w'}\} \tag{12.4b}$$

$$\rho^* = \frac{(\delta s/\Delta)\{w' + k\sqrt{w'}\}[E^* - 1] - \{1 + D + g_i[1 + |E^*_{in}|]\}}{b[1 + |E^*_k|]} \tag{12.4c}$$

$$\rho^* = \frac{w' + k\sqrt{w'} - w_i^{*\prime}}{\beta/a} \tag{12.4d}$$

As in the last two chapters, s is the magnitude of conductance change associated with a single pulse, and is the interval over which the synapse is active. Note that since X is a random variable, ge is a function of a random variable, and therefore a random variable itself. That means simply that, although we may prescribe its likely distribution of values in terms of a probability density function and mean over some number of trials, we cannot determine exactly what value it will have in any given single trial.

It is instructive and useful to rewrite Eq. (12.4a) in terms of the mean and standard deviation as shown in (12.4b). In this formulation we are stating that X will have a value on any given trial at some distance of k standard deviations from its mean. We understand that k will have an individual value on any given trial, that will generally vary from trial to trial. In this sense then we use the quantity k to carry the implication of the random nature of ge and its dependence on the particular probability density function for X, and so on.

Since we can write ge as in (12.4a) and (12.4b), we can also write the expression for the input activation function, ρ^*, as in Eq. (12.4c). This expression is obtained by using Eq. (12.4b) in the general equation for ρ^* given in Eq. (9.3c). We note that in this case, ρ^* is a function of the random variable ge (or X or k) and therefore is a random variable itself. Equation (12.4c) can be rewritten in terms of the characteristic nondimensional parameters, w'_{ich} and β, defined in the last two chapters.

Relation of Output Firing Rate to Probabilistic Input Parameters

Equations (12.4) refer to the input activity in any given representative time window, Δ. The quantity k is expected to vary from window to window according to constraints of the probability density function defined in Eq. (12.2). Let us now utilize the probability density function to relate k

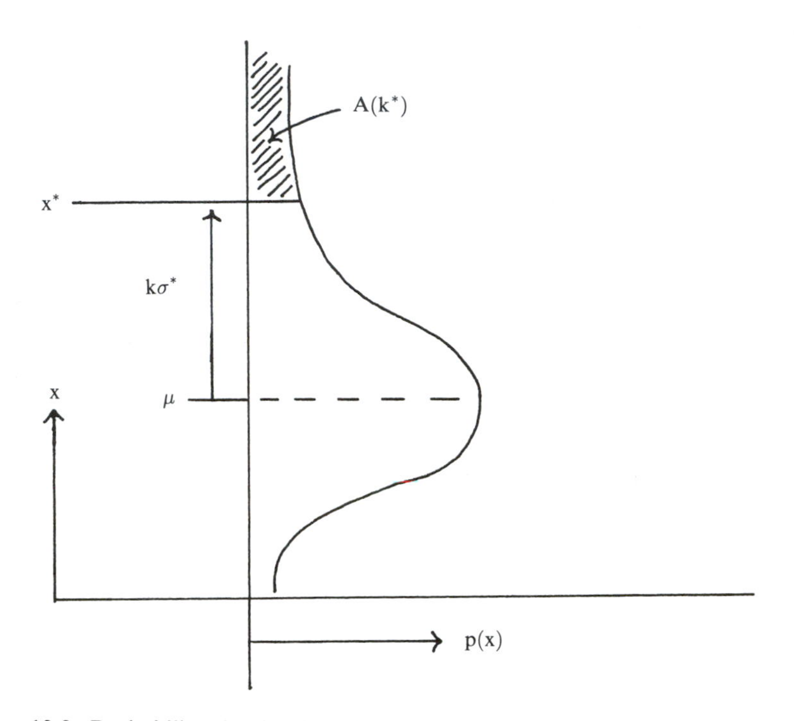

FIGURE 12.2. Probability density function for x PSPs arriving in one integration interval.

explicitly to the output firing rate. We can do this nicely by fixing the number of input pulses in a given Δ to the threshold value.

This process is illustrated in Figure 12.2. If we stipulate that in a given representative window, the number of input pulses, X, is to equal the threshold for firing during that interval, this fixes a particular value of k, say k^*, for that interval. This is shown in Figure 12.2. Now, the total area under the probability density curve is 1. Therefore, the area within any range of k measures the probability of getting a k in this range on any given trial. The graphical interpretation of this figure for our present case is that the area in the probability density function above k^* gives the proportion of times that k will have a value of k^* or above.

The physical interpretation of this is that every time k has such a value it will fire an output pulse. If the neuron would fire at every time window, Δ, its output rate would be $1/\Delta$. Since by Figure 12.2 it is expected to fire at $A(k)$ percent of such intervals, the output firing rate is given by Eq. (12.5).

$$w_o = (1/\Delta)A(k^*) \tag{12.5}$$

This fundamental result tells us how often threshold will be exceeded (and therefore output spikes produced) when a value of k standard deviations above the mean is required to cross the threshold.

Equation (12.5) is not sufficient in itself to solve our problem because we still lack an explicit expression for k. Equation (12.5) is essentially one equation for two unknowns, k and w_o. To determine the value of k necessary for firing, we turn to the membrane equation.

First, consider the case where output firing is so low that refractoriness from g_k can be neglected. For this case we can use Eq. (12.4b) in Eq. (9-1c) to get Equation (12.6a), from which Eq. (6.b) follows immediately.

$$0 = (\delta s/\Delta)\{w' + k\sqrt{w'}\}[E^* - 1] - \{1 + D + g_i[1 + |E^*_{in}|]\} \qquad (12.6a)$$

$$k^* = \frac{\{1 + g_i[1 + |E^*_{in}|]\} - w'}{\sqrt{w'}} = \frac{w_i^{*\prime} - w'}{\sqrt{w'}} \qquad (12.6b)$$

$$w' = \frac{k^* + 2w_i^{*\prime}}{2}\left\{\sqrt{1 + \frac{w_i^{*\prime}}{\left(\frac{k^* + 2w_i^{*\prime}}{2}\right)^2}} - 1\right\} \qquad (12.6c)$$

Equation (12.6b) together with Eq. (12.5), then complete the solution for low output firing rates. For a given input firing rate, w_i', activating a point neuron whose parameters are collected in a characteristic number, w_{ich}', firing occurs at k^* standard deviations from w_i' where k^* is given by Eq. (12.6b). This value of k^* in turn can be take from taken to the distribution tables for the binomial distribution* to determine the area function A for this k^*, and thereby the output firing rate, w_o. Equation (12.6b) can be solved for w_i' as given in Eq. (12.6c)

Let us consider now the case where output firing is sufficiently high that refractoriness from postfiring g_k elevation is significant. We are considering a given representative window in a situation where input and output firing levels are stationary at w_i and w_o, respectively, but both vary over any given time range. In fact, a given time window, Δ, may experience a variable amount of refractoriness depending on the particulars of firing over the recent history. We can nonetheless make an intelligent estimate of this effect by taking the amount of refractoriness experienced in the case of regular output firing. This should not be too different from the average experienced in the irregular case by many representative windows taken at various places in the firing pattern.

* The binomial model can be approximated by the normal distribution for large values of w_i'.

For this case, then, we can write the expression shown in Eq. (12.7) for the output rate:

$$w'_o = \frac{1}{\tau'_k \ln[1 + 1/\rho^*]} \tag{12.7a}$$

$$W'_o = \frac{1}{\tau'_K \ln\left\{1 + \dfrac{\beta/a}{w' + k\sqrt{w'} - w_i^{*\prime}}\right\}} \tag{12.7b}$$

Equation (12.7a) is the general expression for output firing rate in terms of the input activation function, ρ^*. Equation (12.7b) is then obtained by using Eq. (12.4d) in Eq. (10.7a).

Where refractoriness is significant, then, the Eqs. (12.5) and (12.7b) are to be solved simultaneously for k^* and w_o. Physically, we interpret this as follows: neglecting refractoriness, we need k^* from Eq. (12.6b) to get to the threshold; this happens $A(k^*)$ percent of the time, so that output firing rate w_o is given by Eq. (12.5). However, this output rate, in fact, produces refractoriness so that the value of k^* must be increased from our original estimate to the value given by Eq. (12.7b) for this w_o; this means that we must determine a lower w_o by (12.5) for this new k^*; which means we must again reevaluate k^* for this new w_o from (12.7b), and so forth.

Outer Limits of the Significance of Irregular Firing

It is of interest to define a lower level of random irregular activity that produces some just noticeable output rate, w_{ol}. This can be found directly from Eqs. (12.5) and (12.6b). Thus, Eq. (12.5) can be used to determine a (high) value of k^* for any given nominal w_{ol}, say 1/sec. This value of k^* can then be used in Eq. (12.6c) to find the corresponding input frequency, w_{il}, necessary to attain this value of w_{ol}. (Presumably w_{ol} will be such that refractoriness is not a factor.)

The characteristic input frequency can be placed in the random irregular model. Note from Eq. (12.6b) that if there is no refractoriness, then k^* is equal to 0 when w'_i is equal to the input characteristic frequency, w'_{ich}. Now, if the distribution function is symmetric, then $A(k^* = 0) = 1/2$. This means that, with no refractoriness, applying irregular input at the characteristic input rate produces output firing at the rate $w_o = (1/2\Delta)$, which is what we have defined as the output firing rate. Note further that regular input applied at this input rate produces no output firing. (This in fact is the threshold for output with regular input firing.) Therefore, irregular input is significantly more effective in producing output than regular input.

Furthermore, if $k^* = 0$, this means that the mean input rate alone is adequate to attain the threshold for output firing at this rate. This is equivalent to the observation that this point is the threshold for firing for regular input.

In fact, refractoriness will normally be non negligible when the output rate approaches the characteristic output rate, $(1/2\Delta)$, so that output at the input characteristic rate will be somewhat less than this value, as determined by Eq. (12.5), and (12.7b). It is therefore useful and instructive to determine the input firing rate necessary to drive output firing to the characteristic output rate, $(1/2\Delta)$, when refractoriness is present. This can be done by solving Eq. (12.7b) for w_i' when setting w_o equal to $(1/2\Delta)$ and k^* equal to 0, (thereby satisfying Eq. (12.5). The result is given in Eq. (12.8):

$$w_{il}^{*\prime} = (k_l/2)\left\{\sqrt{1 + \frac{w_{iu}^*}{(k_l/2)^2}} - 1\right\} \tag{12.8a}$$

$$w_{ih}^{*\prime} = w_i^{*\prime} + \frac{\beta/a}{\exp\left[2/\tau_k'\right] - 1} \tag{12.8b}$$

The characteristic output rate is in fact a relatively high normal "on" rate for a neuron. The neuron is firing every two integration periods. If the neuron is, for example, a cortical pyramidal cell with membrane time constant about 15 msec, then the characteristic output rate, $(1/2\Delta)$, is about 33/sec. Moreover, at this point, the input firing rate equals w_{ih}', and the mean input rate is itself producing output at the characteristic output rate. It is reasonable to suggest that at some point not too far from w_{ih}' the detailed temporal structure of the input train is no longer significant, and the model for regular firing should be approximately valid. At the point w_{ih}' in fact the irregular and regular firing models give the same result. For input rates greater than w_{ih}', output firing rates are theoretically predicted by the regular model.

Summary

The complete universal transfer curve, shown as Figure 8.2, is now complete. The response for temporally unstructured input is theoretically described by coupling the Poisson representation developed in this chapter for lower input rates and the average conductance model (or summed regular PSP trains) developed in Chapter 11 for higher input rates. The two branches of the curve join at the point $w_i' = w_{ih}^{*\prime}$, $w_o' = w_o^{*\prime} = 1/2$. The response for temporally structured input is theoretically described by the curve $w_o' = w_i'/n$, modified at higher rates by refractoriness. (Here n is the number of simultaneous input pulses needed to trigger an output spike.) This input pattern is also discussed in Chapter 11.

The value of the universal firing rate curve is that it allows one to relate the comparative input operating range of any neuron according to its anatomical and physiological parameters. The characteristic output rate, w_o^*, which is one-half the reciprocal of the neuron integration interval (approximated by the membrane time constant) is a key ingredient in the approach. This rate is taken as representative of a strongly "on" state of the neuron. The characteristic input rates, w_{is}^*, w_{il}^*, w_{iu}^*, and w_{ih}^*, are all useful in relating neuronal parameters to predicted output firing sensitivities and making comparisons across types of neurons.

Comments on the Scope of the Theory

As stated in Chapter 8, the theory developed here can be applied universally to most types of neurons when properly interpreted. Further, the theory can be intelligently modified to apply to situations involving biological factors that are not explicitly considered here.

Dendritic Action Potentials, Accommodation

For example, the firing of action potentials in dendrites can be represented by a simplification of the neuron. Firing in any local dendritic region can be treated by applying the theory to the reduced structure of this region as if it were a self-contained, spike-generating neuron (with appropriate leakage). The entire neuron could be seen as a device that responds either to activation of any of a number of such local components or to conventional summation of their locally subthreshold inputs to the soma.

The influence of accommodation can be characterized by using the value for resting threshold (for transient responses) or adapted threshold (for steady response) in the characteristic numbers.

Target Values, Steady and Transient Responses

In many respects the theory transcends its various simplifying mathematical devices. For example, even though time-varying terms are not explicitly considered in the analysis, the results nonetheless, should apply to the central trends of influence of parameters on transient responses. The idea of "target values" introduced in Chapter 7 is useful here. The time-varying system is driven toward its target potential values; these target values in turn are defined by the governing equations when the time-varying terms are removed. Therefore, the steady-state equations can be a meaningful measure of the parametric influences and a meaningful foundation for cross-neuron comparisons.

A similar undercurrent applies to the interpretation on the output side. The characteristic numbers and the trends of their dependence on the parameters are meant to apply to output firing sensitivities and propensities generally; for example, to peaks of transient responses, and not only to the steady firing levels for which they are explicitly derived.

Average Conductance Representation

The equations are developed in terms of expressions for input synaptic conductances of the form $g = (\delta^* s/\Delta)^* w'$. This in itself does not restrict one to a steady average conductance assumption since the term w' can include any temporal variations one wants to impose. Nonetheless, it is instructive to note that an average conductance representation of activation over regions of a dendritic tree might be a plausible representation in at least some typical normal situations. For example, hippocampal and cortical neurons typically require the equivalent of 200 or 300 simultaneous PSPs to fire. This would correspond to about 200 or 300 PSPs or more in about a millisecond if this is to produce firing. If they are spread over time, steady rates of 50 per millisecond are required to produce steady firing. If inhibitory or accommodative processes are active, even higher numbers are involved. Alternatively, tens of thousands of inputs firing at rates of, say 1 per sec (typical noise level), produce tens of PSPs per millisecond on the neuron. Higher input rates (normal on levels might be 15 to 30 per second) produce more. For a neuron with, say 30 dendritic branches, one might imagine that a PSP per millisecond (which would corresponds to a more or less continuous synaptic activation) is likely a typical activation rate. In a prototypical neuron model with, say, four levels of branching, the average conductance model seems even more strongly justified.

Status of Application of the Theory

The theory as developed in this part is essentially formally complete. Input–output behavior of computer-simulated neurons have been described very well by the theory. However, it remains for future work to apply it to naturally occuring biological neurons.

III

Coordinated Dynamic Firing Patterns in Local Networks

Chapter 13 Introduction to Coordinated Firing Patterns in Neural Networks

The purpose of this part of the book is to spell out an explicit model for the organization of coordinated dynamic firing patterns in neural networks. These patterns are intended to embody the means by which neuronal populations represent meaning and communicate with each other. This third part describes such a model with particular reference to local neural networks and their recurrent connections. Chapter 14 defines the sequential configuration model for this representation of coordinated network firing patterns and discusses its central properties. Chapter 15 presents a mathematical model for the tendency for unorganized activity (in contrast with the highly structured sequential configurations) to internally sustain itself by recurrent connections in local networks. Chapter 16 estimates the memory storage capacity of local modular networks of the cerebral cortex in terms of the sequential configuration model. The following fourth part discusses the broader significance of the sequential configuration model with regard to communications within composite networks of interacting populations and in neuronal junctions generally.

Introduction

The first two parts of this book have dealt with the mechanics of neuroelectric signals and the characteristic parameters of neuronal firing sensitivities. We now turn to the much larger and much less clearly understood world of neural networks. This is a centrally important area in the neural and brain sciences because neural networks (and not neurons) are the fundamental unit of functional organization in the brain.

It is also an area of great mystery because the fine-grained tools of experimental anatomical and electrophysiological study cannot deal with its vastness, and the global tools of these disciplines are not sufficiently refined to reveal the secrets of its intricate operations.

Moreover, our ways of thinking have not yet caught up with the recognition of the intrinsic importance of networks as functional units. We tend to think too much in terms of single neuron signals and single neuron interconnections.

What is needed for a mechanics of neural networks is description of the natural language spoken by neural networks—the coordinated dynamic firing patterns by which meanings are represented in entire neuronal populations and communicated from one population to another.

Such a description should allow us to envision the fundamental functional operations of neuronal populations, networks, and systems in their proper and natural perspective. It should lead us to see into the mechanisms of interaction of the supportive, competitive, or integrative patterns falling on particular networks from their several input activation systems. It should lead us to search into the recurrent connections in local networks for the means by which patterns are formed and sculpted. It should allow us to begin to formulate a mechanics for the processes and properties of neuroelectric signalling in neural networks and systems in their intrinsic language.

An explicit model in this framework is a central necessity for both the development of a theoretical mechanics of neuroelectrical signalling in neural networks and systems and for bringing broad theoretical theories of the type proposed by, say, Lashley, Hebb, and John into explicit comparison with experimental electrophysiological observations. Moreover, explicit definition lays the basis for revealing analysis. It allows, for example, the working out of specific predictions and design constraints.

These third and fourth parts of this book present and study just such a model for just such purposes. Fundamentally, we will assert simply that meanings are represented in neural networks by particular coordinated firing patterns across entire neuronal populations. These patterns are called *sequential configurations*. In them, particular sets of neurons fire in particular temporal sequences. We will see these sequential configurations as embedded in the synaptic structure of the network according to particular rules defined by the particular spatiotemporal structure of each individual dynamic pattern.

Local and Composite Neural Networks

A fundamental preliminary question concerns how a neural network should be defined within the context of normal brain interconnective structure. The brain stem reticular formation and the cerebral cortex, for example, representing two extremes of organization, both seem to be composed of numbers of interacting functional populations. Probably they should be considered as composite, multimembered neural networks. The particularly difficult question in these two cases is just how one might define the parcellation of subnetworks within the large whole.

On the other hand, the anatomical design structure of much of the brain and nervous system seems clearly based on a partitioning of local recurrently connected networks within larger composite neural networks and even larger neural systems.

Local networks consist of large numbers of neurons that are all members of a much smaller number of types or classes. Some types are highly individualistic for particular networks; other types (such as local inhibitory neurons) are quite similar in many networks. Cells within such local networks are usually very highly interconnected. Local networks are activated by one or more input fiber tracts originating in other local networks. In turn, axons of one or more types of cell within the local network project externally in a collected tract or tracts to another network or networks.

This third part of the book will define the sequential configuration model with particular regard to recurrently connected local neural networks and discuss its possible application to the question of memory capacity in local recurrently connected modules of the cerebral cortex. The fourth part will consider the sequential configuration model more broadly in terms of interactions within composite networks and discuss an interpretation it suggests for the organization of composite networks in the cerebral cortex.

Functional Anatomy of Local Neural Networks

The local recurrent connections within local neural networks are likely central devices by which the brain and nervous system defines and effects its representations of meaning in coordinated patterns of neuroelectric signals across its populations. They embody the means by which the dynamic patterns in local networks can transcend the particular spatial and temporal limitations of their input activation signals.

Recurrent inhibition likely serves as the primary device for defining and sculpting coordinated network patterns through the means of selectively distributed spatiotemporal synaptic inhibition of the excitatory fields provided by input activation systems and reverberated by the local recurrent excitation system. The salient example of this is the operation of "lateral inhibition," which has been shown to mediate the process of "feature extraction," which seems to be the terms in which the visual system identifies objects in the visual field. Lateral inhibition is the one intrinsic mechanism of neural network operation whose properties and function in the definition and representation are clearly known and established. The more ubiquitous and presumably further reaching and more ramified processes of recurrent inhibition in sculpturing coordinated patterns more generally are only our best speculation. Specific examples and a detailed understanding of these forms are lacking.

Recurrent excitation is a means by which local neural networks can capture input signals over time and thus can serve as a primary device for maintaining an ongoing representation of meaning in coordinated population firing patterns. In an extreme case, recurrent excitation can be adjusted with recurrent inhibition to provide ongoing activity that is self-sustaining, even in the absence of continued input activation, if once initiated by an input signal. In this case the local network has captured (and is projecting) a lasting internal representation from a passing stimulation pattern. Broader cases of such capturing of internal representations in time certainly involve composite as well as local networks.

The particular development of the sequential configuration model given in Chapter 14 is based on a local neural network with both excitatory and inhibitory recurrent connections. The sequential configurations are internally sustaining, once initiated. For background comparison to these highly structured sequential configuration patterns, Chapter 15 presents an analysis of the tendencies of recurrently connected networks to exhibit internally sustained activity when that activity is unorganized.

Some Miscellaneous Basic Questions

Several basic questions relate to the idea of neural networks representing meaning in more or less standing patterns in neural networks. One such question has to do with the significance of duration in a syntax or a code. How long does a standing pattern have to stand to establish its identity? Is it more instructive to consider the temporal coincidences of pulses across the family of input fibers rather than the individual sequences of pulses in constituent single neurons? In this context, is the frequency of firing in individual neurons a directly meaningful entity in the syntax and coding of neural nets? Firing rate is usually considered as a neuron's main output signal. However, the relative timing of action potentials in single neurons are almost always statistically indistinguishable from chance levels at a given rate when they are separated by more than 50 msec or so. Perhaps in most cases the individual spikes in any given neuron are not significant with respect to each other, except to indicate its general propensity to fire over given time windows. However, its syntactical and coding secrets, which determine the precise instants of firing, may depend rather on its coordinations with specific other neurons.

A second question is the extent to which the microstructure (single neurons with their often exquisite anatomical detail) and the microphysiology (single action potentials with their sharply defined times of occurrence) of the brain are indeed as important in the syntax of the brain as most of us believe them to be. Is it possible that the brain averages in much more global ways over the activities of its vast numbers of cells? Are its principles of operation (any or all) based more on statistical averages as are the principles of thermodynamic based on statistical averages over vast numbers of interacting molecules?

Given the richness and diversity of the brain and its networks, it is likely that such questions will have various and different answers in its different regions. In the sequential configuration model, timing of individual action potentials is significant, durations of syntax and code are necessary, and coincidences of firing across cells is central.

A third question has to do with the universality of the language developed in this section. It may very well be that different networks with different functions and at different levels of development may utilize different languages and different codes. Probably, however, a common principle will be the temporal correlation and coincidence of signals in separate input channels at each network junction.

Another question asks how much tolerance is acceptable in the different recalls of a given dynamic pattern. Must all recalls be essentially identical in cell composition and timing? Alternatively, might there be sufficient tolerance to variability that large numbers of participating neurons may be quite different in the recall of a single memory trace from one recall to another? The approach taken here defines an anatomically based "bed" for any given pattern, against which any of its recalls may be measured. We assume that these recalls are fundamentally stochastic and, in fact, may differ considerably from one recall to another.

The mechanics developed here leads to some intriguing questions of an entirely different order. For example, how might degrees of overlap or compatibility of distinct dynamic patterns correspond to psychological overlap and compatibility of the represented elements? Further, might one search for the mechanics of certain cognitive or psychological processes in the neurophysiological mechanics of the coordinated dynamic patterns by which they are represented? Here rises the central question of exactly how the rich and nebulous world of psychological entities will eventually map onto such coordinated dynamic firing patterns, if, indeed, such firing patterns are elemental units of representation. The point of view taken here is that the process of defining such dynamic patterns explicitly leads us to a mechanics wherein this level of questioning is at least within sight, if not yet within reach.

Finally, in an operational context, note that the approach taken here to the mechanics of neuroelectric signalling in neural networks bypasses a learning stage. We stipulate that the patterns are embedded in the selective distribution of synaptic strengths within the network according to prescribed rules, reflecting the individuality of the patterns, and then imagine that these patterns represent equilibrium configurations toward which any learning process will converge. The idea here is that the general principle of a structured population pattern is more useful for the development of a large-scale mechanics of neural network operations than are the various refinements to be found in the processes by which such patterns may be learned by the net. These learning processes will likely relate meaningfully to the questions raised earlier regarding compatibility of different representations, but are seen as secondary in usefulness and significance to the idea of the final patterns in themselves.

Chapter 14 The Sequential Configuration Model for Firing Patterns in Neural Networks

This chapter presents a theoretical model (called the *sequential configuration model*) for the coordinated firing patterns in neural networks. The explicit formulation of such a model is seen as a central necessity both for the development of a theoretical mechanics of neural networks and systems and for bringing broad neuropsychological theories of the type proposed by, say, Lashley, Hebb, and John into explicit comparison with electrophysiological observations.

The model is developed in this chapter explicitly for local, recurrently connected networks having one excitatory and one inhibitory population. The concept, however, is applicable to any population or combinations of populations.

The model, in fact, is very simple and straightforward. One basically is saying simply that particular coordinated network firing patterns consist of the firing of particular sets of neurons in particular temporal sequences. However, several ingredients of the model are sufficiently uncommon to warrant comment. First, and perhaps most significantly the patterns defined here are conceptually partitioned into beds and realizations. The former consists of a firm and constant theoretical base against which its various variable and context-dependent manifestations may be clearly considered.

Second, this model jumps past a learning stage to define a final target stage of synaptic organization. We suggest that the sequential configurations and their underlying synaptology as presented here represent equilibrium configurations toward which any learning processes can be envisaged as gravitating. Various models for the learning processes might be compared in terms of how they approach such target final states.

Third, the patterns here are intrinsically internally sustained. That is, once initiated, they maintain themselves in the absence of any further reinforcing stimulation and, indeed, even in the face of some significant levels of competitive stimulation.

Fourth, the sequential configurations, as indicated by their name, are intrinsically temporal in structure as well as spatial. Therefore, patterns that utilize exactly the same neurons may be simultaneously embedded in a given net and selectively recalled.

Theory and Methods*

The essential feature of this model is to specifically delineate (1) the nature of coordinated firing patterns used by local neural networks to represent items of information, (2) the enhanced synaptic connections underlying these patterns, and (3) the sources of variability in the representation that allow it to correspond favorably with the character of experimental observations.

The Nature of Coordinated Firing Patterns: Sequential Configurations

The theory posits that items of information are represented by the firing of specific sequences of sets of neurons. That is, the representation of item A consists of the firing of specific sets of cells $A1, A2, A3, \ldots$, at the times $t1, t2, t3, \ldots$. Item B is represented by the firing of different specific sets of cells, $B1, B2, B3, \ldots$, at times $t1, t2, t3, \ldots$. The individual sets of cells, $A1, A2, A3, \ldots, B1, B2, B3, \ldots$ are referred to as *links*. A complete sequence ($\{A1, A2, A3, \ldots\}$, or $\{B1, B2, B3, \ldots\}$), is referred to formally as a *sequential configuration*. Any particular sequential configuration is referred to more casually as a *trace*. Sequential configurations are imagined to

* This theory was first presented in *Biological Cybernetics* **65** (1991), 339–349.

exhibit some finite length, L. They may then terminate, or they may repeat cyclically by closing back on themselves.

The theory further posits that any item of information is represented by a pair of simultaneous coordinated sequences, one member in an excitatory pool and a parallel member in an inhibitory pool of neurons. This situation is illustrated in Figure 14.1. This figure shows the cell number in each pool on the vertical axes and time position in the sequence on the horizontal axes. The specific positions of the dots indicate the firing of specific individual cells at any one time position.

The Nature of Synaptic Connections Within a Sequential Configuration

The theory further postulates that once initiated these sequences are maintained by enhanced synapses between the cells of links close together in the sequence. Specifically, the synapses from cells of any given link of excitatory cells to all cells in all those links forward in time position up to a time-position distance of n_L are enhanced above the initial level. The synaptic increase is greater for links closer together and less for links further apart. Excitatory synapses to cells in both excitatory and inhibitory links are increased.

The specific quantitative rule for this synaptic increase is given in Eq. (14.1).

$$\text{Synaptic strength factor} = d_l = \exp\left(-dc^*(l-1)/n_L\right), \quad l = 1, 2, ...n_L \quad (14.1)$$

In this equation, dc is a model parameter representing the decay of synaptic enhancement with position distance between links, and n_L is a model parameter representing the number of links synaptically related to a given link. If n_L is set equal to 1, then the sequence is Markovian in the sense that any one link projects enhanced synapses only to the next or contiguous link. In the more general case, n_L is 2 or more, indicating that synapses are enhanced to more than one downstream link.

The purpose of the cells in the inhibitory links is to suppress firing by neurons not in the trace at hand and firing of other neurons in the trace except at the appropriate time. Therefore, inhibitory synapses in the net are initially set to a nominal maximum value (each inhibitory neuron projecting maximum inhibition to every network neuron). Then, inhibitory synapses among trace neurons are diminished in precisely the same manner as excitatory trace neurons are increased. That is, inhibitory synapses are decreased to 0 for all synapses from all cells in a given inhibitory link to all cells in the next excitatory and inhibitory

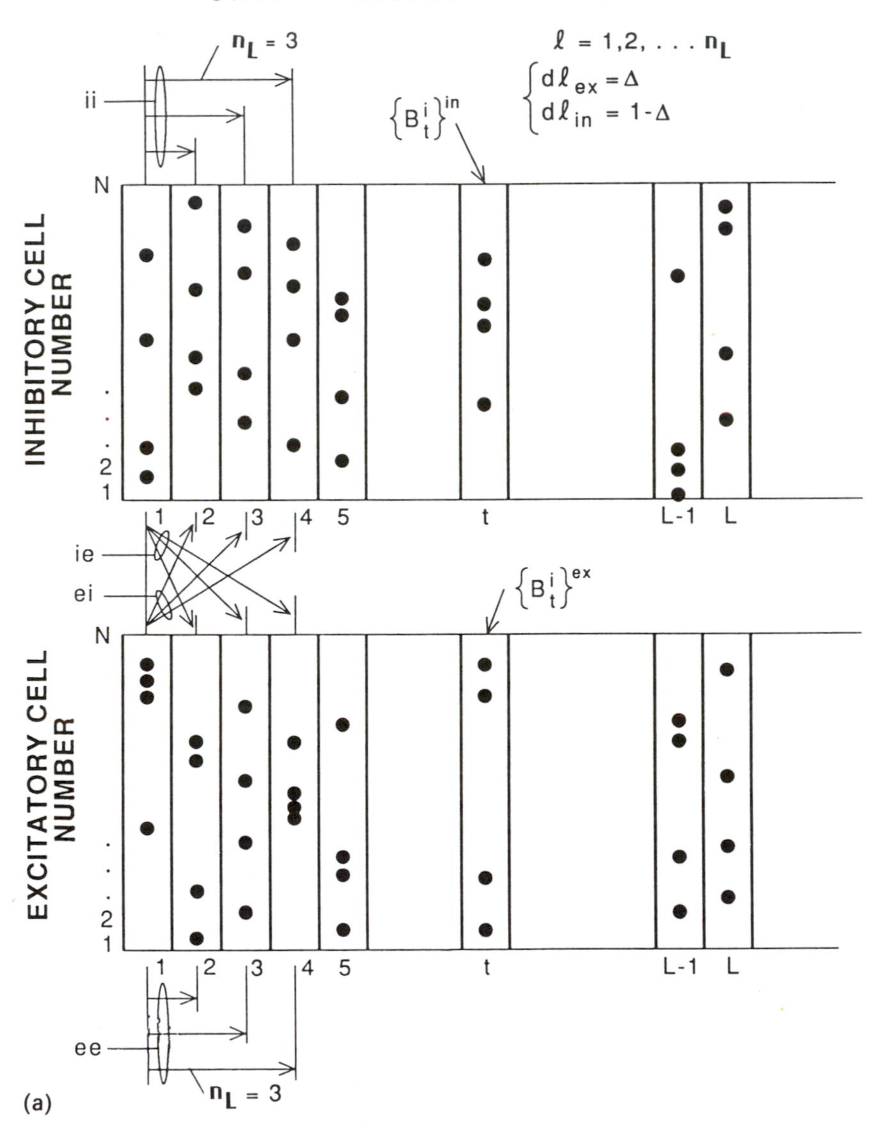

FIGURE 14.1 (a) Schematic of beds and their synaptic underpinnings in a two-population network; (b) synaptic activations within sequential configuration traces ("anticipatory ramps") (reprinted with permission from R. J. MacGregor, Sequential Configuration Model for Firing Patterns in Local Neural Networks, *Biological Cybernetics*, Springer-Verlag, 1991, **65**, 339–349).

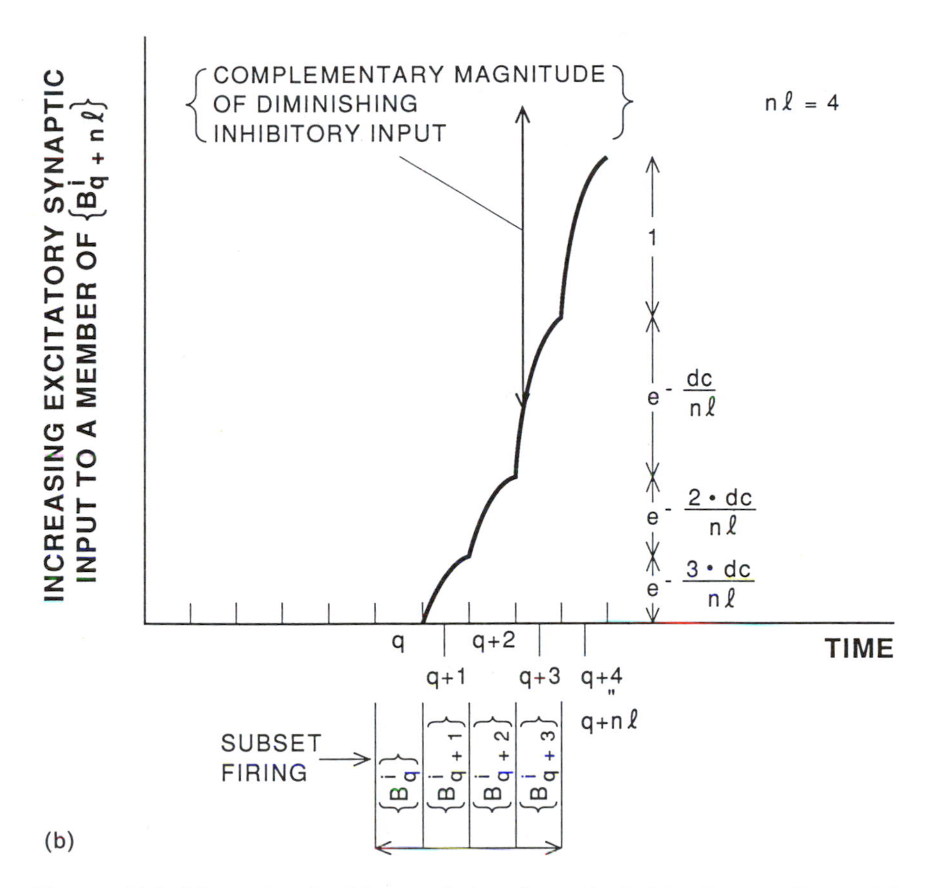

FIGURE 14.1 (b) reprinted with permission from R. J. MacGregor, Sequential Configuration Model for Firing Patterns in Local Neural Networks, *Biological Cybernetics*, Springer-Verlag, 1991.

link. Inhibitory synapses to cells in the second downstream links are decreased but a bit less and so on. Specifically, the inhibitory synaptic strengths are given by Eq. (14.2):

$$\text{inhibitory synaptic strength factors} = d_l = 1 - \exp(-dc^*(l-1)/n_L), \quad l = 1, 2, \ldots. n_L \qquad (14.2)$$

Thus, this procedure provides a progressively increasing ramp of excitatory input and a corresponding temporal "window" of decreasing inhibition, both delivered to trace neurons in the n_L time units prior to their prescribed times of firing. Both effects reach their maximum effectiveness (maximum excitation and minimum inhibition) at the prescribed time of firing for the receiving links.

Beds and Realizations: Representation of Physiological Variations

Sequential configurations are posited to be mediated by "beds" and to manifest "realizations." The bed of a trace consists of an ordered sequence of sets of neurons (links—in both the excitatory and inhibitory populations) are shown in Eq. (14.3):

$$\text{Bed} = \{B_t^i\}, \qquad i = 1, 2, \ldots n, \qquad t = 1, 2, \ldots L \tag{14.3}$$

where each set, B_t^i, constitutes *all* the cells associated with that link. Equivalently, a bed may be defined as all the synaptic modifications defined by Eqs. (14.1) and (14.22) for the trace. Here n is the number of cells in a link, t is time, and, L is the number of links in a trace. A bed is primarily anatomical and moreover, may be fundamentally an idealization or an abstraction in multiply- embedded nets; this can be better appreciated by comparing a bed to a realization.

A realization of a trace consists of an ordered sequence of sets of neurons, such as shown in Eq. (14.4):

$$\begin{aligned} \text{Realization} = \{F_t^i + R_t^j\}, \qquad & i = 1, 2, \ldots n_f \leq n \qquad & j = 1, 2, \ldots n_r \\ & t = t_a,\ t_a + 1, \ldots t_b, \qquad & 1 \leq t_a \leq t_b \leq 1 \end{aligned} \tag{14.4}$$

some of which, F_t^i, compose a subset of the corresponding set, B_t^i, of the bed of the trace, and other neurons, R_t^j, which are not members of the B_t^i and therefore constitute firings extraneous to the trace, or "noise." Here, n_f represents the number of trace neurons that fire at any given time in a realization; n_r represents the number of nontrace cells that fire at any given time in a realization; t_a and t_b represent the end points of the time window over which a given firing history is considered to be a realization of a given trace.

A realization is primarily physiological. It represents the manifestation of a trace on any single recall or recurrence. Notice that realizations may be markedly different from one recall to another; in fact, it is possible that two different recalls may make use of totally distinct subsets of trace cells.

The distinction between beds and realizations allows us to recognize the underlying unity of a sequential configuration even when its manifestations are fundamentally noise-ridden stochastic samples. Note also that if more than one trace is embedded in a net (a situation we will call *multiply embedded*), a single synapse may be called upon for different values by different traces. Clearly, at most one such call can be satisfied. That means that some synaptic beds will not be the precise ones called for by the original trace. In this case the "bed" of the trace corresponds to the initial, unadulterated request, defined by Eqs. (14.1), (14.2), and (14.3), and not to the actual state of affairs current in the synaptology of the net. In such a case, a "bed" has become an abstraction, an idealization. Conceptually, this is valuable as it speaks to effects of multiple embedding of traces in a net. Such effects are considered in more detail in Chapter 16.

Computer Simulations

The results that follow are obtained from computer simulations of recurrently connected (all-to-all) pools of excitatory and inhibitory neurons. Particular sequential configurations are "embedded" in the pools by explicitly identifying the sets of neurons firing at all times and by setting the interconnecting synaptic strengths according to Eqs. (14.1) and (14.2). When a given synapse is called to more than one value by different traces, the maximum value called is used. The dynamics of the individual neurons are represented by the state variable equations introduced in Chapters 6 and 7. Traces are selectively recalled by stimulating in sequence the first few (usually n_L) links. A copy of the computer program used for these results, ldsys6g, is given in the appendix.

Results

Recall of Singly Embedded Traces

To recall a trace from a net in which it is the only embedded trace, one simply stimulates in sequence the first few (usually n_L) links of the trace and then terminates external stimulation. The trace sustains itself by the forward-projected excitation described previously. To fine tune this behavior, one adjusts, relative to cell thresholds, four global synaptic strength parameters, s_{ee}, s_{ei}, s_{ie}, and s_{ii} (representing, respectively, excitatory to excitatory, excitatory to inhibitory, inhibitory to excitatory, and inhibitory to inhibitory). The effective synaptic strength for a given single synapse of type xy is then the product of s_{xy} and its individual trace weighting factor, d_l, given by Eq. (14.1) or (14.2). Recalls (i.e., realizations) of singly embedded traces are complete and noise free; that is, the realization is equal to the bed.

Selective Recall of Traces in Multiply-Embedded Nets

We have run several dozens cases where two or more traces have been embedded in the same net and the individual traces selectively recalled. The results described here are representative of this large sample. The beds referred to here are shown in Table 14-I.

When more than one trace is embedded in the net and there are cells shared by two or more traces, realizations tend to deviate from their beds even in the absence of other sources of noise. This is because synaptic connections from, say, cell 1 to cell 2 embedded for trace *A*, may be activated inappropriately when trace B is active, or vice versa. Moreover, it is possible that a given single synapse is targeted for a different value in both traces, with the result that at least one trace has a synapse with a different value from that called for by the bed.

The dynamic manifestations of this cross talk in realizations, consist of what one may call *jiggle* and *double firings* of cells, or indeed of whole links. That is, cells or links may tend to come on early or late by one or two time units.

The simulations show that individually recalled realizations of sequential configurations are remarkably easy to recall selectively and remarkably robust when small numbers of traces are embedded. For example, Figure 14.2 shows a representative case of selective recall of two traces with 100 percent overlap of cells. That is, traces *D* and *F* (see Table 14-I) contain exactly the same cells, but order them differently in time. This example illustrates the central character of sequential configurations. Suppose one recorded experimentally from cells 12 and 32 when trace *D* was realized and then again when trace *F* was realized. If one simply noted the firing rates of the cells, one could not discriminate between the two cases, for these would be the same in both. However, the relative timing of the firings differ in the two cases. Firing in cell 32 follows firing in cell 12 by two time units in trace *D*, but by six time units in trace *F*.

Jiggle and double firings become pronounced and present problems when larger numbers of traces are embedded in a net. The associated desynchronization and increased amounts of inhibition may cause termination of the trace, or a "derailing" to a competitor trace. Moreover, there may be increased jiggle to the point of explosive mixing of traces with no selectivity in response to either stimulus.

These disruptive effects can be modulated by adjusting the four global synaptic strength parameters, s_{ee}, s_{ei}, s_{ie}, s_{ii}. The main features are shown in Figure 14.3. The key factor is the degree of inhibition.

TABLE 14.I. Beds Referred to in This Chapter

BED A												
	91	81	71	61	51	41	31	21	11	1	15	55
	92	82	72	62	52	42	32	22	12	2	25	65
	93	83	74	63	53	43	33	23	13	3	35	75
	94	84	74	64	54	44	34	24	14	4	45	85
BED B												
	7	27	47	67	87	96	97	77	57	37	17	6
	8	28	48	68	88	86	98	78	58	38	18	16
	9	29	49	69	89	76	99	79	59	39	19	26
	10	30	50	70	90	66	100	80	60	40	20	36
BED C												
	21	23	25	27	29	20	30	28	26	24	22	31
	41	42	45	47	49	40	50	48	46	44	42	51
	61	63	65	67	69	60	70	68	66	64	62	71
	81	83	85	87	89	80	90	88	86	84	82	91
BED D												
	2	10	18	26	34	42	50	58	66	74	82	90
	4	12	20	28	36	44	52	60	68	76	84	92
	6	14	22	30	38	46	54	62	70	78	86	94
	8	16	24	32	40	48	56	64	72	80	88	96
BED E												
	1	9	17	25	33	41	49	57	65	73	81	89
	3	11	19	27	35	43	51	59	67	75	83	91
	5	13	21	29	37	45	53	61	69	77	85	93
	6	12	18	24	30	36	42	48	54	60	66	72
BED F												
	2	34	66	4	36	68	6	38	70	8	40	72
	10	42	74	12	44	76	14	46	78	16	48	80
	18	50	82	20	52	84	22	54	86	24	56	88
	26	58	90	28	60	92	30	62	94	32	64	96

Note: Cells in individual links are read vertically. Successive links are read horizontally from left to right. For example. Bed A has cells 91, 92, 93, 94 as its first link; cells 81, 82, 83, 84 as its second link; cells 55, 65, 75, 85 as its twelfth link. All beds use $n = 4$, $dc = 2$, $n_L = 4$, and $L = 12$.
Reprinted with permission from R. J. MacGregor, Sequential Configuration Model for Firing Patterns in Local Neural Networks, *Biological Cybernetics*, Springer-Verlag, 1991.

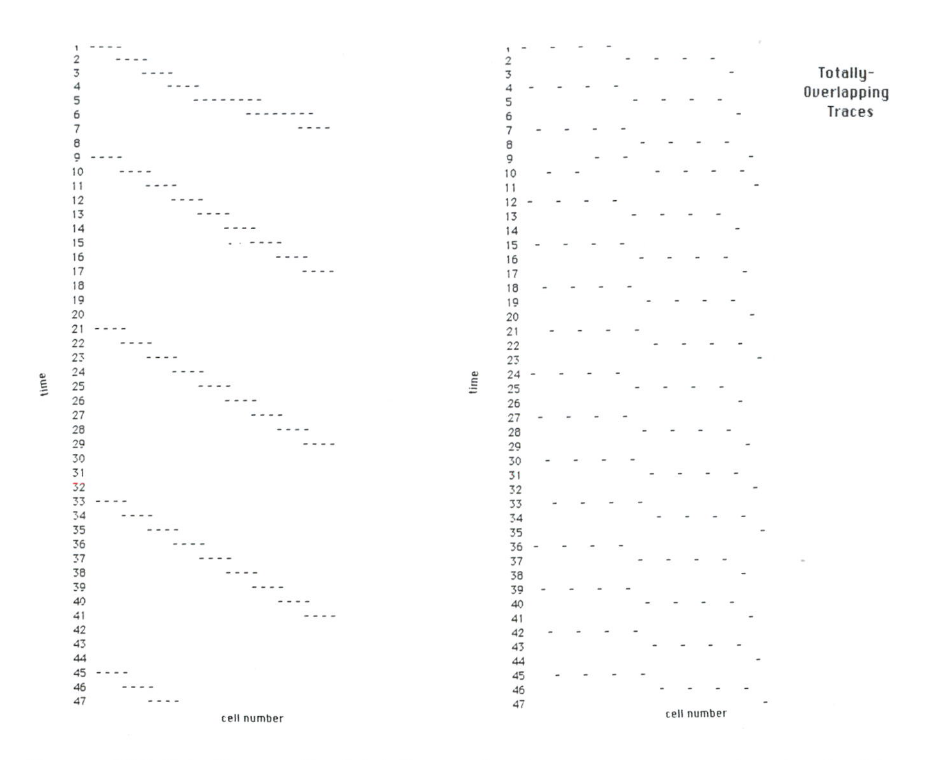

FIGURE 14.2 Selective recall of totally overlapping memory traces (reprinted with permission from R. J. MacGregor, Sequential Configuration Model for Firing Patterns in Local Neural Networks, *Biological Cybernetics*, Springer-Verlag, 1991, **65,** 339–349).

If s_{ie} is increased, the selectivity improves, with decreasing jiggle and noise and increasing fidelity of realizations to beds. This is a pervasive general characteristic.

Moreover, by taking s_{ee} slightly greater than s_{ei} and the inhibitory strengths somewhat higher than both, one produces higher tenacity and stability of realizations. Here recurrent excitation outweighs recurrent inhibition at low levels of excitatory firing (because the low s_{ei} results in fewer active inhibitory cells) and therefore increases tenacity. However, at higher levels of excitatory firing, when the s_{ei} do produce some significant number of active inhibitory cells, the recurrent inhibition comes on relatively strong to push the firing levels back down. (A typical effective set of values is $s_{ee} = 2$, $s_{ei} = 1.5$, $_{sie} = s_{ii} = 4$.) Following this procedure, we have repeatedly successfully recalled traces for long periods of time even when numbers of network cells randomly activated by extrinsic input exceeded the number, n_f, of trace cells firing each time.

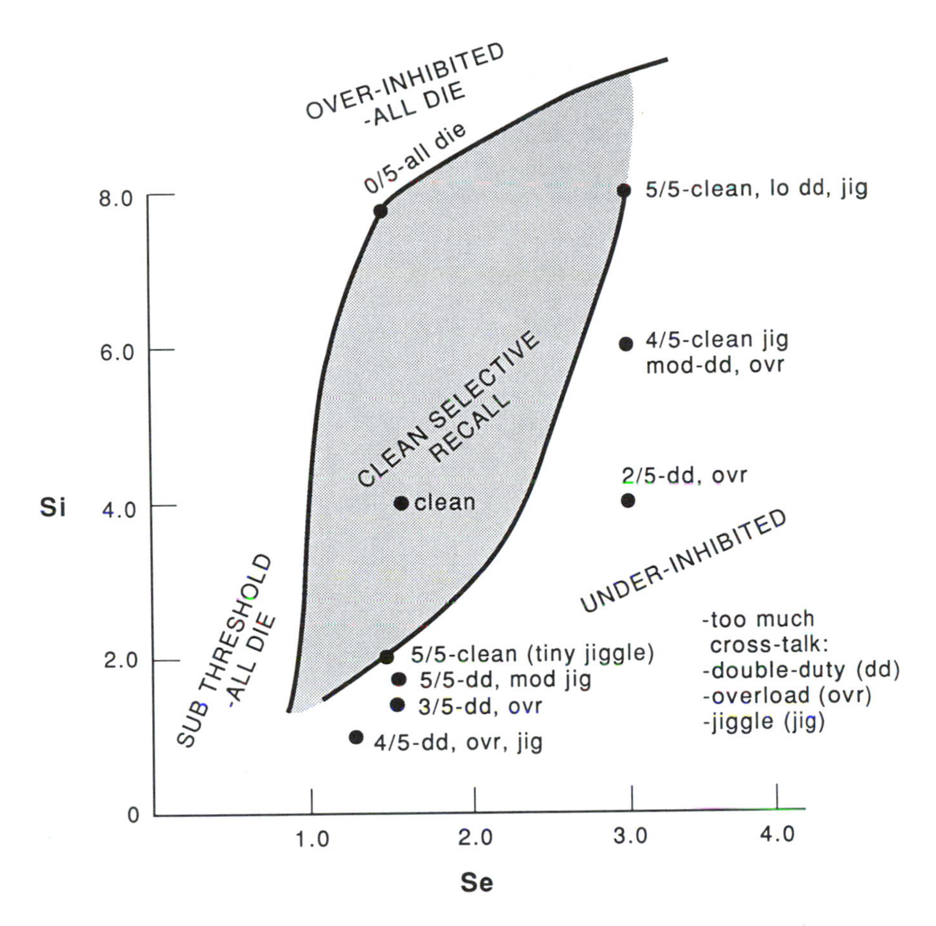

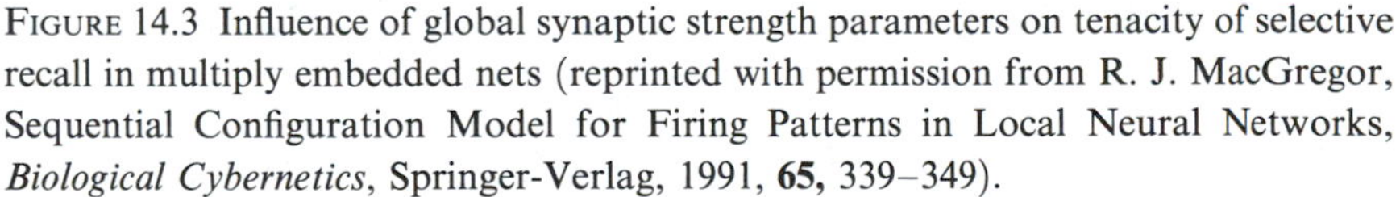
FIGURE 14.3 Influence of global synaptic strength parameters on tenacity of selective recall in multiply embedded nets (reprinted with permission from R. J. MacGregor, Sequential Configuration Model for Firing Patterns in Local Neural Networks, *Biological Cybernetics*, Springer-Verlag, 1991, **65,** 339–349).

Temporal Correlations and Histograms in Sequential Configurations

Several sets of simulations were undertaken to compare the general character of multineuronal activity in sequential configurations to that observed in microelectrode studies. In one such situation three patterns, *A, C,* and *D*, were embedded in a net. Network cells were randomly activated by extrinsic stimulation at a mean rate of about 3 per msec. The simulation was run for a total of 5000 msec. Throughout this duration there were short time periods of random activity in the net separated by much longer periods during which one or another of the three traces was clearly identifiable. The

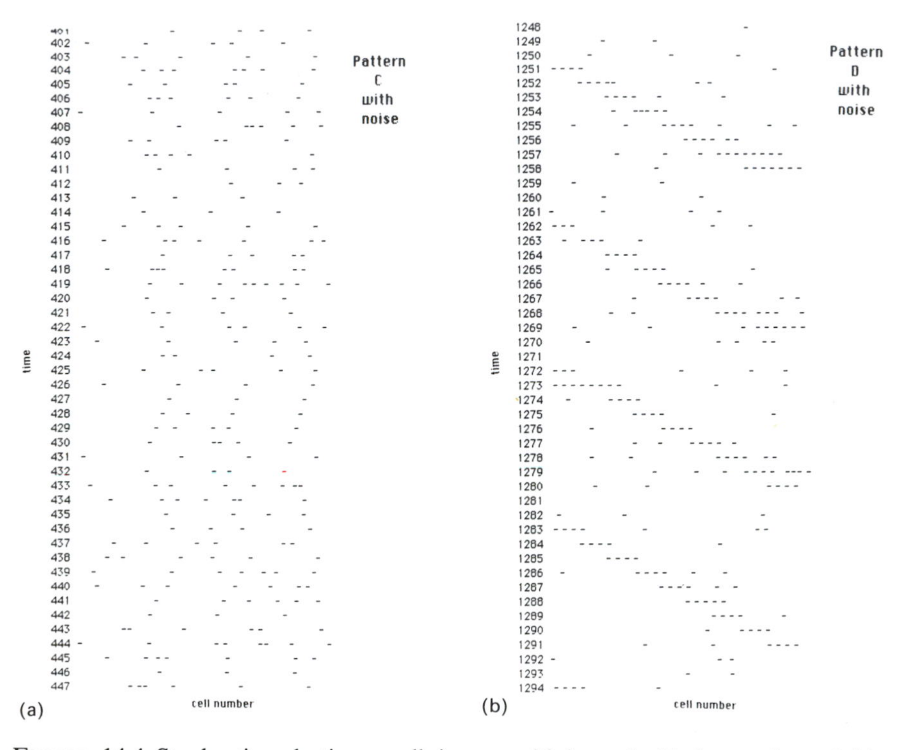

FIGURE 14.4 Stochastic selective recall in a multiply embedded net: (a and b) stochastic realizations of patterns *C* and *D*; (c and d) intracellular and global firing records in patterns *C* and *D*; (e and f) autocorrelation histograms in patterns *C* and *D*; (g and h) crosscorrelation histograms in patterns *C* and *D* (reprinted with permission from R. J. MacGregor, Sequential Configuration Model for Firing Patterns in Local Neural Networks, *Biological Cybernetics*, Springer-Verlag, 1991, **65,** 339–349).

longest runs for traces *D* and *C* were 1300 and 535 msec, respectively. The main results from this simulation are shown in Figure 14-4. (Parts a, c, e, and g show net activity from these runs when pattern *C* is on, and parts b, d, f, and h when pattern *D* is on). To interpret these results, note that cell 96 (1) is in pattern *D* only; cell 44 (2) is in patterns *D, C,* and *A;* cell 60 (3) is in patterns *D* and *C*; and cell 9 (4) is in no pattern.

Figures 14.4a and 14.4b show the network spike train displays for segments of realizations of patterns *C* and *D*, respectively. Note the high degree of noise, the temporal jiggles, and the high tenacity of the traces in the presence of this noise.

Figures 14.4c and 14.4d show the numbers of excitatory and inhibitory cells firing as functions of time and samples of generator potentials from

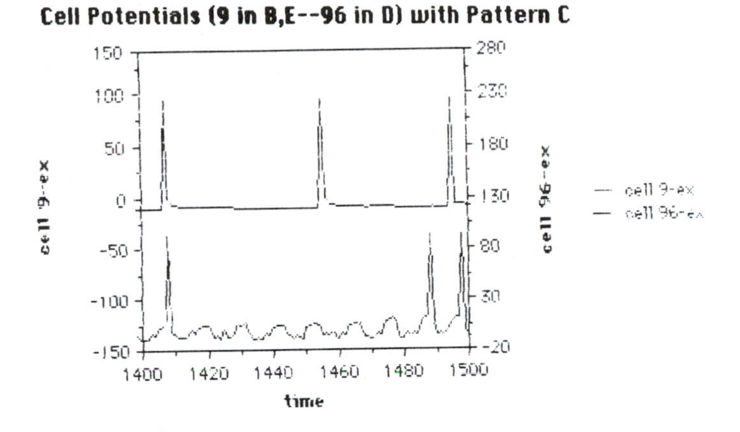

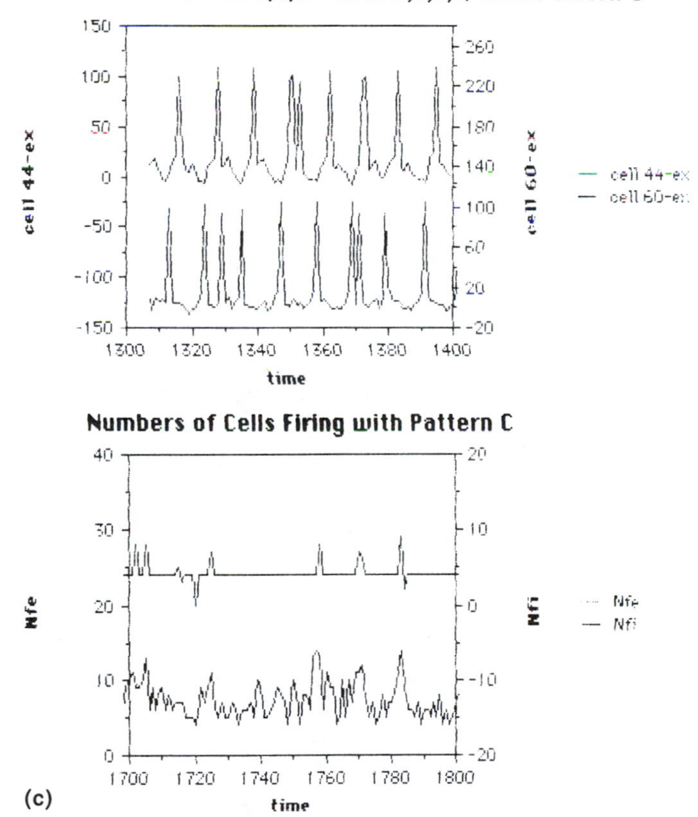

(c)

FIGURE 14.4 (c) Reprinted with permission from R. J. MacGregor, Sequential Configuration Model for Firing Patterns in Local Neural Networks, *Biological Cybernetics*, Springer-Verlag, 1991.

Cell Potentials (9 in B,E--96 in D) with Pattern D

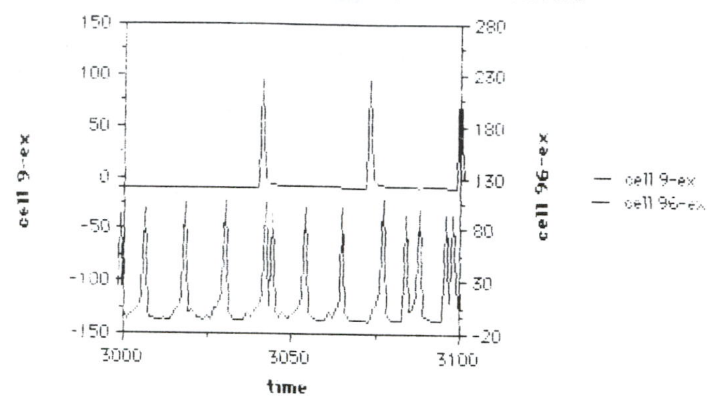

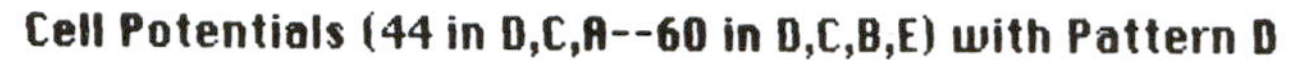

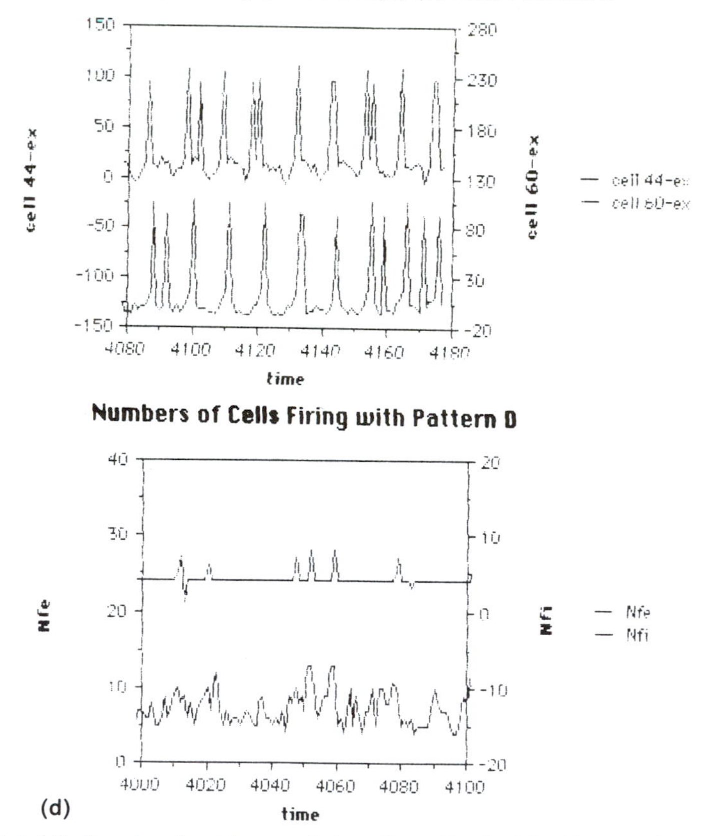

FIGURE 14.4 (d) Reprinted with permission from R. J. MacGregor, Sequential Configuration Model for Firing Patterns in Local Neural Networks, *Biological Cybernetics*, Springer-Verlag, 1991.

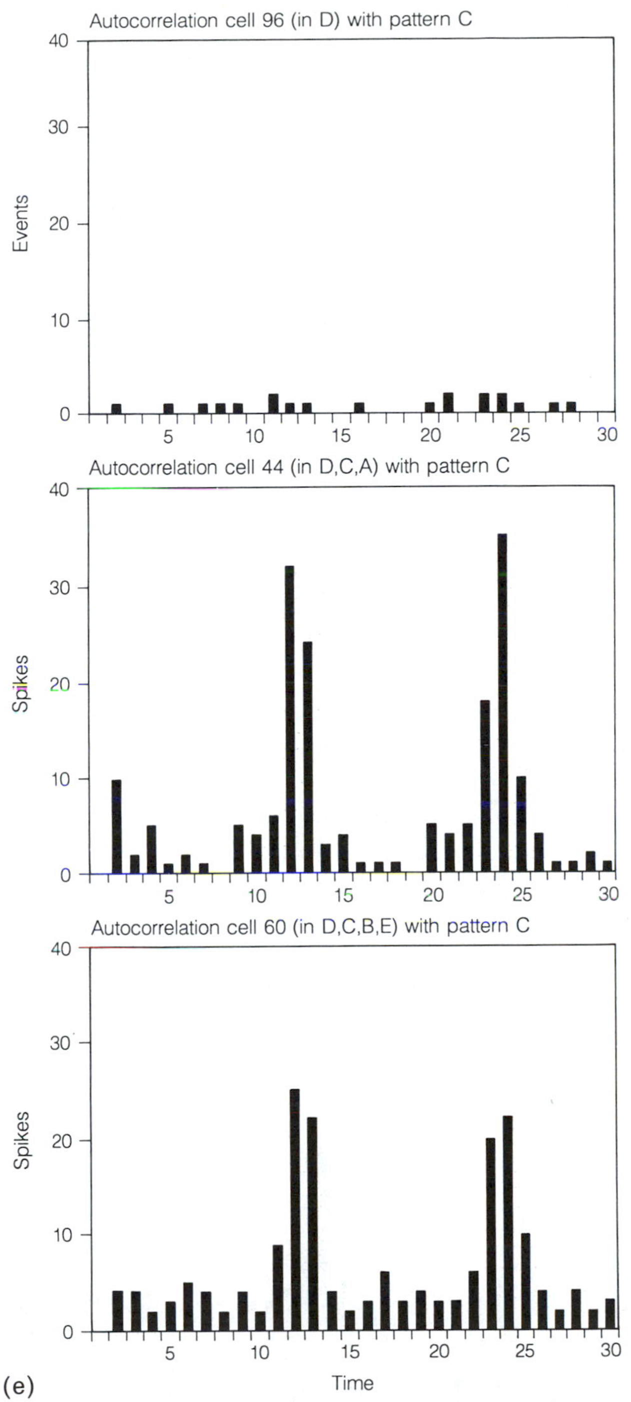

FIGURE 14.4 (e) Reprinted with permission from R. J. MacGregor, Sequential Configuration Model for Firing Patterns in Local Neural Networks, *Biological Cybernetics*, Springer-Verlag, 1991.

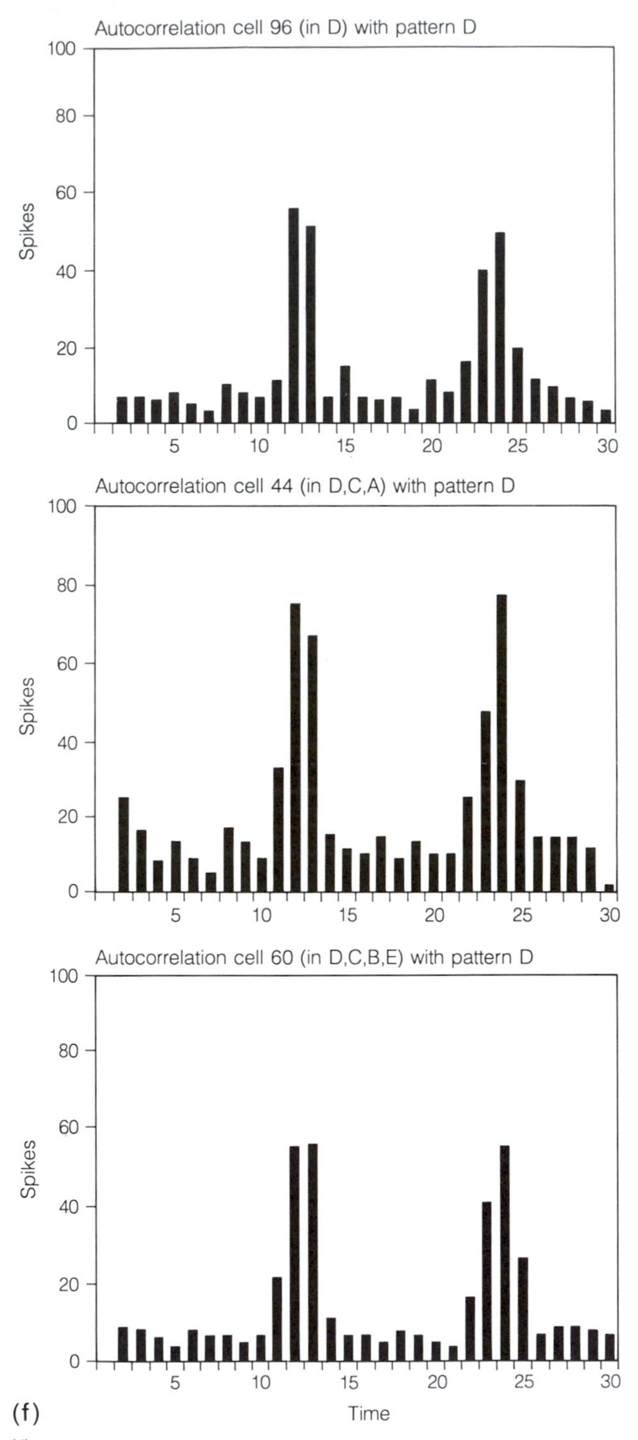

FIGURE 14.4 (f) Reprinted with permission from R. J. MacGregor, Sequential Configuration Model for Firing Patterns in Local Neural Networks, *Biological Cybernetics*, Springer-Verlag, 1991.

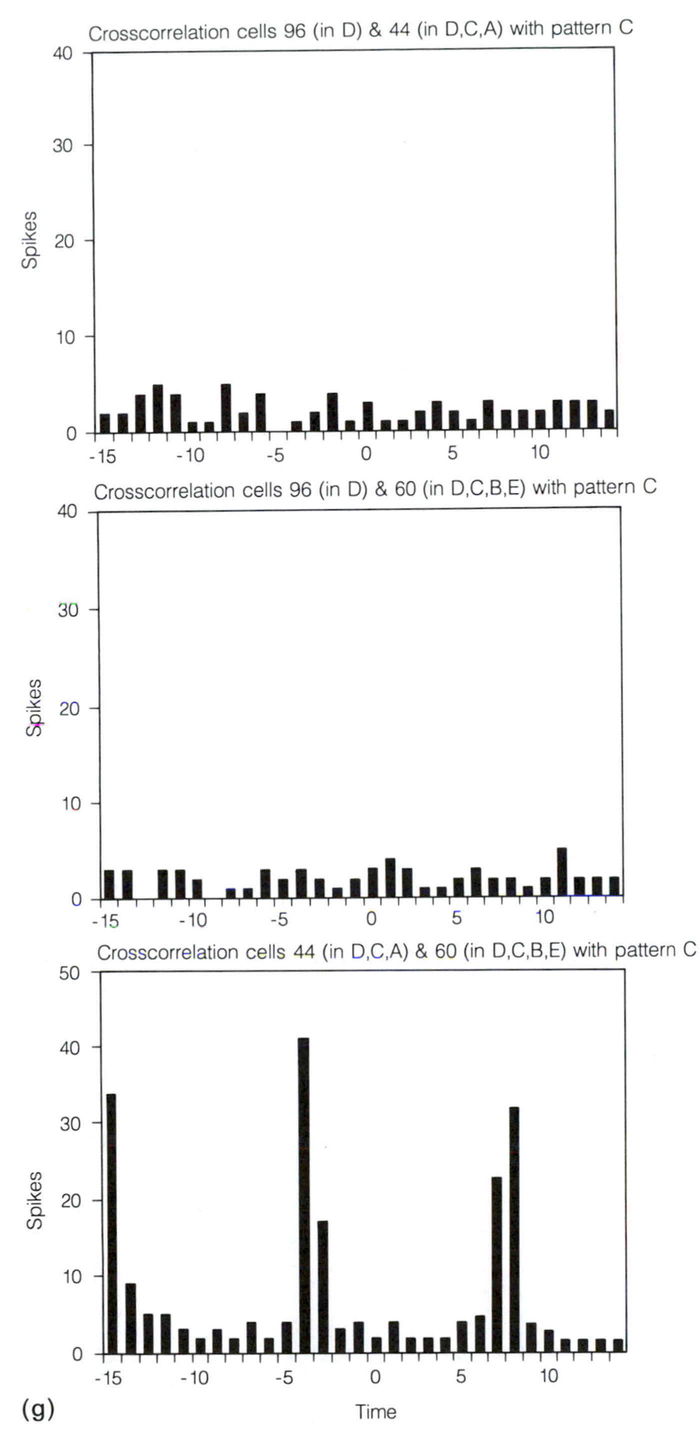

FIGURE 14.4 (g) Reprinted with permission from R. J. MacGregor, Sequential Configuration Model for Firing Patterns in Local Neural Networks, *Biological Cybernetics*, Springer-Verlag, 1991.

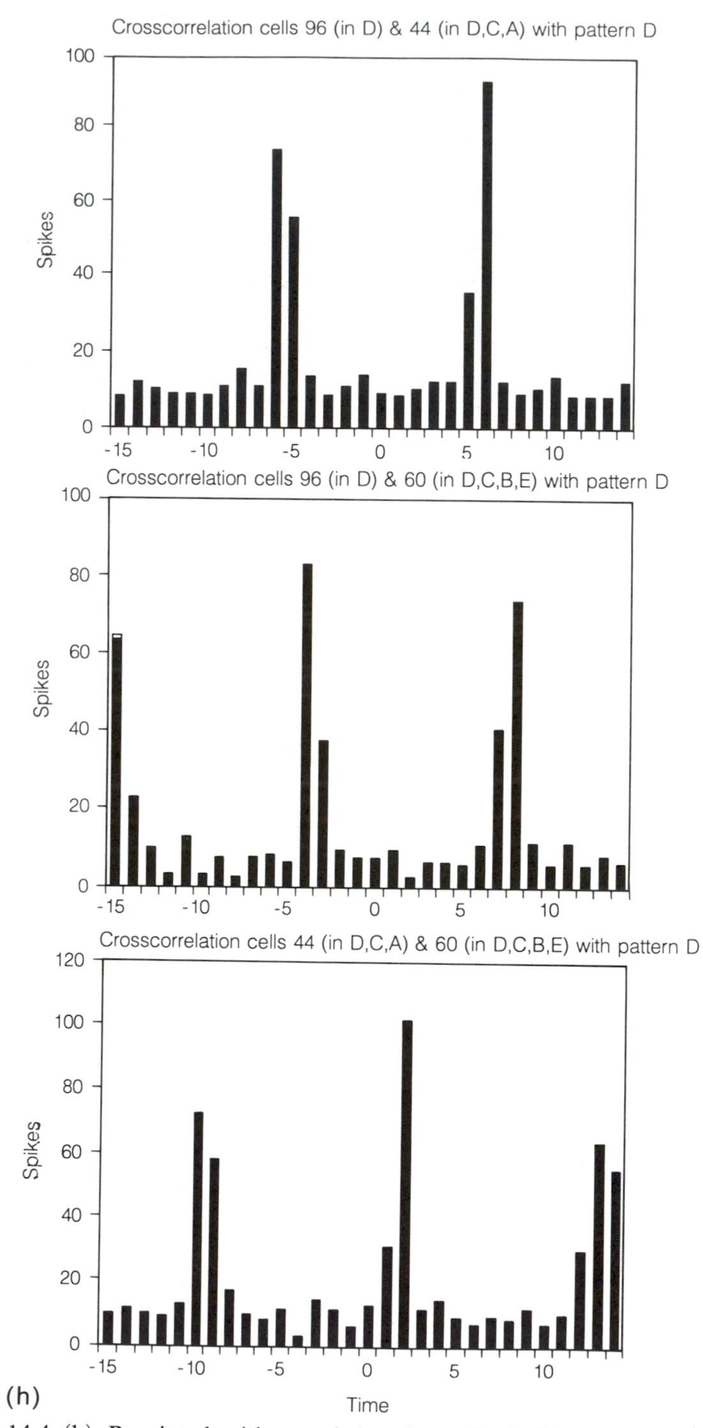

FIGURE 14.4 (h) Reprinted with permission from R. J. MacGregor, Sequential Configuration Model for Firing Patterns in Local Neural Networks, *Biological Cybernetics*, Springer-Verlag, 1991.

these cases. These curves showing total numbers of cells firing illustrate the global feedback mechanisms maintaining relatively stable firing levels. If the sequential configurations are running cleanly, only four inhibitory and four excitatory cells will fire at each time interval. This is approximately the case for the inhibitory cells, as can be seen in Figures 14.4c and 14.4d. The excitatory populations, however, are firing at much higher rates due primarily to extrinsic random activations and, to a lesser extent, to internally generated cross talk. Note that excitatory firing levels are typically around seven or eight, indicating a signal to noise ratio of about 1. The patterns are therefore remarkably robust in the presence of this noise level. Note further that when, and only when, the number of excitatory cells firing reaches dangerously high levels (about ten), a few extra inhibitory cells come on. This is followed immediately by a turn-down or damping in the number of excitatory cells firing. The excess inhibition, however, turns off immediately and is not activated by the lower levels of excessive excitatory activity, thus safeguarding against overinhibition and termination of the sequence.

Figures 14.4c and 14.4d also show representative samples of generator potentials for cells 9, 44, 60, and 96, and the number of excitatory and inhibitory cells firing as functions of time. Note in Figure 14.4c that while pattern C is on cells 44 and 60 (both of which are in C) are firing noisily but clearly rhythmically with period equal to the cycle length of the trace. Note also that cell 96 (which is not in C) is firing only sporadically. Its generator potential, however, exhibits fluctuations due to cross talk since a number of C cells are within n_l steps of 96 in pattern D. Note further that cell 9 (which is not in C and does not have such cross talk connections with C cells) fires only sporadically and exhibits virtually no generator potential fluctuations corresponding to activity in C.

Figures 14.4c and 14.4d show clearly that the firing rates of these model cells go up when the sequential configurations in which they are embedded are activated. Thus, cells 96, 44, and 60 are all firing at high rates when pattern D is on, and only 44 and 60 are so firing when pattern C is on. However, these rates are directly dependent on the cycle times of the pattern, which should likely be longer in networks with larger numbers of cells. The significant point is rather the relative timings in firing between participating pattern cells. Therefore, as can be seen in Figure 14.4c, when pattern C is on, firing in cell 60 leads firing in cell 44 by 3 or 4 msec. However, when pattern C is on (Figure 14.4c), Cell 44 leads cell 60 by 1 or 2 msec.

These correlations in spike train activity are shown more exactly by the spike train autocorrelation histograms shown in Figure 14.4e. Cells 44 and 60 are shown to fire periodically while cell 96 is shown to fire sporadically. The spread in the peaks of the autocorrelations of cells 44 and 60 illustrate the temporal jiggle in the firings of these cells due both to cross talk and to the extrinsically applied random bombardment.

Finally, Figure 14.4g shows the spike train cross correlation histograms between cells 96–44, 96–60, and 44–60, respectively. Note that the firings of cell 96 (1) are not correlated in this example to either cell 44 (2) or 60 (3) even though cell 60 is synaptically linked to 96 by a fourth order synaptic modification by virtue of pattern *D*. Note also that cells 44 and 60 are clearly correlated with cell 60 firing about 4 time units before cell 44 as predicted by the bed of pattern *C*.

Figures 14.4d, Fig. 14.4f, and Fig. 14.4h show neuron behavior when pattern *D* is on in the same net. Note in Figure 14.4d that cells 44 and 60 (both in pattern *D*) are again firing noisily but with clear rhythmicity corresponding to the pattern cycle period, and that now cell 96 (also in pattern *D*) has also assumed this firing pattern. Note further from Figure 14.4d that cell 44 now tends to lead cell 60 in firing by a short time whereas when *C* was on it tended to lag behind cell 60. Cell 9 fires sporadically as it did when *C* was on.

The spike train autocorrelation histograms of Figure 14.4f show that all three cells—96, 44, and 60—are firing rhythmically with cycle period of the trace.

The spike train cross-correlation histograms of Figure 14-4h show that firings in all three cells are time related, with cell 96 out of phase with cell 44 by 6 msec and lagging cell 60 by 4 msec, and cell 44 firing 2 msec before cell 60. Again the spread in the autocorrelations and cross correlations is due to temporal jiggle resulting from cross talk and extrinsic random bombardment. Although we do not reproduce the figure here, cell 9 shows flat autocorrelations and cross correlations with both cells 44 and 60 in both realizations.

Influence of a Weak Reinforcing Stimulation

Table 14-IIa shows a similar summary of simulations from a network in which the five traces, *A, B, C, D,* and *E,* were embedded and the network was bombarded with random activation with a mean level of one random firing per millisecond. As in the previous example, reducing the level of background noise resulted in longer, more stable realizations, and individual traces could be selective retrieved by selective triggering stimulation for

TABLE 14-II. Recalls of Traces in Multiply Embedded Nets

Trace	Time (msec)	% Time	# Periods	Period mean (msec)	Longest period (msec)
Part a:					
Random	239	5	10	24	—
C	862	17	3	287	535*
A	1597	32	4	399	762
D	2302	46	3	767	1300*
Part b:					
Random	339	7	9	38	—
A	836	17	2	418	570
C	1252	25	2	627	1083
D	2571	51	6	429	1119
B	0	0	0	0	—
E	0	0	0	0	—
Part c:					
Random	488	27	18	27	—
A	0	0	0	0	—
C	1328	27	6	221	630
D	3126	63	11	284	1099
B	19	0.5	1	19	19
E	39	1	1	39	39

*Runs used for multi-unit analysis shown in Figure 14-4.

substantial periods, which were longer for decreasing levels of random bombardment. Patterns *B* and *E* did not occur with this random bombardment, but could be selectively recalled for periods ranging from 62 to 300 msec by selective stimulation.

We then stimulated the same five-trace net continuously with the .01 random bombardment of excitatory cells, but now added a weak reinforcing stimulation for pattern *D* consisting of activation of the four cells of its first subset—2, 4, 6, and 8—every 24 time units. As shown in Table 14-IIb, this partial reinforcing tended to increase the dominance of pattern *D* in the net rather remarkably, mostly by disrupting competing patterns. The presence of the weak reinforcing stimulation brought about many more changes of pattern (37 as compared to 19); the mean durations went down from about 450 to 260 msec; the number of changes to *D* went up dramatically (6 to 11); and the total time spent in *D* increased (51 to 63%). These effects seem due largely to the action of the activated inhibitory cells.

Storage Capacities for Multiply-Embedded Traces

We determined the storage capacities of multiply embedded nets for several combinations of bed parameters. The procedure was to embed first one, then two, then three, and so on traces, ensuring at each step that all the individual traces were successfully retrievable. During the process we would adjust as appropriate the four synaptic strength parameters, but keep all other parameters fixed. The storage capacity for the embedding was the maximum number of traces that could be packed in by this procedure wherein all traces were successfully retrievable. For reasonable values of strength parameters (about 2 to 4), we could embed up to eight traces with $n_f = 4$, $L = 12$, and $n_L = 4$; up to five traces with $n_f = 7$, $L = 7$, and $n_L = 4$; and only one trace with $n_f = 10$, $L = 10$, and $n_L = 4$. When more than these numbers of traces were embedded, cross talk prevented successful recall of all the traces. By increasing the inhibitory strengths to unreasonably high values (10), it was possible to pack in more traces.

Discussion

These results show that the essential dynamic properties of firing patterns determined by the sequential configuration model are consistent with the observations of multimicroelectrode experiments: neurons exhibit different temporal correlations when jointly participating in different patterns. The results show further that the spread or variation in cross correlation histograms is a measure of the temporal jiggle in the system, which in turn reflects cross talk and noise. Cross correlations should show increased spread when these levels increase; for example, nets with larger numbers of embedded traces should exhibit more cross-talk and therefore more cross-correlational jiggle.

Operational Design Considerations for Neural Networks

Moreover, this work has shown how to design the basic synaptic strength parameters of networks to minimize the disruptive influences of cross talk and noise on the basic ingredients of multiple embedding of traces—selective recall and storage capacity. Specifically, jiggle, overload, and self-extinction or derailing by overload—which are the basic manifestations of cross talk and noise—are systematically related to the relative balances of excitatory and inhibitory strengths in the network. Tenacity of selective recall is optimal with s_{ee} and s_{ei} somewhat above a minimum threshold level and s_{ie} and s_{ii} rather strong (as illustrated in Figure 14.3), and with $s_{ee} > s_{ei}$ to favor recurrent excitation over recurrent inhibition at low levels of firing.

Chapter 16 shows further how the parameters of the sequential configurations, n_f, n, n_L, and L, determine the storage capacities of networks by determining nature and degree of cross talk.

These central findings show how network parameters determine essential network properties (selective recall and storage capacity) and thereby provide the basis for interpreting the design of naturally occuring nets or designing artificial nets to perform certain tasks.

Essential Stochastic Nature of Firing Patterns

The distinction between beds and realizations is fundamentally significant for the theory. For example, dynamic firing patterns determined by the sequential configuration model are essentially stochastic in large part because of cross-talk influences. The beds provide a clean template against which the stochastic nature of realizations can be clearly seen.

The distinction between beds and realizations allows one to define four measures of the quality of realizations, as shown in Eq. (14.5):

$$\text{fidelity} = \text{avg}\,(n_f/n) \tag{14.5a}$$

$$\text{signal-to-noise ratio} = \text{avg}\,(n_r/n_f) \tag{14.5b}$$

$$\text{completeness} = (t_b - t_a)/L \tag{14.5c}$$

$$\text{jiggle} \tag{14.5d}$$

Fidelity can be defined as the percentage of bed cells that fire (instantaneous or averaged over some time interval) as shown in Eq. (14.5a). A second measure is the signal-to-noise ratio as given by Eq. (14.5b). Third, a measure of the temporal completeness of a temporally finite trace can be defined as the percentage of the bed's duration over which the realization maintains any given degree of fidelity, as shown in Eq. (14.5c). If the trace is cyclic, this measure could be reduced to simply the number of cycles over which the realization survives.

Note that all of these first three measures require the definition of an underlying bed as well as information on the realization, which alone is immediately available from recorded data.

A fourth useful dimension of clarity and noise involves the concept of temporal "jiggle" in realizations. We observe in stimulations that frequently a set $\{F_t^{i^*}\}$ at some t^* is displaced one to two or three time units with respect to $\{F^i_{t^*} - 1\}$. Particularly when cross talk is high in heavily embedded nets or is not adequately constrained by synaptic strength parameters, certain trace links or some of their members are driven to come on slightly before their targeted time by cross talk. Similarly, certain strategically placed inhibition sometimes delays the firing of some members of a realization set. Therefore, instead of the cleanly temporally discrete pictures of realizations (and beds) described in Eq. (14.1), one sometimes has a more jumbled occurrence where multiple sets of trace cells occur simultaneously at various times throughout the realization. As indicated, jiggle maps directly onto the observable spread in cross-correlation histograms.

Organized versus Unorganized Activity

An explicit definition of a model for organized firing patterns, such as the sequential configurations discussed here, thus allows the deduction of considerable hypothetical insights into network operations. Further, as we will establish in Chapter 16, it allows the explicit mathematical analysis of the capacity of neural nets to store such traces in terms of the constituent structural parameters of the nets. That is, one is in a position to consider the design of natural and artificial nets from the standpoint of memory capacity. We will take up these topics in Chapter 16.

First, however, we turn in the next chapter to an explicit consideration of the interaction between recurrent excitation and recurrent inhibition as they relate to internally sustained activity in the case of random or disorganized activity.

Chapter 15 Properties of Unorganized Activity in Recurrently Connected Networks

Recurrent Excitation and Inhibition in Local Neural Networks

Most central neural networks exhibit extensive recurrent connections, both excitatory and inhibitory. A simple model that composes at least part of the structure of virtually all local networks is that of a so-called parent projection population whose axons project systematically to target populations beyond the local network, supplemented by a secondary population or populations whose axons project entirely locally. The parent population cells are usually larger and usually, although not always, excitatory. The intrinsic or local populations often are primarily inhibitory, but often there are local excitatory populations as well. Extrinsic input activates the parent cells. The intrinsic populations are typically activated extensively by collaterals from the axons of the parent cells as these exit the local network toward their most distant targets. The intrinsic cells may be activated by some of the input that activate the parent cells or by separate input fiber populations.

Typically, one tends to conceive that the primary input–output thrust of such networks is from the main input fibers activating the parent projection neurons, and then through these neurons by means of their externally projecting axons. The influence of the local intrinsic neurons is imagined to mediate "integration" or "synthesis" of input and network patterns, the details of which we are, in fact, usually quite ignorant.

One very significant network mechanism in this context is what is called *lateral inhibition*. Lateral inhibition is the distribution of inhibitory signals laterally across a local network by the lateral spread of neuronal axon terminals. These effects can be thought to be pervasive in central networks by virtue of the widely spread generic connection schemes given earlier. One well-known property of lateral inhibition is that the spatial localization of excitatory activation of a laterally projecting inhibitory population produces a field of inhibition surrounding the localized pattern. This mechanism has been shown to produce an enhancement of the representation of boundaries in the visual system and thereby contributes quintessentially to the fundamental operation of "feature extraction" in vision. More generally, it is widely and reasonably thought that local inhibitory populations in central networks are the primary mediators of both spatial and temporal sculpting of information by their actions of the coordinated firing pattern of the parent projection populations in the response of the latter to extrinsic input. The sequential configuration model of the last chapter, for example, uses such lateral inhibition to help select the firing of specific subsets of trace neurons and to inhibit the firing of competitor or inappropriate neurons. Generally, lateral inhibition projects locally, but may be activated directly by extrinsic input, or locally by recurrent collaterals of projection cells, or both.

Therefore, a significant pervasive feature of central networks is recurrent connections. These can be either excitatory or inhibitory. Either type may be mediated directly by collaterals or by intervening interneurons of the appropriate type. The example of lateral inhibition symbolizes the very important, but still largely unknown, involvement of these connections in the encoding of meanings and information in coordinated network firing patterns. The recurrent connections also contribute centrally to a few important global properties of local networks. Recurrent excitation, for example, may on occasion drive a network to abnormal and undesirable high levels of activity. Such uncontrolled recurrent excitation is thought to be centrally involved in epileptic seizures. The likelihood of global control of recurrent excitation by recurrent inhibition is an important question area in this context.

A further refinement of this picture is the concept that a local network might mediate internally sustained activity by a careful balance of recurrent excitation and recurrent inhibition. Indeed, it might well be that the potentially explosive properties of unbalanced recurrent excitation and the involvement of local recurrent connections in fine-grained sculpturing of information are interrelated in the human brain. Thus, the hippocampus is both the primary site of human seizures and strongly implicated in the formulating and storing of permanent memory traces.

The present chapter presents a model for the global iterations of recurrent excitation and inhibition in a representative generic local network. The purpose is to show how the central physiological and structural parameters of networks influence the dynamic properties of recurrently connected nets. The results may be applied to these basic questions of stability and explosive responses in local networks throughout the nervous system.

In the context of the present part of this book we are particularly interested in the propensity of local networks of the cerebral cortex to exhibit internally sustained activity. The sequential configuration model developed in the previous chapter has shown clearly that a recurrently connected net may be adjusted to produce internally sustained activity as long as that activity involves carefully defined configurational organization. On the other hand, what about irregular or nonorganized activity? Should we expect cerebral cortex to be thrown into modes of unorganized, internally sustained activity by irregular random bombardment? Or, relatedly, by the noise associated with the cross talk among competing traces? How stable is the system? Should it be easily thrown into seizure like activity?

These questions are addressed here in the framework of a simple probabilistic model based on the assumption of random connections that is consistent with the focus on unorganized activity. The primary product of the model is a Markovian transfer function that predicts the number of network cells that will be on at time $i+1$ given the number that are on at time i. We will then define the character of reverberatory activity determined by different shapes of the transfer curve and the influence of different constituent parameters on the curve. This allows us to consider how the different parameters influence the dynamic properties of the network; that is, its engineering design.

The highly simplified structure of the model allows to produce explicit equations for the basic transfer curve. Computer simulations show that the broad general influences of parameters predicted by the model hold for computer simulations wherein these assumptions are not made. Therefore, this model, first presented by Harth, *et al.*, 1970 (see p. 362), is an excellent example of a simplified theoretical model whose predictions and conceptual value transcend the limits of its assumptions.

Probabilistic Model for Recurrent Activations

The model considered here is illustrated in Figure 15.1. Consider a parent population of N excitatory cells, and an intrinsic population of M inhibitory cells. Suppose there are three intrinsic junctions in the system, excitatory to excitatory—*ee*, excitatory to inhibitory—*ei*, and inhibitory to excitatory—*ie*. Hence, for the parent excitatory population there is a recurrent excitatory path, *ee*, and a recurrent inhibitory path through the inhibitory cells, *ei* and *ie*. Suppose that the number of output terminals (and output synapses) per output axon for each of these junctions is respectively, n_{ee}, n_{ei}, and n_{ie}. Suppose that the strengths of all the synapses are represented by the average strength for the junction. Let the average strength of *e*PSPS at the *ee* junction be $1/\theta_e$, where θ_e is the threshold of the excitatory cells measured as the number of EPSPS necessary to trigger an excitatory cell. Similarly, let the average strength of EPSPS at the *ei* junction be $1/\theta_i$, where θ_i is the threshold of the inhibitory cells measured as the number of *e*PSPS necessary

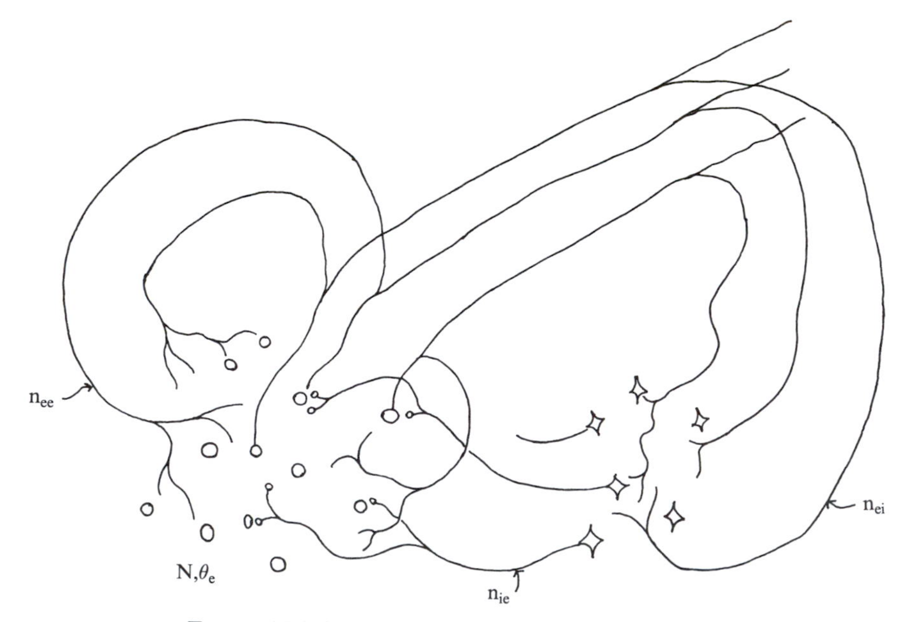

FIGURE 15.1 Recurrently connected local network.

to trigger an inhibitory cell. Let the average strength of the *i*PSPS at the *ie* junction be a factor, a , times the strength of the average *ee* *e*PSP. That is, if a, is, for example, 3, then each *i*PSP will wipe out 3 *e*PSPS.

Let us partition time into windows of a length Δ, where Δ is again an integration interval for the neurons; as for the model of irregular firing in Part II, this interval will correspond to the life times of PSPS, and be roughly equal to the membrane time constant. Suppose that at time i, N_i of the excitatory cells are active. It follows that $N_i{}^*n_{ei}$ *e*PSPS will fall on the inhibitory cells one time unit later; that is, at time $i + 1$. Now, neglecting any spatial or temporal organization of activity, let us imagine that all the individual pulses fall at random on one or another of the M neurons in the inhibitory population. In particular, assume that any single one of these *e*PSPS has a probability of $1/M$ of falling on a given single inhibitory cell, and that the placements of all the individual *e*PSPS are independent of each other.

Collectively, these assumptions mean that this situation of the transfer of N_i excitatory cells firing at i to the number of inhibitory cells firing at $i + 1$, M_{i+1}, can be approached from the standpoint of a Bernoulli trial situation, as seen by a single representative inhibitory neuron. There is a trial for each active *e*PSP: will the *e*PSP fall on this cell or not. The probability of success for each PSP is $1/M$; correspondingly, the probability of failure is $1 - 1/M$. There are as many trials as there are *e*PSPS, namely, $N_i^* n_{ei}$. We know that the probability density function for the number of successes, x, in a Bernoulli trial situation is the binomial distribution, and we know its mean and standard deviation in terms of the parameters of the situation. These are given for this case in Eq. (15.1).

$$\mu_M = N_i n_{ei}/M \tag{15.1a}$$

$$\sigma_M = \sqrt{\mu_M(1 - 1/M)} \tag{15.1b}$$

For a typical neural net, $1/M$ will be very small compared to 1. This means that $1 - 1/M$ can be approximated by 1, and that the binomial distribution can be approximated by the Poisson distribution as shown in Eq. (15.2):

$$p_M(x) = (\mu_M^x/x!)\exp[-\mu_M] \tag{15.2a}$$

$$\mu_M = N_i n_{ei}/M, \quad \sigma_M x \approx \sqrt{\mu_M} \tag{15.2b}$$

Note now that the representative inhibitory neuron will fire provided that the number, x, of *e*PSPS it receives is equal to or greater than its threshold, θ_i. That means we can write Eq. (15.3a) for the probability that the representative cell fires:

$$P_M(x \geq \theta_i) = \sum_{x=\theta_i}^{N_i n_{ei}} p_M(x) = 1 - \sum_{x=0}^{\theta_i - 1} p_M(x) \tag{15.3a}$$

$$\bar{M}_{i+1} = (M - M_i)P(x \geq \theta_i) = (M - M_i)\left\{1 - \sum_{x=0}^{\theta_i - 1} (\mu_M^x / x!) \exp\left[-\mu_M\right]\right\} \tag{15.3b}$$

We can then estimate the number of inhibitory cells expected to fire as the product of this probability times the number of inhibitory cells that are available to fire. We take the number available to fire as simply, M; or we might refine it slightly to approximate the influence of refractoriness at high network firing levels. A step in this direction would be to say that if M_i cells fired at i, then about $M - M_i$ cells are not refractory, and therefore available for firing at $i + 1$. This results in the expression for $\bar{M}_{i+1}$ given in Equation (15.3b).

It is instructive to realize that the expression for $\bar{M}_{i+1}$ given in Eq. (15.3b) is an expected or average value. In fact, the value observed on any one trial is a random variable. The distribution is approximately Poisson, so its mean is $\bar{M}_{i+1}$ and its standard deviation is about the square root of $\bar{M}_{i+1}$. This situation is indicated schematically in Figure 15.2.

These results apply to any neural junction, $N \rightarrow M$. That is, if we have random bombardment of a neural junction with parameters, n_{ei}, and θ_i at a given rate, N_i/Δ, we should anticipate output activity at about the rates given by Eq. (15.3b) and illustrated in Figure 15.2. It can be applied without modification to the case of recurrent excitation by simply imagining that the receiving population was the N population rather than the M population, and correspondingly replacing all values of M in the equations with Ns.

Consider now the confluence of the two recurrent pathways back on the excitatory population. Lets neglect the difference in conduction times between the two pathways; that is, the round trip loop from e to i to e is 1, and what we've called $\bar{M}_{i+1}$ are activated at some point inside this time. Thus, at time $i + i$, there are $N_i^* n_{ee}$ ePSPS and $M_{i+1}^* n_{ie}$ iPSPS falling on the excitatory cells. Consistent with the idea of unorganized activity we will imagine that placements of the ePSPS are independent of those of the iPSPS. Then we can say that the probability that a given excitatory cell receives exactly x ePSPS and y iPSPS is equal to the product of the probability that it receives x ePSPS times the probability that it receives y iPSPS. Each of these probabilities reflects a Bernoulli trial situation exactly analogous to that described above for the excitatory to inhibitory junction.

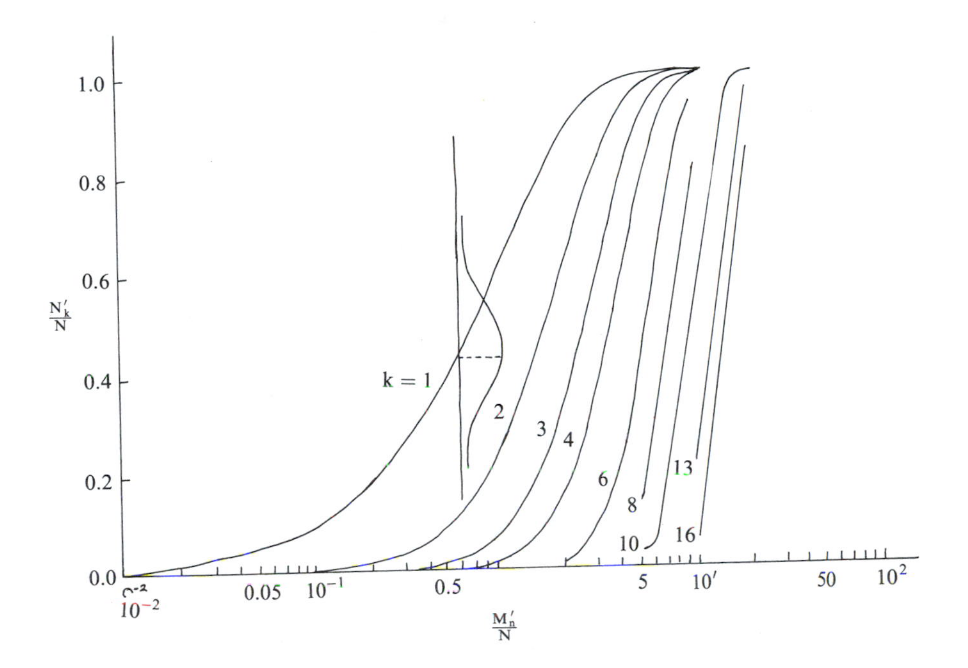

FIGURE 15.2 Population firing transfer for a single feed-forward diffuse synaptic junction (M'_n is the number of input fibers firing; N'_k is number of receiving cells firing; N is total number of receiving cells). Adapted from Theory of monosynaptic transfer between neuron populations, R. J. MacGregor & Teresa McMullen, *Behavioral Science*, **22,** 1977.

Following all the reasoning of the preceding steps then we can write Eq. (15.4) for the probability that a given excitatory cell receives x ePSPS and y iPSPS.

$$p(x,y) = (\mu_e^x/x!)\ \exp[-\mu_e](\mu_i^y/y!)\ \exp[-\mu_i] \tag{15.4}$$

Now recognize that an excitatory cell will fire if the number of ePSPS it receives diminished by the number of iPSPs it receives times a (the relative strength of each iPSP) is greater than its threshold, θ_e, as shown in Eq. (15.5a):

$$x - ay \geq \theta_e$$

$$P(x - ay \geq \theta_e) = \sum_{x=\theta_e}^{N_i n_{ee}} \sum_{y=0}^{y_{\text{up}}} p(x,y)$$

$$y_{\text{up}} = \min\{(x - \theta_e)/a,\ M_{i+1} n_{ie}\}$$

$$\bar{N}_{i+1} = (N - N_i)\ \exp[-(\mu_e + \mu_i)\sum_{x=\theta_e}^{N_i n_{ee}} \sum_{y=0}^{y_{\text{up}}} \{\mu_e^x \mu_i^y/(x!y!)\} \tag{15.5}$$

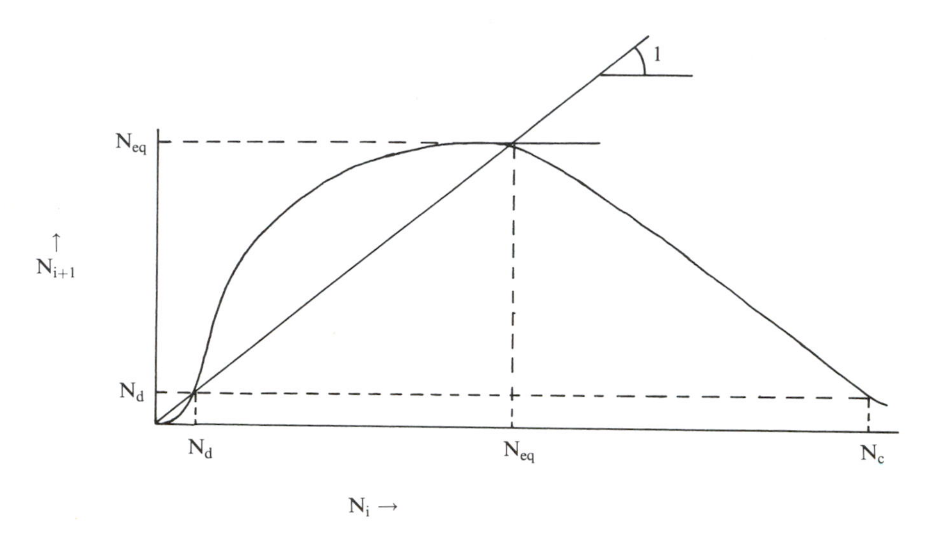

FIGURE 15.3 Recurrent population firing transfer curve (number of cells expected to fire at time $i+1$ (N_{i+1}) as a function of the number of cells that fired at time i (N_i)).

This condition, applied to the density functions of (15.4), provide the results given in (15.5b) for the probability that a given excitatory cell will fire at $i+1$ and for the expected number of excitatory cells to fire at that time.

Equation (15.5c) is obtained by using Eq. (15.4) in (15.5b). It is the desired Markovian transfer curve for these recurrent influences. It estimates the number of network cells to fire in the $i+1$th integration interval, given that N_i fired in the previous interval. The systemic parameters influencing the curve are the numbers of synaptic terminals at the three junctions, n_{ee}, n_{ei}, n_{ie}; the synaptic strengths at the three junctions, $1/\theta_e$, $1/\theta_i$, and a; and the relative number of inhibitory neurons, M/N. The general character of this curve is shown in Figure 15.3.

Dynamic Behavior Predicted by the Probabilistic Model

Most of the important properties of recurrently mediated global activity can be inferred pictorially from the basic transfer curves shown in Figures 15.2 and 15.3. This interpretation is facilitated by noting the lines of slope $+1$ on the curves. On these lines, the number of firings at $i+1$ equal the corresponding number of firings at i; the lines thus represent point of possible equilibrium.

Consider first the concave rising curve for recurrent excitation only shown in Figure 15.2. The behavior of this curve can be interpreted by the notation of Figure 15.3, where the crossing of the curve by the unitary slope line at two points, N_d and N_r, is shown. For the region where N_i is less than N_d, the transfer curve is always below the unitary slope line and therefore the number of expected firings at $i+1$, N_{i+1}, is always less than N_i. This means that any perturbation giving an N_i in this range will be expected to produce successively smaller values of N_{i+1} on successive reverberations and thereby eventually die out; hence the subscript *d* for death. On the other hand, for any N_i greater than N_d (and less than N_r), the transfer curve is always above the unitary slope line and therefore, N_{i+1} is expected to be greater than N_i. Therefore, any perturbation putting N_i into this range is expected to produce successively larger values of N_{i+1} on successive reverberations. That is, activity should be driven progressively upward until the neighborhood of N_r is attained.

In these considerations, we anticipate that there will in fact be noise and fluctuations, since the curves describe average values and indeed there are distributions around these averages. Sometimes there will be N_i to the left of N_d that produce higher values of N_{i+1}, and vice versa. Nonetheless, the trends based on the means are expected to prevail, and the subsequent discussion is based on these predictions for the means. The discussions are borne out by numerous computer simulations.

The value N_d is an equilibrium point of sorts, in that the expected value of N_{i+1} when N_i equals N_d is also equal to N_d. However, N_d is not a stable equilibrium point, but rather a critical point such that lesser activity tends to die out, whereas higher activity tends to drive the net into seizure, limited only by refractoriness at very high rates. This phenomenon was noted early on in simulations of randomly connected nets; it is referred to as *ignition.* The conclusion is clear: recurrent excitation taken alone is a dangerous network structure; such a net is highly prone to seizures.

The higher possible equilibrium point, N_r, is likely not a meaningful one for normal activity levels. At this point, the net is being driven so hard by recurrent excitation that all cells are firing as rapidly as they can consistent with refractory properties. This situation can be representative of nothing but seizure. (The simple rule used here to estimate this effect indicates that approximately half the network cells are firing at every integration interval, meaning that virtually all network cells are firing at what we called in Part II the *characteristic output rate,* a high "on" rate for a neuron. On the other hand, this simple rule surely overstates the refractoriness, and firing rates should be even higher.) We will ignore the regions at extremely high N_i, where this refractory influence comes into play, and consider only regions for much lower N_i where the curve may be brought down not by refractoriness, but by inhibition.

Consider now the case where recurrent inhibition is combined with the excitation as illustrated in Figure 15.3. The unitary slope line crosses this curve in two points also, N_d, and N_{eq}. N_d again represents an unstable equilibrium, partitioning a death region and a regeneratively active region. The higher equilibrium point, N_{eq}, is determined by the recurrent inhibition, not by refractoriness; it can define an equilibrium point in at a meaningful level of normal activity.

Furthermore, the equilibrium point, N_{eq}, is a stable equilibrium point in the sense that if N_i is greater than N_{eq}, then N_{i+1} will be brought lower than N_i (and therefore toward N_{eq}), while if N_i is less than N_{eq}, then N_{i+1} will be brought higher than N_i (and therefore, again towards N_{eq}). Therefore, the value of N_{eq} emerges as a candidate level for internally sustained activity, and networks with transfer curves that produce well-behaved activity in the neighborhood of N_{eq} and reasonable values for N_{eq} emerge as candidates for the exhibition of internally sustained activity.

Let us consider more closely the nature of oscillations around the neighborhood of N_{eq} as determined by transfer curves of various shapes. Specifically consider the two transfer curves shown in Figure 15.4.

Curve A, with a steep negative slope over the region containing N_{eq}, produces oscillatory activity prone to abrupt termination by overinhibition as can be seen as follows. If N_i is greater than N_{eq}, then N_{i+1} is less than N_i because the transfer curve is below the unitary slope line; moreover, since the transfer curve is also below a horizontal line through N_{eq}, then N_{i+1} is also less than N_{eq}. That is, N_i goes from above N_{eq} below N_{eq}. If N_i is less than N_{eq}, then N_{i+1} is greater than N_i because the transfer curve is above the unitary slope line; moreover, since the transfer curve is also above a horizontal line through N_{eq}, then N_{i+1} is greater than N_{eq}. That is, N_i goes from below N_{eq} to above N_{eq}. Therefore, perennial oscillation is implied by this type of crossing. This is indicated in Figure 15.4c.

Further, consider the point N_c (for N critical) defined on curve A by the intersection of a horizontal line at value N_d. Any value of N_i greater than N_c will tend to produce a value of N_{i+1} of less than N_d and therefore through the net back into the death region. This phenomenon can be recognized as a squelching of net activity by overreactive inhibition.

If peak value of the transfer curve, N_p, is higher than the critical value, N_c, the net should be expected to produce this kind of squelching frequently in normal operation. Clearly, a net characterized by a transfer curve of type A is not well adapted to internally sustain recurrent activity.

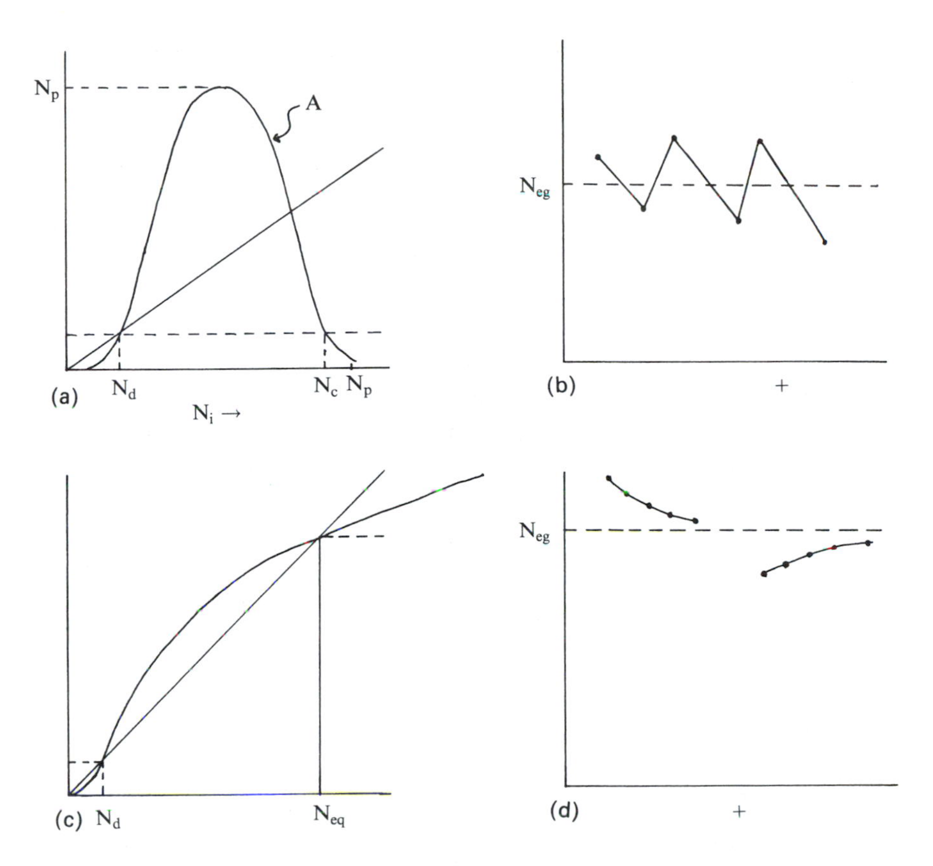

FIGURE 15.4 Comparative shapes of population firing transfer curves: (a and b) poor curve, temporally erratic behavior; (c and d) good curve, temporally smooth returns to equilibrium.

In contrast, consider transfer curve *B* shown in Figure 15.4.b, characterized by a crossing through N_{eq} by a gently positive slope, so that the transfer curve projects between the unitary slope line and the horizontal line through N_{eq} on both sides of N_{eq}. For this curve, if N_i is greater than N_{eq}, then N_{i+1} is less than N_i, but greater than N_{eq}, because the transfer curve is placed between these values. Similarly, if N_i is less than N_{eq}, then N_{i+1} is greater than N_i, but less than N_{eq} for the same reason. Therefore, for any perturbation in N_i around N_{eq} the net will tend to produce a gentle recovery modification back toward N_{eq}, as illustrated in Figure 15.4d. There is no tendency to oscillate and no tendency to dangerous overreactive inhibitory squelchings.

Clearly, then, if one were to design neural networks for the purpose of producing stable sustained reverberations of unorganized activity, one would adjust the design parameters of the net so that it produced a transfer curve of the type *B* shown in Figure 15.4. Further, if one computed from the parameters of a given biological neural net that its transfer curve should exhibit the shape of type *B*, one could anticipate that such a net should be prone to oscillate regeneratively to unorganized input or disturbances. On the other hand, if a transfer curve of type *A* was computed, one would anticipate that, although unorganized inputs or disturbances might trigger momentarily large responses, the net should tend to damp these out.

As we will show and discuss in the last section of this chapter, the circuitry of modules in the cerebral cortex seem to imply a transfer curve of type *A*. That is, this theory predicts that cortical modules should not exhibit sustained responses to irregular stimulation or disturbances.

The proneness of either type of net to seizures would depend on the nature of the curve in the regions beyond N_{eq} out to N_r, on the susceptibility of the network to chemical imbalances or other abnormalities, and perhaps on the details of its fine-grained coding in representing information, which is beyond the scope of this global model.

Design: The Influence of Network Parameters on the Transfer Curve

We can now begin to understand the influence of modulations in network parameters on the dynamical properties of its recurrently mediated activity, by determining how the parameters determine the shape of its basic transfer curve. Specifically, we will project that curves of type *B* are characteristic of nets that will produce internally sustained reverberations to unorganized stimulation, while curves of type *A* will produce oscillations and abrupt terminations.

Note first that the transfer curve is a composite reflecting the two opposing effects of recurrent excitation and recurrent inhibition. The response to excitation alone is illustrated in Figure 15.5. This curve shows that recurrent excitation tends to drive the curve upward, essentially exponentially. The influence of inhibition, then, can be seen from Figure 15.5 to soften this rate of rise and eventually bend the curve back downward.

More specifically, the junctional parameters of the net have the following influences of the transfer curve: increasing n_{ee} or $1/\theta_e$, or both, drives the transfer curve up at all N_i. This effect is clearly a result of increasing the strength of the *ee* junction.

Increasing n_{ei} or $1/\theta_i$, or both, drives the transfer curve down

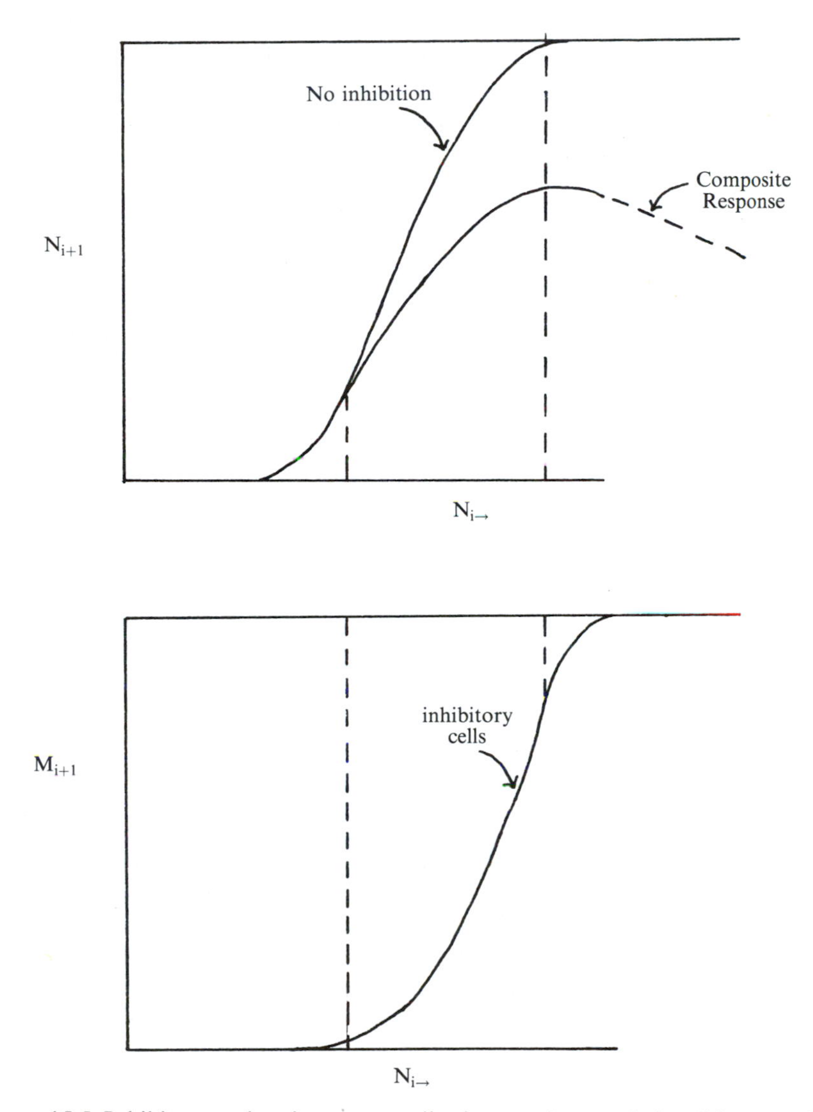

FIGURE 15.5 Inhibitory and excitatory contributions to the population firing transfer curve.

preferentially at low values of N_i. This effect can be understood by picturing the shape of the composite curve as determined by the relative ranges of the N_i axis in which the two effects—recurrent excitation and recurrent inhibition—are active. For example, if the *ee* junction is more sensitive than the *ei* junction, then the rise effects for excitation only will occur over some range of N_i for which the inhibitory cells are not yet activated. Increasing the strength of the *ei* junction (n_{ei} and $1/\theta_i$) decreases the value of N_i necessary for a certain response level in the inhibitory cells and thereby shifts the effectiveness of this level of inhibition to the lower N_i position bringing the curve down.

On the other hand, increasing the strength of the *ie* junction is a little more subtle. Increasing the strength of the individual *i*PSPS—the parameter, a—tends to selectively bring down the curve at higher levels of N_i. The interpretation of this effect is that at lower values of N_i the preponderant number of cells activated to fire at $i+1$ are being fired by input levels close to their threshold, and the numbers of *i*PSPS are relatively small. Thus, firing is blocked in a substantial number of cells by *i*PSPS of any appreciable size; the important factor is the occurrence or nonoccurrence of an *i*PSP, not its size. On the other hand, at higher levels of N_i, there are many more *e*PSPS and *i*PSPS; here increasing the size of *i*PSPS is influential on many cells.

Increasing the number of inhibitory terminals, n_{ie}, on the other hand, brings the curve down more or less uniformly over most of its range. This effect simply increases the amount and distribution of inhibition in a rather continuous graded fashion.

These effects may be observed by comparing transfer curves produced from Eq. (15.5e) derived here for the simple Markovian model. The effects can also be observed in computer simulation models of neural networks, wherein the simplified assumptions of this model are considerably relaxed. One may modify the transfer curves in such artificial neural networks by trial and error, systematically making use of these influences. Some of these cases are illustrated in NBM.

One can see further, although we do not develop it here, that the influence of adding and increasing recurrent inhibition of the inhibitory cells back on themselves (an *ii* junction) is to soften the curvature effect of the inhibition across the entire N_i range.

Propensity to Internally-Sustained Activity in Local Cortical Networks

We have illustrated in Chapter 14 that it is not difficult to produce internally sustained activity in recurrently connected nets when that activity is carefully constructed with respect to the synaptology of the net. The question of whether a net should continue to reverberate in response to unorganized disturbances however depends on the more global properties of the net. One would anticipate that it would be more desirable for a network intended to represent information in fine-grained patterns of standing reverberations to be largely nonresponsive to unorganized disturbances; random reverberations likely would be disruptive and confusing.

It is of some speculative interest then to study the properties of reverberatory activity in recurrent nets with parameters reflecting best estimates of local cortical circuits. Current best estimates for the parameters of local "module" of cerebral cortex ("column" in visual cortex, "barrel" in somatosensory cortex) are* 24, 000 excitatory cells; 6000 inhibitory cells; 4000 locally directed *ee* projections per excitatory cell, resulting in 4000 locally originating excitatory input synapses per excitatory cell; 1000 locally directed *ei* projections per excitatory cell, resulting in 4000 locally originating excitatory input synapses per inhibitory cell; 500 locally directed *ii* projections per inhibitory cell, resulting in 500 locally originating inhibitory input synapses per inhibitory cell; 2000 locally directed *ie* projections per inhibitory cell, resulting in 500 locally originating inhibitory input synapses per excitatory cell; *e*PSPS might be of the order of a tenth of a millivolt; threshold about 24 mV; membrane time constants about 15 msec.

If one uses these values in the approximate theoretical expressions of this chapter, or in computer simulation studies as described in Chapter 18 of NBM, one finds it virtually impossible to produce internally sustained reverberations of random activity. Model nets with these parameters universally exhibit sudden brief explosions of high activity that in turn drive high levels of self-inhibition, invariably causing abrupt termination of activity.

A parallel study, which we have not done, would be to apply these techniques to circuits of the hippocampus for possible relevance to the particular propensity of the temporal lobe to epileptic seizures.

* I am grateful to Professor Moshe Abeles for these anatomical estimates.

Chapter 16 Memory Capacity in Recurrently Connected Networks

Chapter 14 spelled out a theory for the coordinated organization of firing patterns in recurrently connected local networks and showed that the dynamic characteristics of neurons in such firing patterns are compatible with neuroelectric observations. We now want to address the question of storage capacity of such networks.

The picture entertained here is that memory traces consist of the systematic sequential firings of particular sets of neurons; that different traces consist of different sequences of different sets of neurons. Any given neuron may, and generally does, participate in many such traces. The temporal relations among participating neurons define a given trace, not the firing rates of its individual constituent neurons.

The traces are embedded in a given local network by means of selective modulations of the strengths of the synapses among the participating neurons, according to how the latter are arranged in sets in the firing patterns associated with the traces. From the very onset, anatomical connections and associated spatiotemporal firing patterns are two distinct but intrinsically coupled manifestations of a unitary phenomenon. We will formally recognize that distinction by distinguishing and defining anatomically based "beds" and physiologically based "realizations."

The central idea developed here is that, since different memory traces will call for different synaptic adjustments, there will be cross talk mediated by inappropriately activated synapses whenever any single memory trace is activated, except in the most unusual case, when none of the neurons in the trace participates in any other trace. That is, cross talk is an essential and inescapable characteristic of networks that encode overlapping traces by these means.

This cross talk exhibits two manifestations. First, it is possible that two distinct traces call for the same single synapse between two cells that participate in both traces, but each asks for a different value to be assigned to the synapse. In this case, there must be particular value assigned, which is distinct from that requested by the bed of at least one of the traces. When that trace is activated, its behavior will be altered by the fact that at least this synapse is activated in a somewhat different manner than called for by its bed.

Second, it is likely in multiply embedded nets, that a given neuron will project to some subset of cells because of its position in one trace and to another subset of cells because of its position in a second trace. Then, when one of these traces is active, the neuron will project PSPS to cells that are outside the active trace or to cells of the trace but at inappropriate times.

This chapter will spell out a theory of the memory capacity of recurrently connected local nets based on the disruptive effects of such cross talk for the case where multiple memory traces are embedded in a random fashion, each completely independent of the others. The question of selective as compared to random packing of traces will be considered at the end of the chapter.

The theory will provide mathematical expressions to show how the memory capacity of such networks depends on their characteristic structural and physiological parameters. A central result of the theory is the inference that the cerebral cortex may have designed itself in terms of modules of the size of 30,000 neurons at least in part to optimize memory storage capacity. A second central result is that modules of such size might be expected to exhibit memory capacities of about 300 to 900 traces. The broader implications of these results and the theory in general are considered in Part IV.

Before turning to the theory, it is instructive to recognize what very large numbers of terminals are involved in densely interconnected local nets. Imagine for example a very small local network with 40 neurons, each of which projects 25 terminals back into the net. Even such a small net has 1000 synapses. Imagine standing in front of a display board or screen visualizing the 40 neurons. It might seem manageable to visualize ongoing activity in 40 neurons; but to critically visualize the ongoing activity in 1000 terminals is certainly not reasonable.

Further, a local network in the cerebral cortex might have about 30,000 neurons and might send (and receive) about 10,000 terminals locally. Here, the number of (only local) terminals is about 300 million.

If the threshold is 24 mV, and average EPSPS are about 0.1 mV, about 240 simultaneous PSPS are necessary to fire a cell (or perhaps 400, taking into account nonlinear synaptic interactions) and perhaps several thousand individual firings are required to adequately signal a particular memory trace. So, millions of terminals are likely involved in the recall of a single trace.

Therefore, exceedingly large numbers of axon terminals and synapses are fundamentally involved in the interconnection patterns discussed in this chapter. The tremendous complexity of interconnectivity in the brain in underscored by the observation noted in Chapter 2 that if all the axons and axon terminals from a single human brain could be laid out end-to-end, the resultant connective tissue would stretch to the moon and back.

Stochastic Theory of Cross Talk*

Step 1: The Sequential Configuration Model of Memory Traces

The theory developed here is based on the sequential configuration theory of memory traces developed in Chapter 14. Recall that in this theory memory traces consist of a "bed" (primarily anatomical) and "realizations" (primarily physiological). The bed is defined in Eq. (16.1a):

$$\text{Bed} = \{B_t^i\}, \quad i = 1, 2, \ldots, n, \quad t = 1, 2, \ldots, L \quad (16.1\text{a})$$

$$\text{Synaptic strength factor} = d_l = \exp(-d_c^*(l-1)/n_L), \quad l = 1, 2, \ldots, n_L \quad (16.1\text{b})$$

$$i = 1, 2, \ldots, n_f \leq n$$

$$\text{Realization} = \{F_t^i + R_t^j\}, \qquad j = 1, 2, \ldots, n_r \quad (16.1\text{c})$$

$$t = t_a, t_a + 1, \ldots, t_b, \quad 1 \leq t_a \leq t_b \leq L$$

$$n = k_n^* n_f, \quad k_n \geq 1 \quad (16.1\text{d})$$

It may be thought of as both an ordered sequence of sets of active neurons and, equivalently, as we will see, as the sets of synaptic connections among

* This theory was presented in greatly abbreviated form in *Biological Cybernetics* **65** (1991), 351–356.

these neurons corresponding to their sequence as defined* in Eq. (16.1b) and illustrated in Figure 14.1.

Realizations are defined in Eq. (16.1c). A realization is an observable physiological manifestation of an underlying bed. It consists of an ordered sequence of subsets of active neurons, some of which are members of the corresponding sets of the bed and which fire over a given time interval in the same temporal correspondence as exists among the cells of the bed. A single set of active trace cells in either a bed or a realization is called a link.

We imagine that a given memory trace includes coupled sequences such as these in both excitatory and inhibitory populations, and perhaps in additional populations as well, but this need not concern us in this chapter. We will consider nets in which multiple memory traces have been embedded and refer to such nets as *multiply embedded nets*.

Step 2: Probability of an x-Degree Overlap of a Given Realization Link with any Arbitrarily chosen Bed Link

We now want to develop a theory for the memory capacity of recurrently connected neural nets as limited by cross talk within the framework of the sequential configuration theory. It is most convenient to approach the problem in terms of links. A link is a single set $\{F_t^i\}$ of n_f cells of the bed that fire simultaneously at some time, t. Links are both senders and receivers; we will refer to a link as a *sender link* or *receiver link* when thinking of it as a sender or receiver, respectively.

Note that if a given activated sender link, $S1$, shares a cell with another link, $S2$, the shared cell will project to each of the cells in both target links $R1$ and $R2$. $R1$ is part of the trace at hand, but the single PSPS sent to each of the cells in $R2$ constitutes activation extraneous to the trace containing $S1$ and can thereby be labeled *cross talk*. Note further that if $S1$ and $S2$ share x cells, for x between 0 and n_f, then each of these x cells projects to all cells in both $R1$ and $R2$, producing cross talk of x PSPS in all cells of $R2$. It is therefore of interest to estimate the number of links that share x cells with any given sender link.

This situation is described in Figure 16.1. Imagine that a given link of size n_f has been chosen from N cells. This corresponds to a single link activated at a particular time in the active realization of a memory trace. Suppose that independent of this we choose another link of size n from the same population. That is, we make n new choices from N, which has been

* All symbols used in this chapter are defined in the glossary in the appendix.

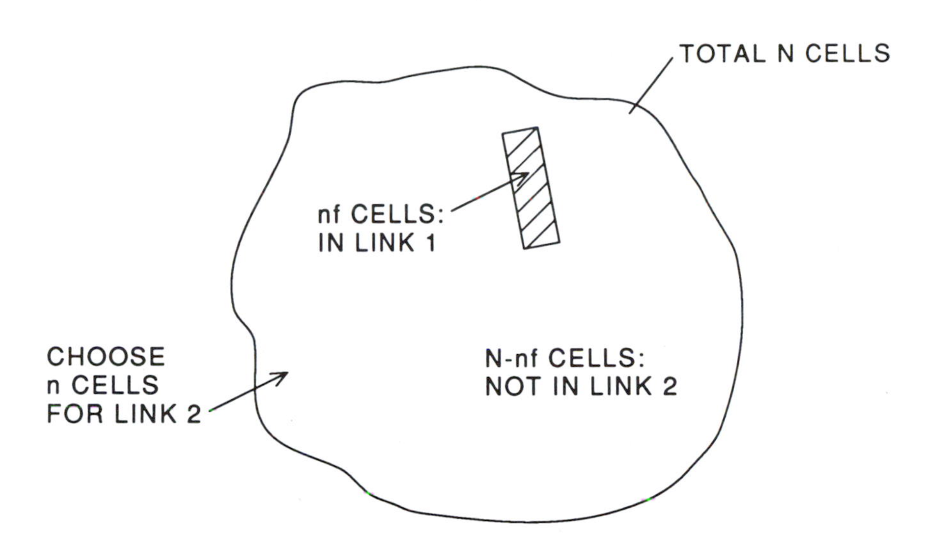

FIGURE 16.1 Combinatoric choices for hypergeometric distribution for overlap of two links (reprinted with permission from R. J. MacGregor & G. Gerstein, Cross-Talk Theory of Memory Capacity in Neural Networks, *Biological Cybernetics*, Springer-Verlag, 1991, **65**, 351–356).

partitioned by link 1 so that n_f of the N are in link 1 and $N - n_f$ are not in link 1. We want to know how many of these n new choices will also be in link 1. This situation is an example of the hypergeometric distribution. The desired probability that x of these choices will be in link 1 is given by Eq. (16.2), which shows also the mean, μ_x, the expected value, x, the standard deviation, σ_x, and the expected value of x^2, $\overline{x^2}$:

$$
\begin{aligned}
p(x) &= \frac{\binom{n_f}{x}\binom{N-n_f}{n-x}}{\binom{N}{n}}, \quad x = 0, 1, 2, \ldots, n_f \\
\mu_x &= x = n^* n_f/N = r^* n_f \\
\sigma_x &= \sqrt{\frac{(1-r)^*(1-n_f/N)^* x}{(1-1/N)}} \\
\sigma x^2 &= \overline{x^2} - \bar{x}^2 \\
r &= n/N
\end{aligned}
\tag{16.2}
$$

Step 3: Expected Number of Links, of y Embedded Links, which Share x Cells with n_f Firers

Consider now the case illustrated in Figure 16.2, where a net contains in addition to the given firing link, $S1$, an arbitrary number, y, of embedded links, each of size n. The number of these links that share x cells with the n_f firing cells of link 1 can be represented by a random variable, Z_x, and the set of such links for each x can be represented by O_x. If the links have been independently chosen then Z_x can be determined by a Bernoulli trial process. Its distribution is then binomial, and its mean and standard deviation are given in Eq. (16.3):

$$\begin{aligned} \mu_{Z_x} &= p(x)^* y = O_x \\ x &= 0, 1, 2, \ldots, n_f \\ \sigma_{Z_x} &= \sqrt{p(x)^*(1 - p(x))^* y} \end{aligned} \tag{16.3}$$

We have used the result that each link has probability $p(x)$ from Eq. (16.2) of sharing x cells with the n_f firing cells of link 1.

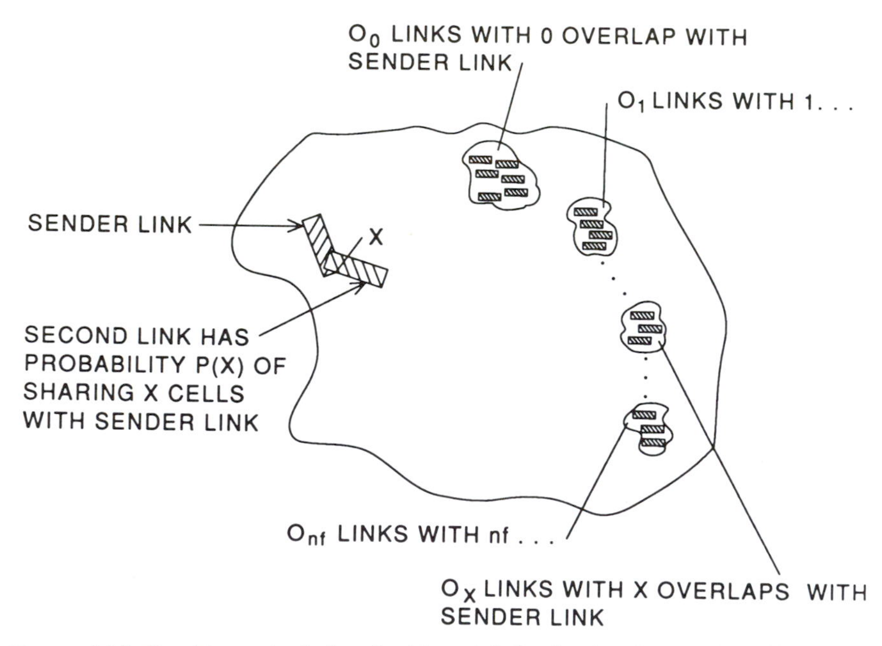

FIGURE 16.2 Combinatoric choices for binomial distribution for number of links, Z_x, sharing x cells with a sender link (reprinted with permission from R. J. MacGregor & G. Gerstein, Cross-Talk Theory of Memory Capacity in Neural Networks, *Biological Cybernetics*, Springer-Verlag, 1991, **65**, 351–356).

Step 4: $(n_f + 1)^* n_L$ Sets of Links, S_{xl}, which Receive x PSPS of Strength d_l

The O_x defines $n_f + 1$ sets ($x = 0, 1, 2, \ldots, n_f$) of sender links, each of whose members share x cells with the given sender link, $S1$. Now, each of the links in each of the $n_f + 1$ sets projects uniquely to n_L receiver links, where n_L is the number of links synaptically connected to a given sender link as indicated in Eq. (16.1b). Moreover, each link projects exactly x cross-talk PSPS to all cells in its corresponding receiving links. Therefore, the O_x also

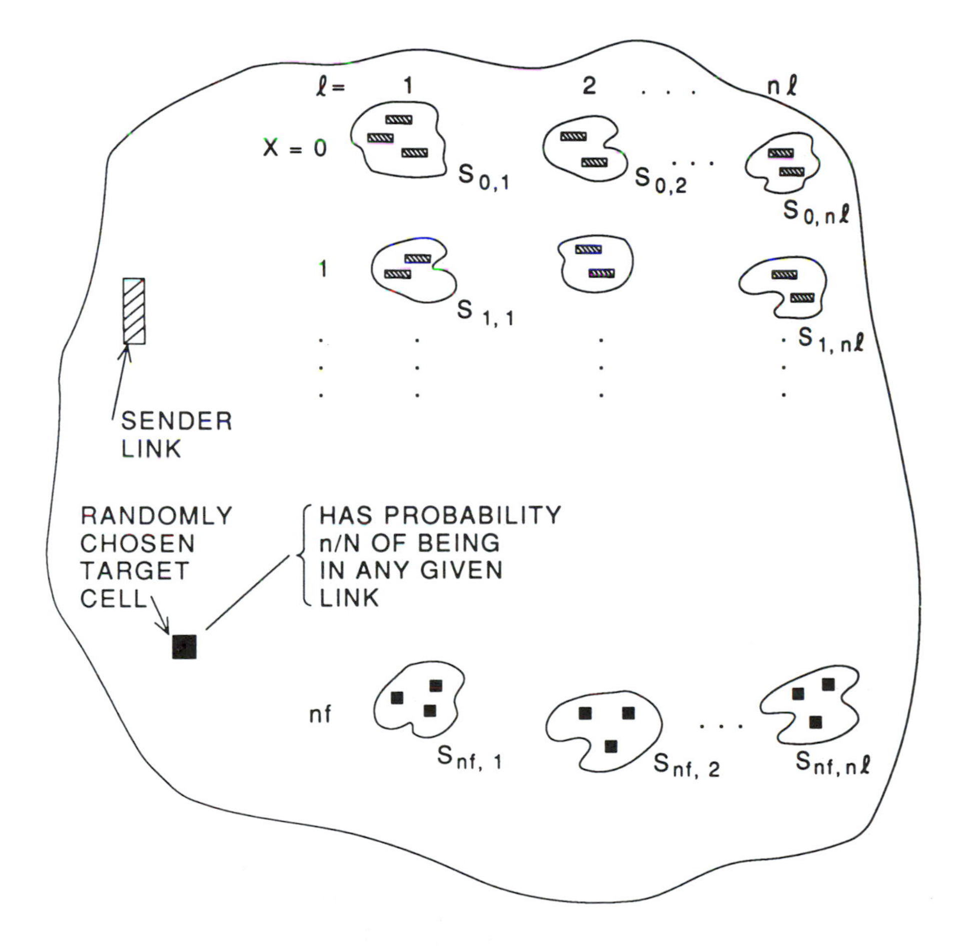

FIGURE 16.3 Combinatoric choices for occurrence of a target cell in z_{xl} of S_{xl} links (reprinted with permission from R. J. MacGregor & G. Gerstein, Cross-Talk Theory of Memory Capacity in Neural Networks, *Biological Cybernetics*, Springer-Verlag, 1991, **65**, 351–356).

determines $(n_f + 1)^* n_l$ sets of receiver links, S_{xl}, each of whose members receives cross talk of x PSPS of strength d_L upon the activation of the initial sender link. This is illustrated in Figure 16.3.

Step 5: Random Variables, x_{xl}, for the Number of Extraneously Activated Links in which a given Network Cell Participates

We now want to know the extent to which a given arbitrarily chosen cell of the population will be activated by this cross talk. We can attain this by defining $(n_f + 1)^* n_L$ random variables, z_{xl}, which give, respectively, the number of S_{xl} links in which the given cell participates. The z_{xl} are given by Bernoulli trial situations with S_{xl} trials and a probability of success on each trial equal to n/N (which is the probability that the cell will be in any particular one of the O_x links) in each case. The probability that the cell will be in exactly z_{xl} of the S_{xl} links is then given by the binomial distribution and has mean and standard deviation as given in Eq. (16.4):

$$\begin{aligned} \mu_{z_{xl}} &= r^* \mu_{Z_x} = r^* p(x)^* y \\ \sigma_{z_{xl}} &= \sqrt{r^*(1-r)^* \mu_{Z_x}} \approx \sqrt{r^*(1-r)^* p(x)^* y} \end{aligned} \tag{16.4}$$

Step 6: Random Variable, ξ, for Total Cross-Talk Activation Received by a given Network Cell

We can now state that the amount of cross-talk activation falling on a given arbitrarily chosen cell due to the activation of a single link embedded among y other independently chosen links is determined by a random variable, ξ, which in turn is a function of x, and the random variables, z_{xl}. Specifically, ξ is given by Eq. (16.5):

$$\xi(x, \{z_{xl}\}) = \sum_{x=0}^{n_f} \sum_{l=1}^{n_L} z_{xl}^* x^* d_l \tag{16.5}$$

Physically, Eq. (16.5) is based on the observation that a given cell receives x cross-talk inputs, each of strength d_l, for each single S_{xl} link it is in. It thereby receives $x^* d_l^* z_{xl}$ input if it is in $z_{xl} S_{xl}$ links; all its cross-talk input is obtained by summing over all possible S_{xl} links.

Since ξ is given in Eq. (16.5) as a linear summation of random variables, z_{xl}, whose means and standard deviations are given in Eq. (16.4), the mean and standard deviation of ξ can be approximated directly as shown in Eq. (16.6)*:

$$\begin{aligned} \mu_\xi &= \sum\sum x^* d_l^* \mu_{z_{xl}} = \sum\sum x^* d_l^* r^* p(x)^* y \\ &= \sum d_l^* r^* y^* \sum x^* p(x) = d^* r^* y^* \bar{x} \\ &= d^* r^{2*} n_f^* y \end{aligned} \tag{16.6a}$$

$$\begin{aligned} \sigma_\xi^2 &= \sum\sum x^{2*} d_l^{2*} \sigma_{z_{xl}}^2 = \sum\sum x^{2*} d_l^{2*} r^* (1-r)^* p(x)^* y \\ &= D^* r^{2*} (1-r)^* n_f^* [(1-r)^* (1 - n_f/N)/(1 - 1/N) + r^* n_f'^* y \end{aligned} \tag{16.6b}$$

$$d = \sum d_l, \qquad d = \sum (d_l^2) \tag{16.6c}$$

Equations (16.6) are a central result of this theory. They allow us to predict the distribution of cross-talk activation over network cells for any given number of embedded links, y, as a function of network parameters.

Step 7: Criterion for Disruption of a Realization by Cross Talk

The variable, ξ, defined by Eqs. (16.5) and (16.6) represents the amount of cross-talk activation received by single cells in the net as a result of the firing of a single link of n_f cells. We now want to determine how significant a given level of ξ might be with respect to disrupting ongoing trace activity in the net. Figure 16.4 indicates schematically the distribution of ξ over network cells. The number of cells which will have greater than any particular number is itself a random variable. Its expected value however can be determined by Eq. (16.7a), which can be understood on the basis of Figure 16.5:

$$\mu_\xi + k(A)^* \sigma_\xi = \xi_{\text{crit}} \tag{16.7a}$$

$$\mu_\xi + [k(A) + m]^* \sigma_\xi = \xi_{\text{crit}} \tag{16.7b}$$

Equation (16.7a) has the property that $A^* N$ cells will be expected to have a $\xi^\#$ such that $\xi^\# > \xi_{\text{crit}}$ on a given trial. If, for example, $A = n_f/N$, and ξ_{crit} = threshold, then the number of cells that would be expected to fire at a single trial due to cross talk alone would be n_f. We can generalize this criterion as shown in Eq. (16.7b). Equation (16.7b) can be interpreted as

* In fact the standard deviation of ξ should be slightly larger than indicated by Eq. (16.6b) because of the variation possible in the values of S_{xl}. Nonetheless the mean value for ξ is correctly given by Eq. (16.6a), and the estimate for the standard deviation is a good first approximation.

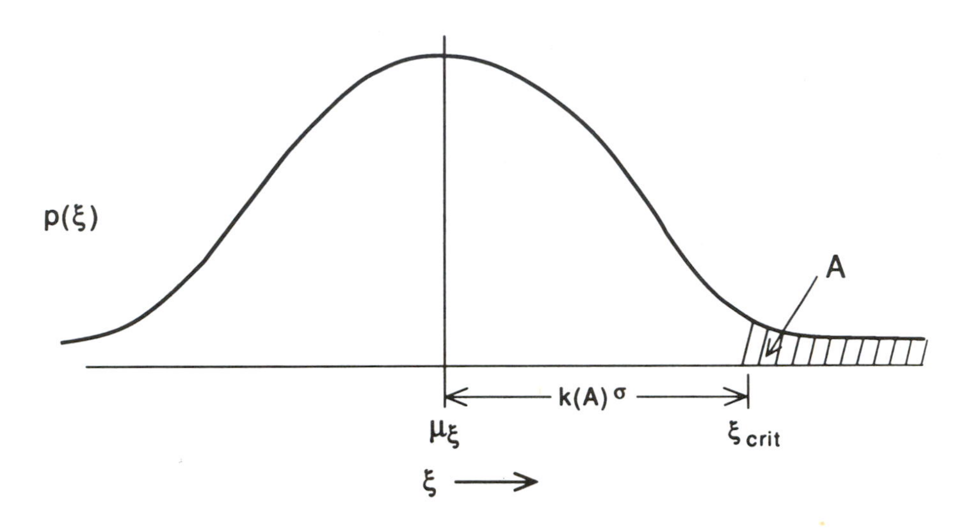

FIGURE 16.4 Probability distribution of ξ (if $\mu_\xi + k(A)\sigma_\xi = \xi_{crit}$, then AN cells will have $\xi > \xi_{crit}$ upon firing of a single link), (reprinted with permission from R. J. MacGregor & G. Gerstein, Cross-Talk Theory of Memory Capacity in Neural Networks, *Biological Cybernetics*, Springer-Verlag, 1991, **65**, 351–356).

determining that the number of cells having $\xi^{\#} > \xi_{crit}$ will be expected to be within m standard deviations of A^*N. This criterion can represent the disruption of activity by a high level of cross talk even if that high level occurs only once in a large number of link firings. In this chapter we will usually apply the criterion in the form of Eq. (16.7b) with $k(A) = 0$, and think of μ as being within m standard deviations of ξ_{crit}.

Step 8: Maximum Number of Links Allowed by Disruption Criterion

If one now uses the expressions for μ_ξ and σ_ξ from Eq. (16.6) in the criterion of Eq. (16.7) and solves the resulting quadratic equation, the expression given in Eq. (16.8) is obtained for the allowable number of links, y_c, that may be embedded in the network before disruption by cross talk should be expected:

$$y_c = B/2 - \sqrt{(B/2)^2 - (\xi_{crit}/(d^*r^{2*}n_f))^2} \tag{16.8a}$$

$$B = \frac{(k+m)^{2*}D^*(1-r)^*}{d^{2*}r^{2*}n_f}\left\{\left[\frac{(1-r)^*(1-n_f/N)}{(1-1/N)} + r^*n_f\right]\right\} + \frac{2^*\xi_{crit}}{d^*r^{2*}n_f} \tag{16.8b}$$

Specific Disruption Criterion: Physiological Estimates of S

We now need to relate the critical cross-talk level, ξ_{crit}, to the physiological parameters involved in the ongoing activation of a memory trace, so that Eq. (16.8) can be evaluated. We anticipate that cross talk will become threateningly disruptive when it attains magnitudes comparable to the signal levels mediated by the sequential configurations. Therefore, let us define the critical level of cross talk as that value which repeated at every time step would drive cells to fire even in the absence of any signal input. Before expressing this mathematically, however, let us first develop an expression for the signal levels found in sequential configurations and apply a constraint to ensure that these are at an adequate level for triggering firings.

Signal Levels in Sequential Configurations: Spatiotemporal Summation of PSPS

Consider now the signal level received by trace cells during the realization of a trace. Each trace cell receives n_f PSPS of strength d_l at each of n_l times according to the arrival times relative to the prescribed firing time. If we assume that each PSP (and hence each composite PSP) decays exponentially with time constant T and that the inputs are spaced by a time interval of Δ, as illustrated in Figure 16.5, then the temporal summation of PSPS can be approximated by Eq. (16.9a). As in Chapters 9 and 10, this sequence of terms can in turn be interpreted as a geometric progression in both the strength and temporal decay terms and thus is concisely represented as in Eq. (16.9b):

$$\begin{aligned} S_S &= d_1 + d_2^* \exp(-\Delta/T) + d_3^* \exp(-2^*\Delta/T) + \cdots) \\ &= \sum \exp(-d_c^*(l-1)/n_L)^* \exp(-\Delta^*(l-1)/T) \qquad (16.9a) \\ &= \sum \exp(-[d_c/n_L + \Delta/T]^*(l-1)) \end{aligned}$$

$$S_S = \frac{1 - \exp(-w^* n_L)}{(1 - \exp(-w))} \qquad (16.9b)$$

$$w = d_c/n_L + \Delta T \qquad (16.9c)$$

$$S = c^* S_S$$

In fact, the summation will not be linear because of nonlinear leakage of current through the activating synaptic conductance channels. This loss can

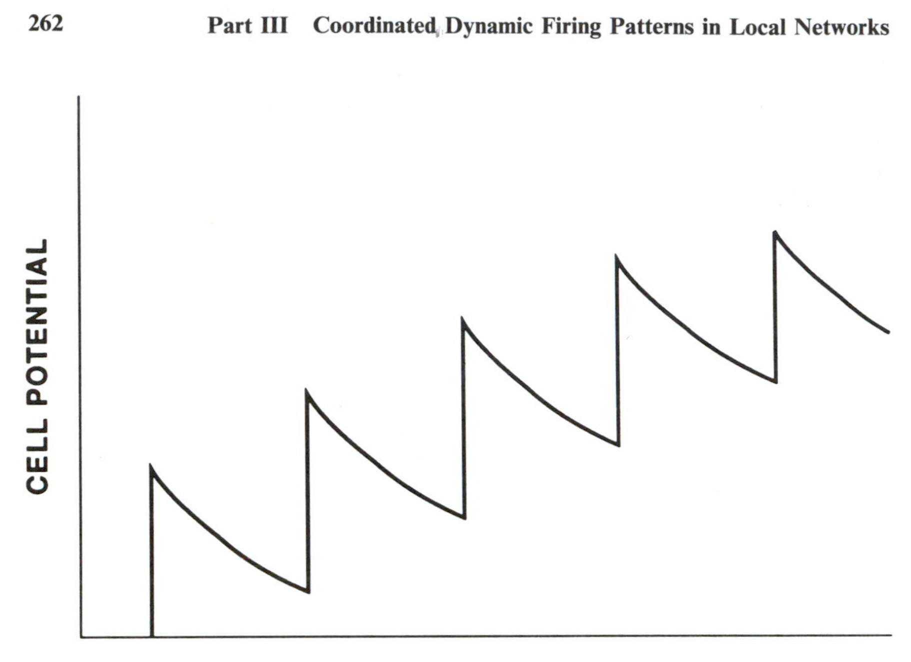

FIGURE 16.5 Linear summation of PSPs (reprinted with permission from R. J. MacGregor & G. Gerstein, Cross-Talk Theory of Memory Capacity in Neural Networks, *Biological Cybernetics*, Springer-Verlag, 1991, **65**, 351–356).

be estimated by the use of an occlusion factor, c, somewhat less than 1.0 (typically about 0.6) as shown in Eq. (16.9d).

We can now estimate that the nondisturbed trace signal level by neurons participating in sequential configurations is given by $n_f^* S$. We can further constrain that for sequential configurations to be comfortably regenerative, this value must be somewhat above the threshold, θ, as given in Eq. (16.10):

$$n_f^* S = b^* \theta \tag{16.10}$$

Here, b is a number somewhat above 1, say about 1.1.

Note that Eq. (16.10) gives us an estimate of the minimum value of n_f required by a self-sustaining trace with summation factor S in a bed of neurons with threshold θ. If we take n_f to be this minimum allowable value, we will predict an upper limit for the number of traces the net can store. Moreover, even though only n_f cells may be firing in a realization, cross talk will be distributed according to the possibly larger number, n, of bed cell connections at each link as shown in Eq. (16.1). Again, we will take the minimum value of $n = n_f$, (corresponding to $k_n = 1$ in Eq. (16.1d) and thereby produce an upper limit for the allowable number of traces.)

Criterion for Firing in Corticallike Modules

We can now use Eq. (16.11) to define the critical level of cross talk, ξ_{crit}, as that value which repeated at every time step would drive cells to fire even in the absence of any signal input:

$$\xi^*_{\text{crit}} S^*_{mx} s\# = \theta \tag{16.11}$$

The summation factor for such a case, S_{mx}, can be obtained from Eq. (16.9) by letting n_L get infinitely large and taking $\Delta = 1$ and $c = 0.6$. (This produces $S_{mx} = 9.30$.) The quantity $s\#$ is the mean synaptic strength; it is applied here because the cross talk should be expected to draw randomly from the entire range of strengths, not the orderly array found in the sequential configurations.

We now can combine Eqs. (16.10) and (16.11) to produce the linear relations among n_f, ξ_{crit}, and θ shown in Eq. (16.12):

$$n_f = b^*\theta/S = \xi_{\text{crit}}/s \tag{16.12a}$$

$$\xi_{\text{crit}} = \theta/(S^*_{mx} s\#) = s^* n_f \tag{16.12b}$$

$$\theta = S^*_{mx}(s\#)^* \xi_{\text{crit}} = n^*_f S/b \tag{16.12c}$$

$$s = S/(b^* S^*_{mx} s\#) \tag{16.12d}$$

Here s is the combination of terms defined in Eq. (16.12d).

Representative Values for Cortical Nets

It is instructive to consider some representative values for cortical nets of the various terms appearing in Eqs. (16.10), (16.11), and (16.12). Let us suppose that the bed parameters in Eqs (16.1) and (16.9) are $n_L = 30$ and $d_c = 2$; that the neurons' time constants, T, are 15 msec; that intralink communication time, Δ, is 1 msec; and that the synaptic occlusion factor, c, is 0.6. We can then determine from Eq. (16.6c) that the quantities d and D are respectively 13.41 and 7.86. This allows us to estimate by Eq. (16.9) the signal spatiotemporal summation factor, S, as 4.72. From Eq. (16.1b), the synaptic strengths for traces with $d_c = 2$ and $n_L = 30$ range from 0.145 to 1.0, with mean, $s\# = \exp(-d_c) = 0.368$.

Now suppose that the threshold of cortical neurons is about 24 mV and that the maximum trace PSP amplitude, 1.0, corresponds to 0.1 mV. This would give the threshold θ as 24 mV/.1mV = 240 control level PSPS. This in turn allows us to project from Eq. (16.12a) that the minimum required n_f is 56. (Note for future comparative purposes that if we took 0.1 mV to correspond to the average trace PSP, $s\#$, then the biggest PSPS, 1.0, would correspond to .267 mV, θ would be 24 mV/.267 mV = 90, and $n_{f\min} = 21$. If we took the biggest model PSPS, 1.0, to correspond to .06mv, then θ would be 24mV/.06mV = 400, and $n_{f\min} = 93$.)

Finally, note that for θ equal to 240, the value of 70.1 is predicted for ξ_{crit}.

Storage Capacities

It is convenient to illustrate the main large-scale properties of the theory by plotting predictions for memory capacity y and NT_{mx} from Eq. (16.8) as a function of network size, N, where the ratio $r = n_f/N$ is held constant, and the threshold θ, is taken as proportional to net size. These results are shown in Figure 16.6.

Three important features are illustrated in this figure. First, notice that the allowable capacities for any relative threshold level levels off at higher values of N. Physically this corresponds to the fact, deducible from Eq. (16.7), that the ratio σ_ξ/μ_ξ becomes progressively smaller as N gets large. This means that one can develop approximate equations by setting $\sigma_\xi = 0$ in Eq. (16.7) for application to larger nets. Second, note that the curve corresponding to $r = .00187$, which corresponds to our best estimate for the cortical module, $\theta = 240$, levels off at a net size of just about the observed module size, $N = 30,000$. We will pursue this point later. Third, note that the trace capacity goes down by a factor of about 16 for every rise of a factor of 4 in the ratio, r.

Approximate Theoretical Equations for Large Nets

Using the assumption that $\sigma_\xi = 0$ for larger nets and using expressions from Eq. (16.6) for μ_ξ and from Eq. (16.12) for η_{crit}, we can write the criterion of Eq. (16.7a) as shown in Eq (16.13a and b). These allow the explicit prediction for trace capacity, y, given in Eq. (16.13c):

$$\mu_\xi = \xi_{crit} \tag{16.13a}$$

$$d^* r^{2*} n_f^* y = s^* n_f \tag{16.13b}$$

$$y = s/(d^* r^2) = s^* N^2/(d^* n^2) \tag{16.13c}$$

$$N_{Tmx} = y/L \tag{16.13d}$$

These equations show us that the memory capacity for large nets increases as the square of net size, N, and inversely as the number of cells per bed link, n, with a proportionality factor of s/d.

Numerical Estimates for Cortical Nets

The general approximate expressions given in Eq. (16.13) can be applied to

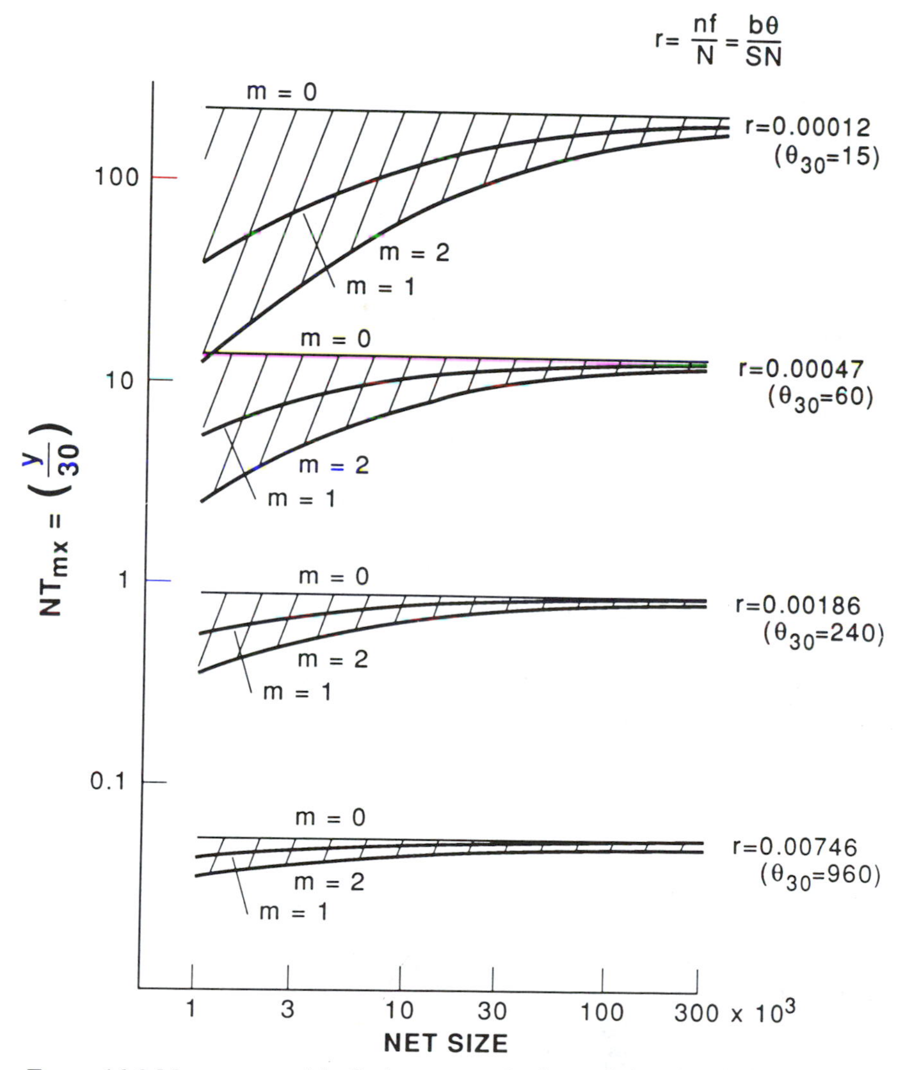

FIGURE 16.6 Memory capacities for large nets—basic trends (reprinted with permission from R. J. MacGregor & G. Gerstein, Cross-Talk Theory of Memory Capacity in Neural Networks, *Biological Cybernetics*, Springer-Verlag, 1991, **65**, 351–356).

estimate local cortical circuits by using the values $s = 1.253$ and $d = 13.41$. This gives the explicit approximate values shown in Eq. (16.14):

$$y = .09346/r^2 \quad (16.14a)$$

$$N_{Tmx} = y/L = .0945/(L^*r^2) \quad (16.14b)$$

If we further take $n = n_f = 56$, and $N = 30,000$, we find $y = 26,880$; if $L = 30$, then $N_{Tmx} = 896$. These estimates all correspond to the best model case, $\theta = 240$.

The more exact solutions shown in Figure 16.6 for k and $m = 0$ agree exactly with these approximate values. The more exact solutions show further that for $m = 1$, $N_{Tmx} = 814$, and for $m = 2$, $N_{Tmx} = 740$.

Corroborative Results from Simulations of Small Nets

As mentioned in Chapter 14, computer simulations show approximate storage capacities of small nets ($N = 100$) for the following three cases: ($n_f = 4$, $L = 12$), ($n_f = 7$, $L = 7$), ($n_f = 10$, $L = 10$). All three cases used $d_c = 2$, and $n_L = 4$. With trace signal levels moderately above threshold (b about 1.1), the 100-cell nets allowed successful retrieval of up to eight 4–12 traces, five 7–7 traces, and only one 10–10 trace). The simulation programs and other representative results are described in Chapter 14.

The theory for this case, taking $d_c = 2$, $n_L = 4$, $c = .6$, and $b = 1.1$, yields from Eq. (16.9), $w = .5667$ and $S = 1.243$. S_{mx} and $s\#$ are again 9.303 and .368. These give $s = .330$ from Eq. 22. The predictions from Eq. (16.8) for this case, with $k >$ and $m = 0$, are: for $n_f = 4$, $L = 12$ traces; $y = 94.8$ and $N_T = 7.9$ (compared with eight observed); for the $n_f = 7$, $L = 7$ traces, $y = 31.0$ and $N_T = 4.4$ (compared with five observed); for the $n_f = 10$, $L = 10$ traces, $y = 15.2$ and $N_T = 1.5$ (compared with one observed).

If one uses the criterion that disruption will occur when n_f cells are expected to fire (i.e., $A = n_f/N$: $k(.04) = 1.75$, $k(.07) = 1.475$, $k(.10) = 1.28$), the predictions are higher: $y = 136$ and $N_T = 11.3$ for $n_f = 4$; $y = 55$ and $N_T = 7.9$ for $n_f = 7$; and $y = 29$ and $N_T = 2.9$ for $n_f = 10$.

Both approaches show the right trends; the first criterion matches the numbers more precisely.

Total Numbers of Cells and Synapses used in Multiply Packed Beds

It is instructive to consider the total numbers of cells and synapses involved in multiply embedded nets. A given bed link uses n cells. It projects n^2 synapses to each of the n_l receiving links it activates. A net of N cells can be estimated to have N^2 synapses in an all-to-all connectivity. Hence, the useful estimates in Eq. (16.15) of the cells used per cell and synapses used per synapse in a bed containing y links:

$$\text{cells used/cell} = n^* y/N = r^* y \tag{16.15a}$$

$$\text{synapses used/synapse} = n^{2^*} n_l^* y/N^2 = r^{2^*} n_l^* y \tag{16.15b}$$

It is interesting to note, for example, that, if one would use the criterion that a net's storage capacity would be attained when all its synapses were used up, Eq. (16.15b) would suggest that the storage capacity would be about $1/(n_l^* r^2)$. Note that this differs from the value given by our more detailed approximate equations in Eq. (16.13c) only by the factor s/d.

Estimates of Number of Memory Traces Corresponding to Number of Links

It is of some interest to consider several ways of estimating the numbers of traces associated with constraints on numbers of links, cells, and cell firing rates.

Suppose that a given trace consists of L links in sequence. Therefore the number of links, y, required for a given number of traces, N_T, is simply $N_T^* L$ as shown in Eq. (16.16):

$$y = N_T^* L \tag{16.16}$$

The maximum N_T will then correspond to the minimum of L.

The minimum of L, however, is obtained by allowing each participating neuron to fire at its maximum allowable rate, w_{mx}. Therefore the maximum number of traces for a given y can be estimated by Eq. (16.17):

$$N_{Tmx} = w_{mx}^* y = y/L_{mn} \tag{16.17}$$

A softer lower limit for the number of traces for a given y can be obtained by restricting the number of cells in a single trace to be equal to the total number of cells in the net as shown in Eq. (16.18a):

$$n^* L_{mx} = N \tag{16.18a}$$

$$N_{Tmn} = y/L_{mx} = n/N^* u = r^* y \tag{16.18b}$$

There is in fact no reason why a given cell might not participate more than once in a trace, but the estimate does provide an instructive reference level.

Storage Capacities in Nets Bombarded with Random Background Activity

Consider now the case illustrated in Figure 16.7, which includes the bombardment of network cells by sporadic random firing of other network cells in addition to the cross-talk activation analyzed earlier. This activation can be represented by a random variable, R, and hence the expression for ξ can be generalized from Eq. (16.5) as shown in Eq. (16.19):

$$\xi(x, \{z_{xl}\}, R) = \sum_{x=0}^{n_f} \sum_{l=1}^{n_L} z_{xl}^* x^* d_l + R \tag{16.19}$$

Suppose that any net cell has probability q of firing in one time period and that each neuron has N input synapses in densely connected nets. If these

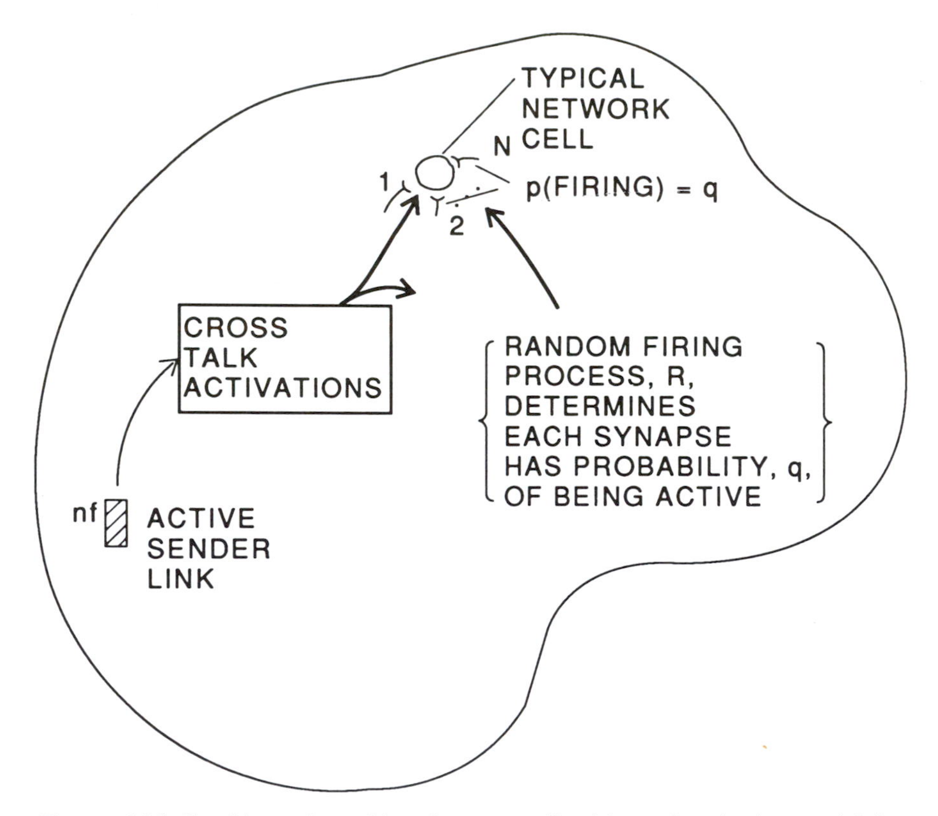

FIGURE 16.7 Combinatoric problem for cross talk with random background firing (reprinted with permission from R. J. MacGregor & G. Gerstein, Cross-Talk Theory of Memory Capacity in Neural Networks, *Biological Cybernetics*, Springer-Verlag, 1991, **65**, 351–356).

random firings are independent, then the number of target cell input synapses activated by this process is a random variable determined by N Bernoulli trials, each with probability of success equal to q. The resulting generalized expression for the mean and standard deviation of ξ are given in Eq. (16.20):

$$\mu_\xi = d^*r^{2*}n_f^* y + q^*N^*s\# \quad (16.20a)$$

$$\sigma_\xi^2 = \sigma_{CT}^2 + q^*(1-q)^*N^*s\# \quad (16.20b)$$

The synaptic weighting factor, $s\#$, is again the average synaptic strength.

The corresponding generalized expressions for the number of allowable links according to Eq. (16.7) is given in Eq. (16.21).

$$y_c = B/2 - \sqrt{B/2 - C/(dr^2n_f)^2}$$

$$C = [\xi_{\text{crit}} - q^*(N-n_f)^*s\#]^2 - (k+m)^{2*}q^*(1-q)^*(N-n_f)^*s\#$$

$$B = (k+m)^{2*}D^*(1-r)^*[(1-r)^*(1-n_f/N)/(1-1/N) + r^*n_f]/(d^{2*}r^{2*}n_f)$$
$$+ 2^*[\xi_{\text{crit}} - q^*(N-n_f)^*s\#]/(d^*r^{2*}n_f)$$

Approximate Theory for Large Nets

Consider first the approximation of neglecting the standard deviation of for large N. Note from Eq. (16.20), that we can also neglect the standard deviation from the random firing at high N, since it also becomes much smaller than the mean of the random firing. We can then write Eq. (16.22):

$$\mu_\xi = \xi_{\text{crit}} \quad (16.22a)$$

$$d^*r^{2*}n_f^* y + q^*N^*s\# = s^*n_f \quad (16.22b)$$

$$y = (s^*n_f^*N^2 - q^*)s\#)^*N^3)/(d^*n_f^3) \quad (16.22c)$$

Notice that we use the weighting factor, $s\#$, on the terms from random firing in Eq. (16.22b), since those inputs should make use of PSPS of all sizes randomly. Equation (16.22c) tells us that the amount of extraneous activation applied to network cells by random firing in densely interconnected nets becomes overpoweringly large as N gets very large (giving a term going as N^3, as compared to N^2 for signal cross talk only). Equation (16.22c) tells us there will be a maximum value of y at some particular value $N = N_p$. These values can be obtained from Eq. (16.22) as shown in Eq. (16.23):

$$\frac{dy}{dN} = \frac{2^* s^* n_f^* N - 3^* q^* (s\#)^* N^2}{d^* n_f^3} = 0 \tag{16.23a}$$

$$N_p = 2^* s^* n_f / (3^* q^* s\#) \tag{16.23b}$$

$$y_p = s/(3^* d^* r_p^2) = (y_{CT}/3) \tag{16.23c}$$

$$NT_{mx} = y_p/30 \tag{16.23d}$$

Equations (16.23) tell us that the storage capacity of large nets experiencing significant amounts of extraneous activation from random firing of constituent neurons should be approximately one-third that of a net of the same size with no random activity and should decrease with the square of the number of cells per link, n, and increase with the square of the optimal network size, N_p. The optimal network size, in turn, increases linearly with the number of cells per link and decreases with the first power of the random firing rate.

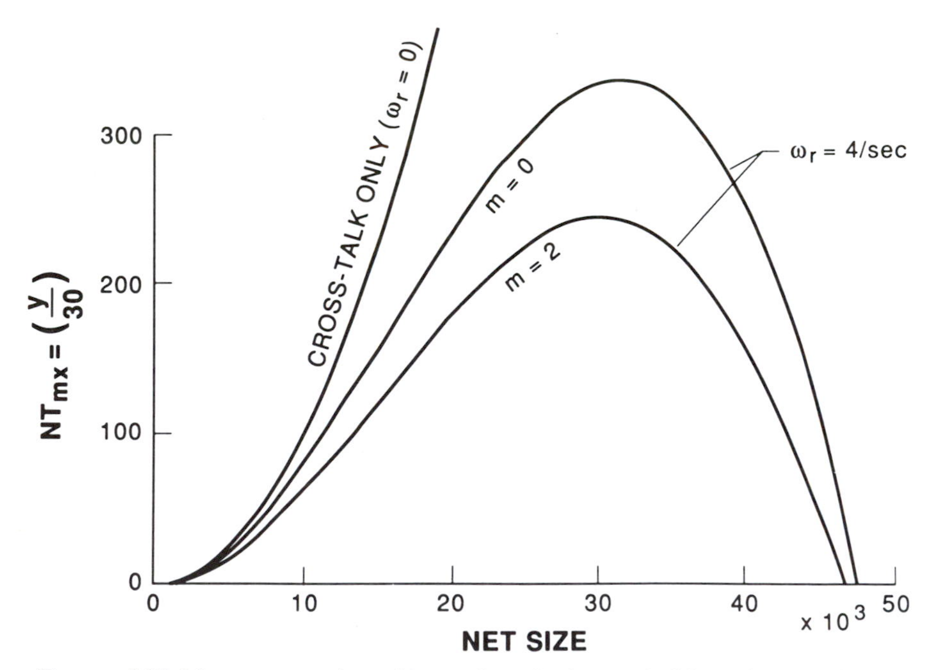

FIGURE 16.8 Memory capacity with random background firings (reprinted with permission from R. J. MacGregor & G. Gerstein, Cross-Talk Theory of Memory Capacity in Neural Networks, *Biological Cybernetics*, Springer-Verlag, 1991, **65**, 351–356).

Numerical Estimates of Storage Capacity with Random Bombardment

Figure 16.8 shows exact solutions for N_{Tmx} based on Eq. (16.21) for this case of random background activity. The central feature is the occurrence of a peak or optimal value of net size with respect to storage capacity.

This figure is based on $\theta = 240$, $n_f = 56$. Note again that the accurate equations reduce to the approximate model $m = 0$. For $m = 1$, $N_p = 31,250$ and $N_{Tmx} = 286$; for $m = 2$, $N_p = 30,500$ and $N_{Tmx} = 244$. All of these values are based on $w_r = 4/\sec$.

Turning again to the approximate model, we find that by using $s = 1.253$, $s\# = .368$, and taking $q = w_r/1000$, where w_r is the random firing rate in spikes per second, we can write Eq. (16.24):

$$N_p = 2,270^* n_f / w_r \tag{16.24}$$

If $n_f = 56$ and $w_r = 4/\sec$, then $N_p = 31,779$ and $N_{Tmx} = 336$.

If we use the more extreme cases introduced previously, we find, for $\theta = 400$, $n_f = 93$, and taking $w_r = 5$, $N_p = 42,331$ and $N_{Tmx} = 426$. For $\theta = 90$, $n_f = 21$, and taking $w_r = 1.5/\sec$, $N_p = 31,755$ and $N_{Tmx} = 2374$.

Concluding Comments

This theory, then, allows us to estimate the number of distinct, overlapping memory traces that can be packed nonselectively into a local recurrently connected net before significant disruption by cross talk. On this basis we can begin to speculate intelligently about the influence of the anatomical and physiological parameters of the networks in this capacity. We begin to approach the principles of engineering structural design of networks (biological and human-made) in this context.

The broadest specific central prediction of the theory is that cortical networks may have designed themselves at modules of about 30,000 neurons at least in part to optimize memory storage capacity.

Another specific central prediction is that cortical modules should exhibit storage capacities of about 300–900 memory traces. This latter specific prediction must immediately be tempered by the observation that it reflects a random placing of traces. Networks in which memory traces are embedded in an interrelated or otherwise rational or nonrandom manner will be able to store larger numbers of traces. This situation is analogous to the comparison of sequential configurations to unorganized firing patterns discussed in Chapters 14 and 15 of this part and to the comparisons of temporally structured to temporally unstructured input patterns discussed in Chapters 11 and 12. Nonetheless, this estimate provides a meaningful preliminary estimate of a minimum level.

One immediate line of inquiry that emerges from this study revolves around the concept that larger areas of cortex might be conceived in terms of collections of coupled modules, the primary dynamical features of which are to reverberate in one or another of some several hundred possible dynamic modes. This concept is discussed explicitly in Chapter 20.

We turn now to Part IV, in which the implications of the sequential configuration model for internetwork interactions in composite networks and neuronal junctions generally are considered.

IV

Theoretical Mechanics of Composite Neural Networks

Chapter 17 Toward a Mechanics of Composite Neural Networks and Synaptic Junctions

This part of the book considers the area of composite neural networks. Our purposes are to point out the conceptual differences between this area and that of simple neural networks, stressing the need for corresponding differences in formulation, and to lay the foundations, however preliminary, for an eventual theoretical mechanics of the area.

Although systems of interacting networks (composite networks) may bear some resemblance to systems of interacting neurons (simple networks), their mechanical operations are fundamentally different and of a higher order. One should not be surprised if their overall organizational and operational characteristics are also fundamentally different and of a corresponding higher order.

There are two large domains here, corresponding respectively to the content and context of the neuroelectric signalling of interacting neural networks. Our mechanics is intended to provide a structure for the content of the area, whereas its essential contextual dimensions are served by the domain of systemic organization of nervous systems. These two faces of the area are in fact symbiotic as is shown by the material of this part.

At these two levels of interacting networks and systemic organization the biological dimension of neural network organization most fully approaches the psychological and behavioral dimensions of brain and nervous system function. The material in this part is necessarily much less developed and much more speculative than that in the preceding three parts. Biologically rooted theoretical guidance is correspondingly more sorely needed here.

Introduction to a Theoretical Mechanics of Interacting Neural Networks

The normal operations of neural networks occur for the most part within the ongoing interactions of multiple interconnected local networks. For example, the operations of the hippocampus involve coordinated interactions among local networks in the entorhinal cortex, the dendate gyrus, the CA3 and CA1 regions, the subiculum, and the septum. Operations in higher motor control include interactions among local circuits in the neocerebellum, the thalamus, the cortex, and the inferior olive. Interactions within various central regions such as the reticular formation and the cerebral cortex likely involve local networks whose structural delimitations are not always clear. This list could be extended indefinitely.

A correct mechanics for the operations of such interacting networks should be based solidly on two ingredients. The first is a clear vision of the coordinated firing patterns by which the networks represent meaning and intercommunicate—a view of their natural language in representing information and meaning. This language is a network language, not a single neuron language. The second ingredient is a corresponding network-level description of the mechanisms and means by which these patterns are passed and interact at the confluent junctional interfaces between networks.

This part of the book strives to build on these two ingredients the cornerstones for an eventual mechanics for interacting (composite) networks. Chapter 14 presented a theoretical representation of coordinated network firing patterns called the *sequential configuration*. The book advocates that this view of sequential configurations or something roughly equivalent should necessarily be a component section of one's thinking about the ongoing operations of interacting composite nets.

The book also advocates that one should think of the multiple-member synaptic junctions between local neural networks as devices for processing these coordinated network patterns and their interactions. To underscore this network view (as opposed to thinking in terms of unitary responses at single synapses) these junctions are referred to as *synaptic junctional matrices*.

Subsequent sections of this chapter develop some initial suggestions for network-based interpretations of the operations of synaptic junctional matrices.

Subsequent chapters in this part apply these ideas in outline to two central composite networks, the hippocampus and the cerebral cortex. An essential ingredient in approaching the understanding of a particular set of interacting neural networks is a guiding concept regarding the overall functioning of those collective networks. For example, the overall concepts that the hippocampus represents and embeds cognitive generalizations in response to affective prompting and that composite cortical networks represent information are vital to the theoretical views of the operations of these composite networks developed here.

Chapter 18 presents a theoretical view of the composite circuitry of the hippocampus. In addition to presenting the vital intrinsic significance of this circuitry, this chapter illustrates, by application to a particular composite network, both the approach of the composite network mechanics outlined here and the significance of systemic contextual information in dealing with composite networks.

A modular view of the organization of classical cortical fields (association cortex, primary and secondary sensory areas, etc.) suggests that these fields also might be considered as involving interactions of numbers of local networks. Chapter 19 develops a tentative theoretical view of modular organization in composite cortical regions based on the signalling of information in component local circuits by sequential configurations.

Chapter 20 provides an overview of experimental approaches to and principles and theories of cortical organization. The chapter also includes a theoretical view of the relationship of the multimodal oscillator theory of Chapter 18 to the processes of cognition. The approach takes us to the point where a neurobiologically based mechanics of cognition can be glimpsed if not yet seen clearly nor reached.

We can label the domain of ideas regarding overall functioning of composite neural networks and multiple composite neural networks as the level of "systemic organization" in the nervous system. This area provides essential contextual guidance concerning the operations of interacting neural networks and their theoretical mechanics. It is a vast area tremendously rich in biological detail and is generally understood only in outline. Moreover, the area is symbiotic with a theoretical mechanics of neural networks, in that full understanding of either is dependent on the other.

Although the area of systemic organization of neural networks is beyond the scope of this book, the book nonetheless dedicates one chapter to it. Chapter 21 concludes the section and the book with a broad theoretical

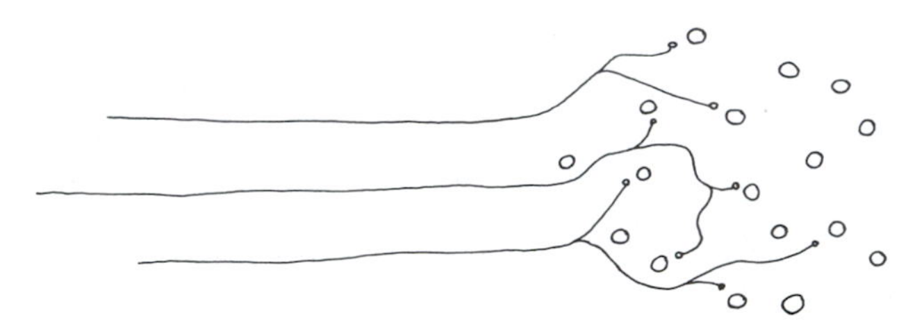

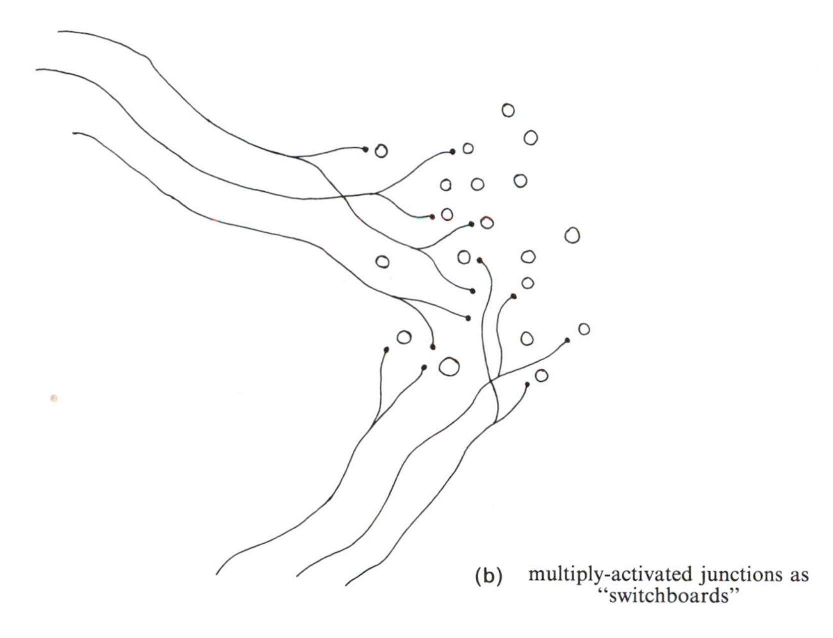

FIGURE 17.1. Main types of excitatory synaptic junctions: (a) singly activated feedforward junction; (b) multiply activated feedforward junction as "switchboard"; (c) recurrently connected junction as "autonomous pattern generator."

interpretation of principles of systemic organization in nervous systems. This chapter is highly speculative. It is included to provide an indication of the types of overall guidance that can be useful in constructing operational models of composite networks.

An Overview of Junctional Operations on Coordinated Network Firing Patterns

The mechanics outlined in the next section suggests the definition of three main classes of excitatory synaptic junctional matrices: through-passes and

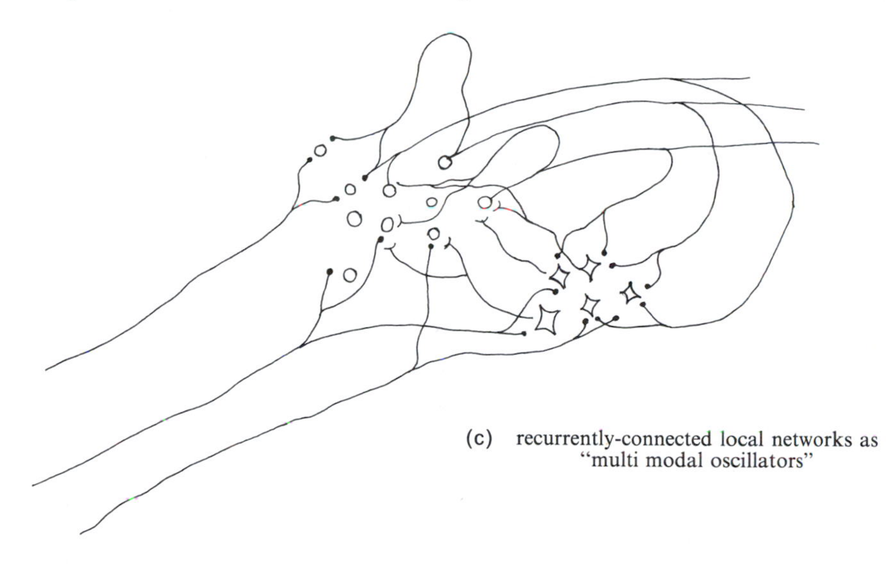

FIGURE 17.1. Continued.

switchboards, representative of feedforward junctions, and autonomous pattern generators, representative of recurrent junctions. The inhibitory components of synaptic junctions, both feedforward and recurrent, are seen as mediating fine-grained spatiotemporal sculpting of coordinated network firing patterns.

Through-passes

Through-passes are operations performed at single neuronal junctions whose input activation is dominated by a single input system. There may or may not be local inhibitory neurons, and these may or may not be activated by either feedforward connections from the input system or recurrent connections from the receiving population. There may not be recurrent excitation, however. In short, such a junction is a prototypical convergent–divergent junction that may or may not house local inhibition. The central operation of such a junction, which is illustrated in Figure (17.1a), is simply to respond to a given sequential configuration in the input system, with a corresponding sequential configuration in the output system. The relations between the input and output sequential configurations are embedded in the synaptic junctional matrix between them. Other input systems may be converging at the junction, but when it is operating in the through-pass mode, these other inputs are in abeyance with respect to single input system that is then dominant or providing input consonant with it.

Switchboards

Individual junctions without recurrent excitation, such as those just defined for through-pass and representational loop operations, may act as switchboards when activated by the confluence of two or more input systems, each carrying a sequential configuration. In such a case, as illustrated in Figure 17.1b, the output is a coordinated network pattern linked to the particular input pattern according to the junctional synaptic matrices for all participating input systems. The guiding idea here is that various combinations of the input patterns selectively couple to the individual members of the output repertoire of patterns and that the component synaptic matrices serve effectively as a composite switchboard in effecting these decisions. This operation is suggested here to be extremely common in composite networks, as is illustrated in Chapters 18 and 19 and discussed more broadly in Chapter 21.

Autonomous Pattern Generators

Autonomous pattern generators are served by local neural networks with recurrent excitatory connections, such as illustrated in Figure (17.1c). These recurrent excitatory connections are the means by which coordinated patterns can be stored and driven locally; they are the means by which local networks can gain autonomy from their input systems. These patterns may exhibit variable degrees of independence of the input and may be dependent or independent of continuing input.

Chapter 14 illustrated the extreme case, where the output repertoire is totally determined by recurrent synapses and where patterns are self-sustaining even in the absence of continuing input. This extreme type of autonomous pattern generator may be referred to as *multi modal oscillator*, which is simply a label for this behavior. Fundamentally, the language of such nets consists of the ability to oscillate in one or another mode; that is, such nets may reverberate in one or another of their repertoires of coordinated firing patterns. Here, the primary factor is the local internal synaptic matrices (excitatory and inhibitory). The input synaptic matrices are significant primarily in how they play upon these.

Outlines of a Mechanics of Neural Junctions

Some of the main characteristics of the interactions of coordinated firing patterns at junctions and within local networks, such as those just indicated,

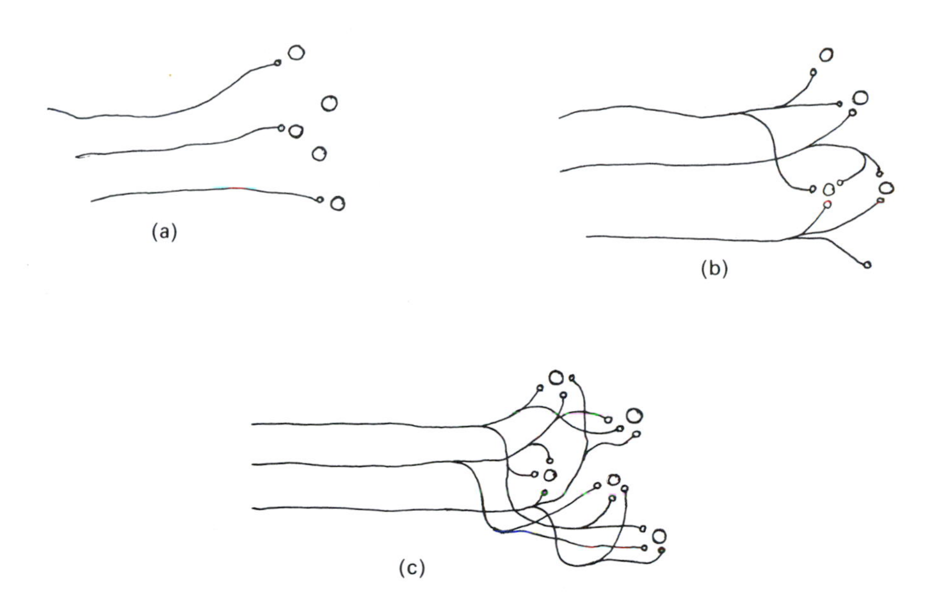

FIGURE 17.2. Prototypical convergent–divergent junctions: (a) Minimal interconnectivity; (b) moderate interconnectivity; (c) Maximal interconnectivity.

can be inferred rather directly. Their detailed properties, and undoubtedly many vitally important characteristics, involve tremendous complexity. Here, we deal only with these former speculative generalizations.

Prototypical Convergent-Divergent Junctions

Degree of Connectivity. A prototypical convergent-divergent junction activated by a single input system is illustrated in Figure 17.2. The significance of the synaptic junctional matrix can be illustrated by considering two extreme cases. Figure 17.2a shows the case of minimal divergence wherein each fiber projects to only one receiving cell. This highly artificial case clearly involves no network integration. Each receiving cell is only a temporal integrator over its particular input fiber.

Figure 17.2c shows the opposite extreme, where every input fiber projects equally to every receiving cell. If we neglect individual differences among the synapses and among receiving cells, all receiving neurons receive exactly the same composite input signal. The receiving population behaves as a unit, wherein all cells either fire or do not. The situation is nonetheless illustrative. Consider for example, the computer simulated network transfer pairs for highly connected junctions shown in Figure 17.3.

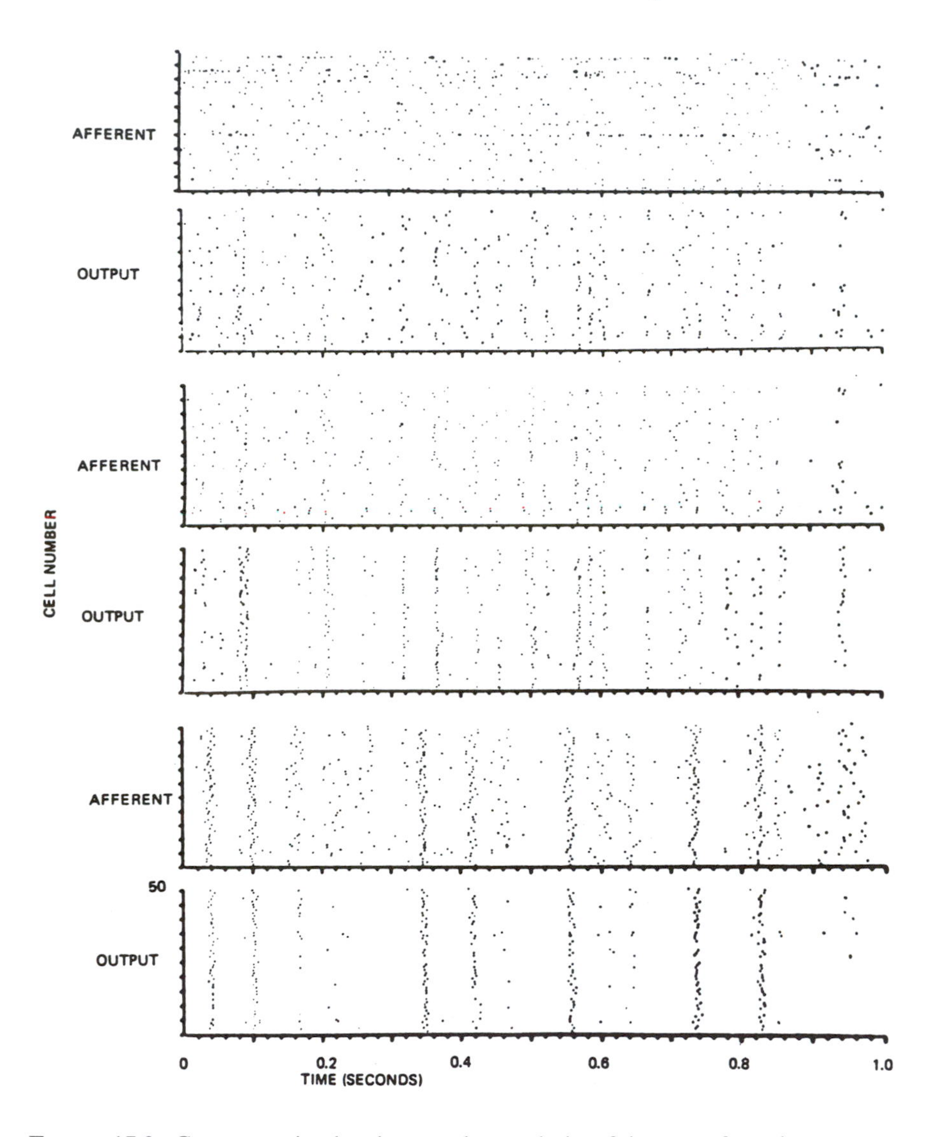

FIGURE 17.3. Computer simulated network population firing transfer pairs.

Such a junction operates essentially as a filter, selectively sensitive to the number of input fibers active over a recent time window. If this number is sufficiently high, all the cells of the network respond. This filter effect is shown in Figure 17.3. If one drives the first of a series of such junctions with a random pattern, there is progressive filtering so that after two or so junctions, only these large clusters of firing remain in the output patterns.

This simple example, even though developed for an extreme case of very high connectivity, illustrates the central significance of large sets of simultaneously firing cells for the basic junctional operations of neural networks.

It seems clear, from considering these extremes of junctional connectivity, that even global network integrations are likely to be best served by intermediate levels of connectivity. These can be illustrated as in Figure 17.2b.

Activation by Individual Coordinated Firing Patterns. Consider the case where the junction shown in Figure 17.2b is activated by a single pattern in its single input system. The successive activations of particular sets of input synapses will be associated with the activation of successive particular sets of receiving neurons. This is a transfer from an input sequential configuration to an output sequential configuration. Generally, there will be little obvious relation between the microstructure of the input and output configurations. These specific relations will be determined by the details of the synaptic junctional matrix.

The presence of a local inhibitory population, activated either forwardly or recurrently, will alter the fine structure of the output pattern, but not the basic concept of a transfer from an input pattern to the output pattern. This, overall behavior is what we have called *through-pass transfer*.

Confluence of Coordinated Firing Patterns with No Local Inhibition. Consider now the case where two patterns converge on a single prototypical junction with no local inhibition. Such a case is illustrated in Figure 17.4. Each input system, if activated individually, will produce a particular corresponding output pattern according to the processes just described. The microstructure of each transfer will be determined by the synaptic junctional matrix for each input system. Suppose, for example, that solo inputs A and B in the two distinct input channels produce corresponding output configurations, A and B.

The activation of both these inputs simultaneously will then produce a composite output pattern dominated by these two patterns, as illustrated graphically in Figure 17.4b. Output cells will be driven by input firings over recent time windows in the combined case, simply summed over both input systems. Some depletion of each of the two responses will occur because of refractoriness of some cells participating across the A and B patterns at times close together, but this effect will not be markedly significant at normal neuronal rates. Moreover, one might expect a small additional shadow region of activation for cells whose subliminal activations by the

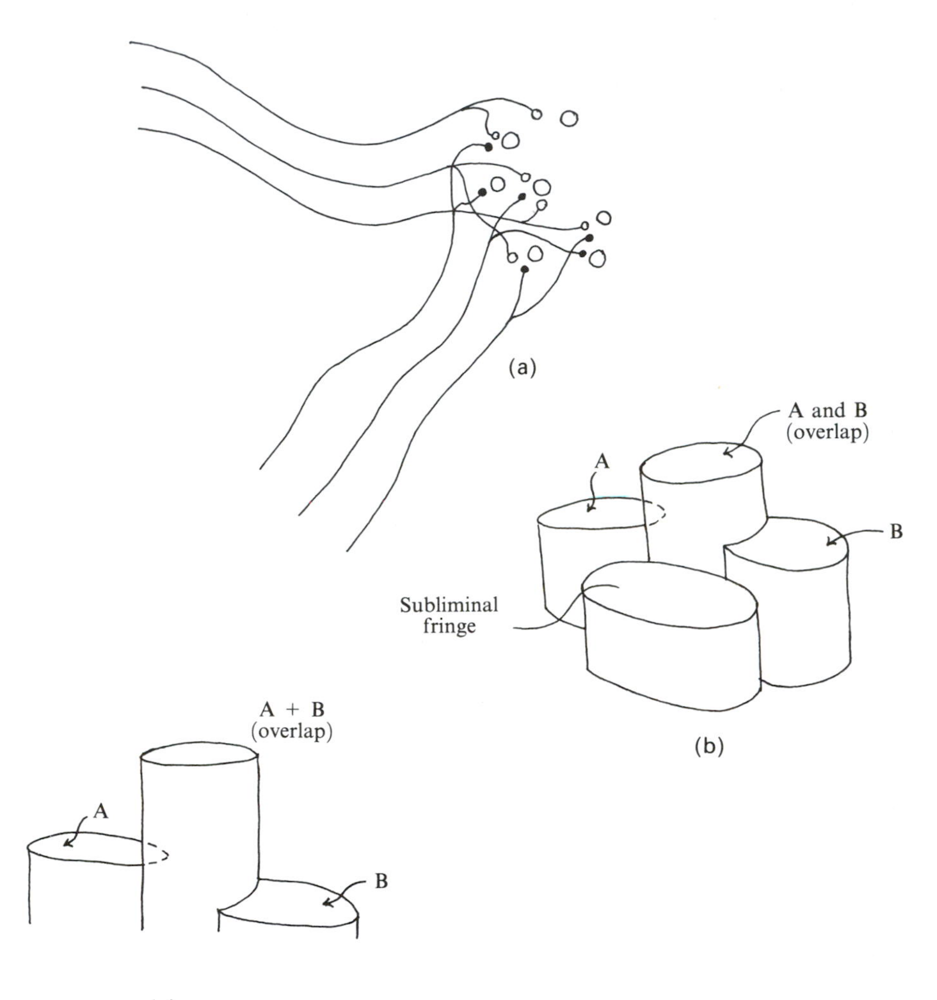

FIGURE 17.4. Confluence of two coordinated firing patterns on a single receiving population: (a) schematic of synaptic junctions—(b) with no inhibition (responses in $A + B$ receivers, in A receivers and in B receivers, and in a subliminal fringe group.); (c) with supportive inhibition for both A and B (responses restricted largely to $A + B$ receivers).

individual inputs sums above threshold for the combined event, as illustrated by region C in Figure 17.4b.

The idea here is that this composite input event triggers an output event that is also a coordinated firing pattern. The transfer can be seen as an instance of a general switchboard effect, where different combinations of input patterns produce distinct patterns, according to the combinations of particular junctional synaptic matrices for the input systems. In this case the composite output configuration closely resembles the sum of the input patterns. Such resemblance may be drawn upon in higher order use of these operations.

Confluence of Coordinated Firing Patterns with Local Inhibition. The presence of local inhibition will modify the response to the composite input *A* and *B* discussed previously to the form shown in Figure 17.4c. Here, the cross-pattern inhibitory influences will tend to diminish all responses except those common to both patterns. Thus, the common or overlap component of the combined response will be the dominant feature of the response pattern. The shadow region will be less likely to appear.

Again, the central effect here is a switchboard transfer from a given input of combined patterns to a particular output pattern, the fine structure of which is determined by the participating junctional synaptic matrices. Also, again, there is a relation between the composite event and individual events that may be drawn upon in higher order use.

Recurrently Connected Local Networks

The central property of recurrent excitatory junctions is to prompt output firing patterns in accordance with the synaptic weightings associated with the recurrent junction. The strength of this influence relative to that of incident feedforward junctions on the same network can be graded. At one extreme the result is simply a through-pass or switchboard; at the other extreme the result is a multimodal oscillator. Intermediate levels of relative strength produce a network that may assert its autonomy or acquiesce to input patterns depending on the structure and level of the input.

Dynamic similarity scaling of neuronal junctions

It is useful to characterize neuronal junctions individually and comparatively in terms of nondimensional numbers according to the dynamic similarity approach developed for single neurons in Part II. The structure of the neuronal junction between any single input fiber system, *i*, and any single population of receiving neurons, *j*, can be characterized by three parameters: (1) a global convergence factor, α_{ij}, which is the ratio of projecting fibers to receiving cells, (2) a microscopic convergence–divergence parameter, γ_{ij},

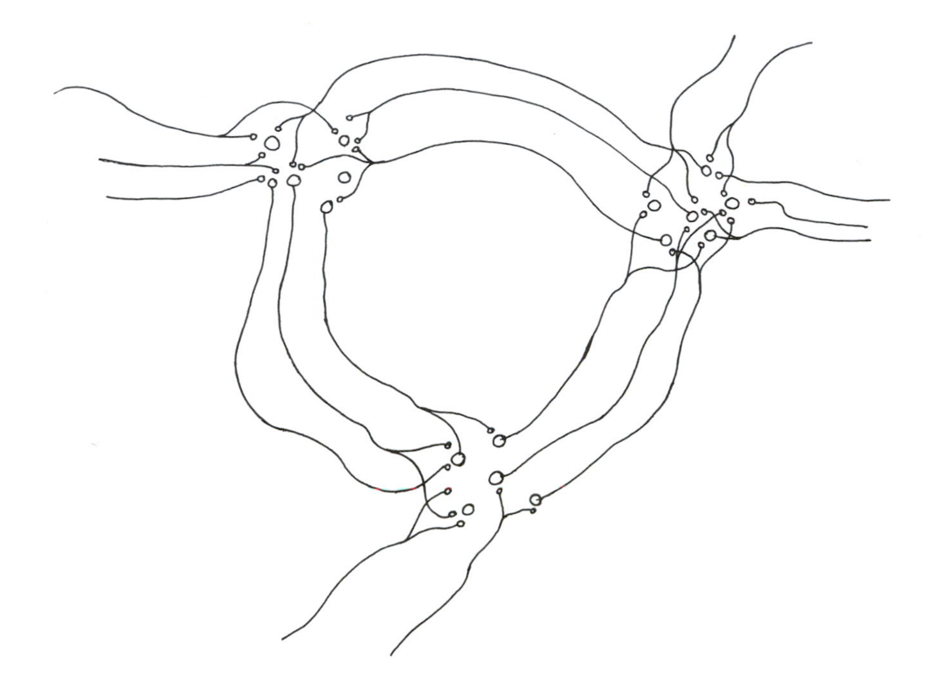

FIGURE 17.5. Recurrently connected series of junctions as representational loops.

which measures both the percentage of cells to which a given sender fiber projects (and the percentage of fibers from which a given cell receives) and the percentage of cell–fiber combinations that are directly connected, and (3) a synaptic sensitivity parameter, σ_{ij}, which measures the representative effective synaptic strength of the junction. This last parameter has been discussed in Chapters 9 and 10. Characterization of junctions in terms of these numbers allows for succinct and useful comparisons of different junctions.

This approach also provides rational guidance for the construction of reduced scaled-down models of neuronal networks involving very large numbers of neurons. If one stipulates that both the percentage of cells to which a given fiber projects and the relative strengths of the different input systems to a given population remain the same, then the rules shown in Eq. (17.1) may be used to determine the parameters of a scaled-down model:

$$x = \text{scaling factor} \tag{17.1a}$$

$$N_i' = N_i/x \tag{17.1b}$$

$$a'_{ij}/\theta_j^{o\prime} = (a_{ij}/\theta_j^o)/x \tag{17.1c}$$

$$S'_{ij} = S_{ij}/x^2 \tag{17.1d}$$

In these equations the primed variables refer to the scaled-down model system and the unprimed variables refer to the original net.

The variable x is the scaling factor. N_i is the number of cells or fibers. The variable a_{ij} is the synaptic strength; θ_j^o is the resting threshold; and S_{ij} is the total number of synapses at the junction.

This topic is treated in more length in the 1988 paper cited on page 361.

Representational Loops

A more speculative theoretical concept is that of representational loops. Representational loops, illustrated in Figure 17.5, consist of recurrently closed loops involving two or more local networks in a sequential feedforward relationship. Representational loops are seen as establishing and maintaining a particular generalized coordinated network firing pattern that manifests in the individual pattern in each of the participating populations. The individual participating patterns are in dynamic capability, the foundations of which are again in the composite synaptic junctional matrices at each junction. Each constituent junction might experience additional input that can influence or modify the representation established when each junction is functioning in the through-pass mode.

This general concept is developed in detail in the next chapter, where the serial recurrent loop connecting five local networks of the mesocortex and hippocampus are interpreted as a representational loop to represent abstractions regarding the external and internal worlds.

The closed loops from the specific sensory nuclei of the thalamus to their corresponding particular sensory cortical regions (lateral geniculate nucleus to and from visual cortex, medial geniculate nucleus to and from auditory cortex, ventroposterior lateral nucleus to and from somatosensory cortex), similarly could be seen to represent particular specific visual, auditory, and somatosensory generalizations according to particular patterns in their representational loops.

The loop involving the motor cortex–inferior olive–cerebellum–ventral lateral thalamus–motor cortex could be seen to represent particular large-scale coordinate movement patterns by generalized representational-loop sequential configurations.

Topics at the Outer Limits of the Network Mechanics Outlined Here

Two important topics are found at the outer limits of the mechanics developed in this part of the book. The first of these points to the central significance of the fine-grained structure of junctional synaptic matrices in carrying out the mechanics of interactions among sequential configurations. The significance of this major topic is clearly illustrated by the examples discussed in Chapters 18 and 19. Its subsequent explicit development will be an essential central component of a mature mechanics of composite networks.

A second major topic resides in the presumed relation between the mechanics of interacting neural networks, on the one hand, and the mechanics of the corresponding psychological and behavioral processes, whose elemental meanings are represented by coordinated firing patterns in these networks. This topic is beyond the scope of the book, but is approached and briefly discussed in the remaining chapters of this part, particularly Chapter 21.

Chapter 18 A Theory of Representation and Embedding of Memory Traces in Hippocampal Networks

The hippocampal circuitry is a vital component in the experiential and cognitive life of human beings. It is the location where significant cognitive representations of the internal and external world are formed, selected, and embedded in the synaptic architecture of our brains. It is the fundamental supportive representational agent in our brains without which any continuing higher cognition is impossible. Bilateral ablation of the hippocampus puts an end to the embedding of new experience. Deficits in its function are a leading candidate source for the cognitive and experiential deficits in schizophrenia.

This chapter presents a theory of the systemic organization and network operations of the hippocampal networks in the representation and embedding of significant memory traces. In addition to the intrinsic value of such a theory, the chapter points out by illustration the value ensuing to the development of understanding of the operations of particular neural networks of having both a general theoretical mechanics of interacting networks and a clear view of the systemic functions of the larger system of which it is a part.

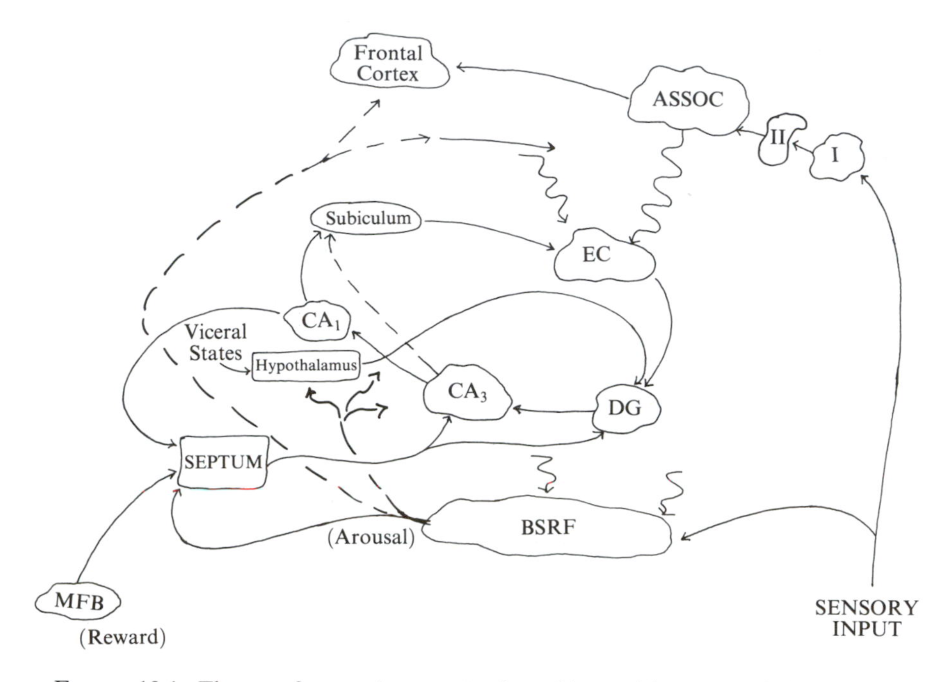

FIGURE 18.1. Theory of systemic organization of larger hippocampal circuitry.

Systemic Operations of the Hippocampal Networks

The larger neural network system governing the hippocampus is shown in Figure 18.1. The main local hippocampal networks are those of the dendate gyrus (DG), the hippocampal regions, CA3 and CA1, the entorhinal cortex (EC), and the subiculum (S). These local circuits are driven by two primary input systems: (1) highly abstracted sensory and cognitive input derived from cortical regions (largely association cortex) and funneled into the DG–CA3–CA1 series through the EC, and (2) input from the medial septum (MS) distributed pervasively to CA3 and DG. This input carries information regarding arousal and attention from the brain stem reticular formation (BSRF), reward states from the medial forebrain bundle, and recurrent cognitive feedback from the CA1 hippocampus. Input to the dendate gyrus is also provided from the hypothalamus. This input carries information regarding internal visceral states.

The hippocampus is necessary for learning new memories, which means it is central to the construction of internal cognitive representation. In humans, new memories are stored in the hippocampus for approximately two weeks before being transferred for longer storage in cortical areas.

Broad Statement of the Theory

The theory advocated here is based on two main tenets. First, the recurrently connected loop from EC–DG–CA3–CA1–S is seen as mediating standing reverberations that represent various candidate abstracted qualities of the external and internal worlds. This loop will be referred to here as the *HML* (for hippocampal–mesocortical loop). These representations are seen as potentially significant generalizations about the worlds formed in part by other higher order cognitive and sensory regions and in part by the hippocampal circuitry itself. The generalizations are represented in the interrelated coordinated network firing patterns of the component local circuits of the HML loop, as sequential configurations or their equivalent.

Second, particular representations are selected for permanent embedding when they are coincident with input delivered to the CA3 and DG circuits from the medial septum. Activity in the septum is seen here as reflecting the occurrence of a significant event in the internal or external world. The level of significance, in turn, is determined by three factors, each delivered to the medial septum by its own input system. Input regarding current levels of reward and aversion are delivered to the septum from the medial forebrain bundle (fundamental reward center) and the septo–hypothalamic–mesenthalamic region of the brain stem (fundamental aversion center). Input regarding current level of overall arousal is delivered to the septum from the arousal portions of the brain stem reticular formation. Information regarding the "newness" of the current pattern in the HML is delivered to the septum from the hippocampal circuits (either CA1 or CA3).* Collectively, these reward, aversion, arousal, and newness inputs define the significance level mediated by the septum for delivery to the HML loop.

When new memories are targeted for embedding by coordination with septal significance input, they must reverberate for some 30 minutes or so for successful embedding to occur. This is assumed to correspond to the time required for the necessary morphological (and perhaps chemical changes) to occur at the synaptic junctional matrices involved in the embedding. As mentioned previously, when new memories are successfully embedded in humans, they are stored in the hippocampus for approximately two weeks and then are somehow transferred to neighboring cortical areas for longer periods of storage.

* Our current inclination is that CA1 mediates this kind of an effect, but there is some justification for considering CA3 in this light. The present discussion is continued in terms of the CA1 option, but this may be adjusted in subsequent developments.

The memories under consideration here are both significant and experiential. It is likely that they constitute the episodic memories described by Penfield and identified by him as occupying specific regions close to the hippocampus in the temporal lobe.

Systemic Engineering of the Interacting Networks

Anatomy

All three local hippocampal circuits exhibit recurrent excitation and local inhibitory cells driven by both feedforward and recurrent connections. Inputs to the CA3 cells from the DG are relatively discrete and very strong (*e*PSPs $\sim$ 2–12 mV, with cell thresholds about 5 mV). Recurrent excitation and inhibition are at high levels in CA3. The drive of CA1 by CA3 inputs, however, is highly diffuse and uses much smaller individual *e*PSPs ($\sim$ 0.13 mV). Recurrent connections are much less prevalent in CA1 than in CA3.

The local circuits of the hippocampus are distributed over one long dimension (along the fornix) and perhaps a second shorter dimension (across the plane of the "interlocking 'C's") in much the same way as the networks of the cerebral cortex are distributed over two dimensions. Further, the hippocampus is bilateral. CA3 circuits on both sides are intensively contralaterally interconnected. These connections parallel the recurrent connections within each CA3 side, so that the two CA3 sides may function as a single distributed system. The CA1 circuits, however, are not contralaterally connected.

The HML loop, then, is basically a serial feedforward loop of individual circuits that may be two-dimensional with one dimension considerably longer than the other. However, the forward projections from the individual regions, particularly, EC, CA3, and CA1, tend to project not only to the next network in series but also further downstream to subsequent networks in the loop. This general arrangement is partially illustrated in Figure 18.2.

Considerable further detailed information exists regarding the numbers of cells, connections, transmitters, and so on for these circuits.

Electrophysiology

Pyramidal cells of the CA3 and CA1 regions fire both complex bursts of action potentials (APs) (seeming to reflect calcium-mediated dendritic APs)

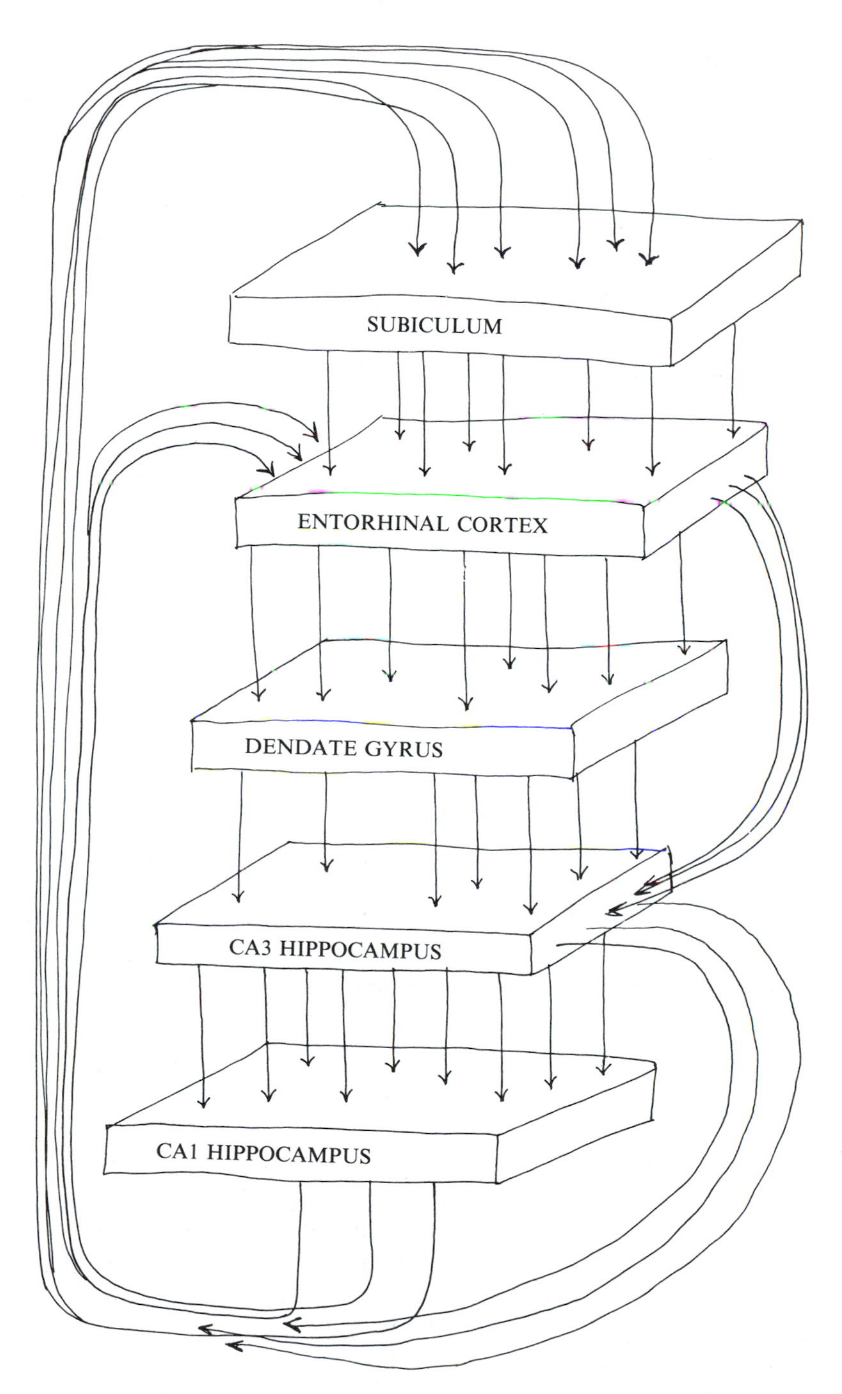

FIGURE 18.2. Multi network recurrent hippocampus–mesocortex loop as a two-dimensional representational loop.

and single, conventional APs. Resting rates are typically less than 2/sec, and maximum rates are about 15–40/sec sustained for less than 2 sec. Firing patterns are irregular and not systematically related to the theta rhythm.

Local basket cells fire only simple action potentials. Firing patterns are regular. Resting rates are high (> 8/sec). Maximum rates are high (~30–150/sec), occur only in conjunction with the theta rhythm to which they are phase related, and are sustained for many seconds.

The complex burst APs accommodate in CA1, but not in CA3. The single conventional APs do not accommodate in either region. Since the complex spikes are presumably of dendritic origin, this would suggest that local dendritic regions may individually accommodate and that these neurons would selectively vary their sensitivities to their different dendritic input regions over time, depending on recent firing history in the local dendritic regions. It is possible that this accommodation produces a network (CA1) particular sensitive to new coordinated input patterns and thereby constitutes a "newness" detector.

The theta rhythm correlates with attention, arousal, and likely memory embedding. It, and these functions, may be mediated largely by the medial septal input to the hippocampal basket cells.

Large-Scale Recurrent Loops

A number of larger scale recurrent loops may be operative in hippocampal function. The Papez loop recurrently connecting the hippocampus with the septum, mammillary bodies, and anterior ventral nucleus of the thalamus has been historically used to emphasize the linkage of visceral and affective components to cognition. The emphasis on the recurrent HML loop is more recent. Chapter 21 provides a broader discussion of these points.

Representation of Patterns by Coordinated Firing Patterns (sequential configurations)

This theory pictures that cognitive representations are embodied in the coordinated network firing patterns of the HML loop. For the sake of specificity, let us imagine these to be sequential configurations as defined in Chapter 14. These coordinated patterns are initiated in these circuits by corresponding patterns in the input delivered to the EC. We picture that some switchboarding is operative at the synaptic junctional matrix between the association cortex and the EC.

Once initiated in the HML loop, the patterns are viewed as at least partially internally sustained by the recurrent connections (both local and loopwide) of the loop and its networks. This picture implies some interesting constraints and coordination among the representation of the patterns as seen in the individual networks and the loop as a whole. One constraint reflects on the relation of the whole loop conduction time to the cyclic duration of the sequential configuration. Since five serial synapses (composing one loop time) would not be expected to last longer than about 10 msec (say 5–25 msec), it would seem likely that the sequential configuration (or its equivalent) would consist of stochastically repeatable cycles, each of which would involve a significant number of loop transit times. Harmonious support of such patterns would involve local supportive reverberations reflected back within a millisecond or so, and supportive whole loop reverberations reflected back at about 10 msec intervals. Since the two recurrent channels (for a given net) involve two distinct synaptic junctional matrices, it is clear that this supportive structure can be embedded in each, in the same way such was embedded in one in Chapter 14.

Therefore the embedding of traces in this multiple network loop involves at least two degrees of freedom associated with the defining of traces locally within local recurrent synaptic junctional matrices and with the definition of their interrelations (supportive, compatible, incompatible, or neutral) within the feedforward junctional matrices of the loop. In this latter light, the multiple downstream projections of the loop may take on a possible new significance.

The feedforward junctions of the loop thus can be anticipated to mediate both switchboarding and supportive sequential configuration definition in some to-be-determined mix. The local recurrent excitatory junctions likely mediate supportive sequential configuration definition, but this might vary among the networks according to the strength of these junctions. CA3, for example, with particularly pervasive and strong local recurrent excitation should be expected to be saliently involved in the definition of the structure of sequential configurations. Weaker or more diffuse recurrent excitation in other nets might serve more diffuse, generally biasing effects.

Local feedforward and recurrent inhibition throughout the loop should be expected to be involved in fine-grained sculpting of local patterns. These local inhibitory populations are the key ingredient in mediating conflict and possible harmonization of competing or different ongoing sequential configurations in the feedforward and recurrent inputs of the net. Thus, if the instantaneous groups of cells fired in the net itself and in its excitatory input are both projecting to the same or a compatible set of firers for the next instant, the two patterns at least instantaneously are highly supportive or at least compatible. If, on the other hand, these two groups of instantaneous firers are projecting to incompatible groups of successors, then conflict is mediated on the cells of the local net that, for resolution, may require multiple reverberations, alterations in either or both of the patterns, and adjustments in the strength levels of local synaptic matrices. All these questions point to the need for guidance from constructive computer simulation.

The situation is further complicated, but not in principle, by the input regarding internal state delivered to the dendate gyrus by the hypothalamus. This input probably carries some significant temporal structuring of it own. It likely intermeshes with the cortically defined input by junctional interactions like those described earlier to produce a combined or experientially colored representational event.

The Mediation of Significance by the Medial Septum

The *inputs to the medial septum* from the medial forebrain bundle and the brain stem reflecting reward, aversion, and arousal are straightforward. There may be a minimal degree of pattern variation in these individual inputs, a simple level of intensity may suffice for each of them. But one may not rule out the possibility of significant variation of meaning here and resultant switchboarding at these junctions. For the present at least, the theory here will take these inputs into the septum as one-dimensional graded signals.

The CA1 and CA3 regions carry significant representations. These are likely projected to the medial septum by their connections to it. That these particular patterns are reflected in the septum's projections back to the CA3 and DG is a definite possibility.

Thus, the theory sees the *septal input projecting to the HML loop* as carrying the representation of significance in the intensity of its firing level and perhaps also coloring this representation with some individuality of particular pattern. If this is the case, then that pattern might be largely reflective of the new signal currently being signalled by the CA1 population. It might, however, also carry some individuality associated with the particular reward, aversion, or arousal context.

The septal input is projected as excitatory synaptic activation on the local inhibitory populations in CA3 and DG. This input can be seen as a bias supporting the responses of inhibitory neurons to the feedforward and recurrent excitatory patterns incident on the CA3 population. The synaptic junctional matrices onto these inhibitory cells from the latter two input systems and that from the inhibitory population onto the local excitatory population have been adjusted or are coming into the process of being adjusted to serve these two patterns and perhaps accommodate them to each other as discussed earlier.

Suppose the sepal input is simply a uniform bias across the inhibitory population, lowering its effective threshold for responses to its two loop-generated inputs. Then the effect of the septal input is to increase the local interactive inhibitory sculpting so as to sharpen the definition and increase the intensity of comparison and perhaps agreement between the two loop-generated inputs. If the septal input is biased in favor of the signal currently generated in the CA1 cells of the loop, that factor simply increases the effectiveness of this biasing action: the bias is favorably biased, so to speak. If the septal input is further biased in terms of selective coloring of its reward, aversion, or arousal inputs, then that coloring would be in turn reflected in the biasing passed on to the CA3 pattern definition. (The possibility cannot be ruled out that the arousal input to the septum reflects a more differentiated, attention like signal than the monolithic arousal concept connotes; such a differentiation could add a new dimension of pattern variability to the hippocampal system largely through this septal pathway, although perhaps also through higher levels of the input feeding initially into EC.)

At a clearer experimental level we know that activity in the medial septum is highly correlated with, and the likely driver of, the theta rhythm. We know further that the occurrence of the theta rhythm is associated with significant experiential situations, and with learning. This couples very directly with the present theory of the septum as a mediator of experiential significance and prompter of embedding.

The fact that firing of the individual inhibitory neurons are phase related to the theta rhythm suggests that the rhythm could be determined by the recovery time of these cells, in the same way that the alpha rhythm is thought by Andersen to be determined by the recovery time of the inhibitory cells in thalamic nuclei. However, experimental evidence does not seem clearly in support of this conjecture.

Breakdown of Hippocampal Integration in Schizophrenia

A considerable amount of experimental evidence from humans and rats has suggested that schizophrenia may be mediated by the disruption of the input from the medial septum into the hippocampus. The typical experimental paradigm for this is called *sensory gating*; it is based on the diminution of the so-called *P* wave. The *P* wave is a positive global electrical response to a brief stimulus generated in the CA3 hippocampus, clearly recordable with macroelectrodes in either humans (latency 50 msec) or rats (latency 50 msec).

According to this measure and subjective assessments, normal humans typically suppress the second of two paired stimuli to under 30 percent. Schizophrenics, however, suppress the second stimulus only to about 80 percent. Normal rats also suppress the second response. Rats whose septal input to the hippocampus is blocked show reduced but approximately equal responses to both stimuli. The following section provides a plausible interpretation of these experimental findings in terms of the general theory of hippocampal operations presented previously.

Theoretical Interpretation of Sensory Gating Breakdown in Schizophrenia

Suppose the *P* wave reflects the summation of local inhibitory synaptic currents in the population of CA3 pyramidal cells. According to the theory just presented, a significant new extrinsic stimulation provided to normal humans and normal rats will produce an arousal response in the BSRF and a new cognitive signal to the HML loop. Both these signals will drive the septum to produce synaptic inhibition onto the CA3 pyramids; hence, the P wave. A repeat of the same stimulation will not produce an arousal response in the BSRF because one is already aroused and oriented for this stimulus; it will also not produce a representational cognitive response in CA1 because CA1 is accommodated to this stimulus; hence, greatly reduced input to the septum, reduced input to CA3 inhibitory cells, and a diminished P wave, or sensory gating. In rats whose cholinergic septal output synapses are blocked with aBT, there is marked diminution of input to the local CA3 inhibitory cells; hence, diminished P wave response to both first and second stimuli.

In this picture, schizophrenics are seen as having a chronic deficit in septal driving of CA3 inhibitory cells. Suppose these cells compensate for this deficit by increasing their sensitivity, to respond more fully to feedforward and recurrent input from the dendate and CA3, respectively. (Disuse sensitization in neurons is well established by Sharpless and others.) In this case, any inputs, new or recycled, to the CA3, will produce similar levels of drive on the local inhibitory cells. All stimulation produces comparable levels of synaptic inhibition, and hence comparable P waves. The root of the phenomenon is that the significance and newness dimensions of input are diminished or lost by chronic blocking of the septal input to the CA3 inhibitory cells.

In this picture, the significant neurophysiological mechanisms of action include the composite inhibitory synaptic current applied to the CA3 pyramids, the biasing of inhibitory cell firing by septal input in normals, and the disuse sensitization of these cells in schizophrenics. The significant network operations include the fine-grained representation of cognitive information across hippocampal populations and the interactions of these across the constituent junctional synaptic matrices of the network. The important functions here are cognitive representation, arousal, reward, newness, and significance.

Closing Comments

The theoretical development presented in this chapter has brought us to the point where intelligently guided computer simulation studies and companion electrophysiological experimentation can be formulated and undertaken. It is instructive to bring out clearly the various levels of theoretical effort that have brought us to this point.

First, note that the complexity and richness of fine-grained detail necessary for a satisfactorily realistic representation of hippocampal operations clearly necessitate both large-scale realistic computer simulation and multiple-microelectrode investigation. Clear or reliable pictures of the processing and interactions of coordinated network firing patterns, and the operations of complex synaptic junctional matrices are impossible without such tools. They are sufficiently difficult even with them.

Second, the preceding discussion shows clearly the value of a specific theory of particular operations in a specific network for guiding the use of computer simulations. Computer simulations can easily be completely meaningless unless guided by intelligent theorizing. The view of this book is that computer simulations simply help clarify the interface between theory and experiment. The most deeply successful science depends on both. We should have theory guiding and experiment deciding.

Third, note how the specific theory of operations for the composite network developed in this chapter is dependent on both the general theoretical mechanics of interacting networks developed in Chapters 13 through 17 and on the general theoretical view of the functional systemic organization of the hippocampal circuits presented in the first section of the chapter and discussed in broader terms in Chapter 21.

It may be that any of these levels of theorizing in this particular example is a little off center. However, none of them is likely to be far off, and the continued cultivation of the combined theoretical–computational–experimental approach advocated here will show up the shortcomings and suggest how to adjust for them.

Chapter 19 Composite Cortical Networks as Systems of Multimodal Oscillators

This chapter presents a tentative speculative theory for the functional organization of regions of cerebral cortex based on the idea that such regions consist of a hierarchical arrangement of interacting local neural networks. As discussed in the next chapter, this modular view of cortical organization is not new in itself. What is new is the explicit and biologically based theoretical description of the patterns and mechanics of neuroelectric signalling in the composite system. This explicit description allows us to formulate a view of how information is represented in the overall composite system (as well as in the individual local networks) and to define clearly the significance of the confluent junctional matrices in mediating the requisite interactions of competing or compatible sequential configurations.

A primary and larger significance of this chapter resides in the broad thrust of the overall theoretical mechanics developed in this book, as indicated in previous chapters. Nonetheless, the specific theory of the cortex presented here is plausible and of interest in its own right. The area is extremely difficult, as is discussed in Chapter 20. Satisfactory understanding seems decades away. Theorizing at this level, although considerably more speculative than at lower levels, is more sorely needed to help form an interface between neuropsychological theories of cortical function with neurobiological experimentation. This theory, for example, suggests new types of experimental approaches, particularly in the area of multiple microelectrode studies.

A Theoretical View of Functional Organization in Composite Networks of the Cerebral Cortex

The sequential configuration theory for firing patterns in local cortical networks suggests various overall views of the operations of local networks. For example, one is led by the sequential configuration theory to see local cortical networks as signalling the activation of one or another dynamic state according to which particular sequential configuration is being mediated by its internal reverberations over a particular time period. In this sense, one may conceive of a unitary local cortical network as an "oscillator" and its repertoire of dynamic states (sequential configurations) as a collection of possible "modes." A local cortical network can thus be conceived as a multimodal oscillator, each possible mode of oscillation corresponding directly to a given specific sequential configuration.

The suggestion follows immediately that some larger order regions of cortex might operate functionally in terms of an interacting group of local networks, each seen as such a multimodal oscillator. In this view, the dynamic configuration of a composite cortical area are determined by, and represented and interpreted in terms of the set of particular sequential configurations existing within its constituent local networks and their interactions and equilibrium configuration.

These composite regions could be thought to represent an entire classical association area, or classical primary or secondary sensory region, or perhaps some significant fraction of such regions. A posterior association area for example would contain about 10,000 local networks, each of which would contain about 30,000 cells.

The overall configuration of such a composite network would be a broadly integrated result of the entire system of local networks, mediated by extensive lateral intracortical connections, both excitatory and inhibitory and both proximal and distal. These lateral connections would operate in conjunction with the locally mediated recurrent connections and any active input patterns to bring about the particular overall configurations and determine their quality, cohesion, and tenacity.

The remainder of this chapter spells out a theoretical mechanics for a particular realization of this theory. The theory suggests that two main laterally mediated intracortical activation systems determine the systemic organizational properties of the composite system. Fine-grained neurophysiological interactions at the synaptic junctional matrices between interconnected local networks are the key substrate of the fine-grained mechanics of the theory.

This theoretical view suggests dynamical and organizational properties consistent with the implications of experimental observations and generalizations reviewed in the next chapter that even elemental psychological units in cognition are likely to be represented diffusely across composite regions of cerebral cortex rather than tightly within local networks.

The Mechanics of Interacting Multimodal Oscillators

States in a Composite Cortical Network

Suppose that a composite region of cerebral cortex (say a posterior association area) consists of N local recurrent networks as indicated in Figure 19.1. Each local network can go into any of, say, M, dynamic modes, each of which is a particular sequential configuration firing pattern.

The state of the entire composite network at any given time is definable in terms of the collection of individual dynamic modes of its constituent local networks. A particular state of the composite system would be a particular set of N modes, each describing the current sequential configuration of one local network. The number of such particular microstates allowable to the system would be a very large number, $(M)^N$. Only some much smaller subset of the total states, however, will be stable because the local networks will interact among themselves to drive the system toward certain equilibrium configurations determined by the intrinsic connections of the network and its current extrinsic stimulation patterns. The nature of these equilibrium configurations as determined by a computational model are discussed here.

The first view of the theory, then, is that the overall state of the composite net is represented by the equilibrium distribution of sequential configurations in its constituent local networks. An immediate generalization of this concept follows from recognizing that individual sequential configurations

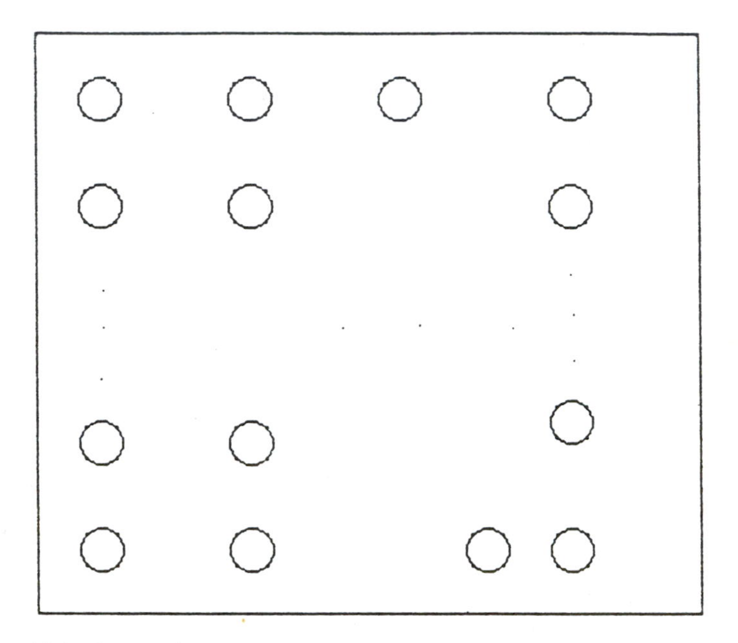

FIGURE 19.1. Composite cortical network as a system of recurrently active local networks.

may be more or less compatible with each other. For example, two sequential configurations may be more compatible if they share some significant number of neurons or exhibit more corresponding temporal similarity of shared neurons. One may picture sequential configurations as falling into any number of higher order groupings determined by this kind of dynamic compatibility. Although the present theory does not explicitly commit itself as to what level of cognitive element corresponds to a sequential configuration in a local network of association cortex, the analogy to compatibility among cognitive elements is apt. Thus, the elements dog, cat, lion, and so on stand in a higher order compatibility with each other than with bicycle, coat, telephone, ball, and so forth. This higher order compatibility could be called *animal,* for example.

Since this kind of compatibility is built into the synaptic structure and overlapping of the sequential configurations, it follows that the collective occurrence of distinct but highly compatible sequential configurations distributed across a number of individual local networks can be interpreted as a higher order event corresponding to this compatibility in the same way that "animal" corresponds to dog, cat, lion, and so on. This significantly generalizes the interpretation of state for the composite network.

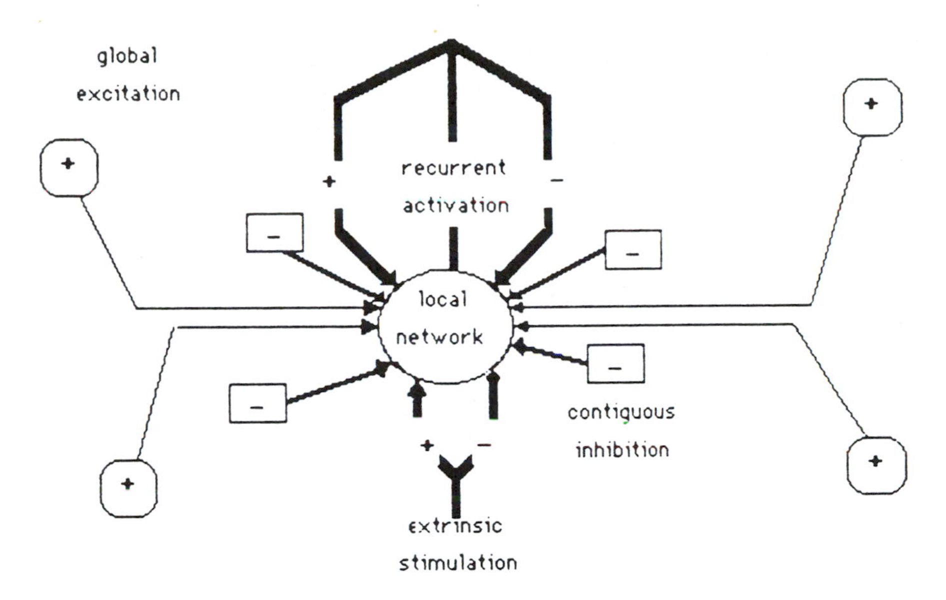

FIGURE 19.2. Four activation systems for driving the sequential configurations of local networks.

Interconnections in the Composite Cortical Network

The dynamic state of an individual local network is influenced by four main activating systems* as shown in Figure 19.2. First, there is extrinsic input directed specifically to the local network in question; this extrinsic input may be signalling supportively for one or another of the patterns in the repertoire of the local network or sending in a disruptive activity pattern. Second, the internal recurrent connections of the network are returning the current pattern of the local network back on itself; these connections are selectively promoting the continued unfolding of that pattern. Both of these extrinsic and recurrent projection systems have both excitatory and inhibitory components.

The last two activating systems are laterally mediated cortico-cortical connections. The third system consists of a massive system of only excitatory input derived from other local networks within the composite network; these projections arise more or less uniformly from throughout the entire composite network. We will call this system the *global excitatory system*, and assume it is mediated by myelinated fibers passing laterally near the basal surface of the cortex. The fourth activating system is a collection of

* The local and lateral interconnections used here are based on descriptions provided by Abeles.

inhibitory input fibers derived from the cortical regions very close to the local network in question; we will call these *contiguous* inputs and assume they are mediated from contiguous local networks by short inhibitory fibers crossing near the apical surface of the cortex.

Modal Interactions at Confluent Synaptic Junctional Matrices

It should be clear that the interactions between local neural networks are extremely intricate, complicated, and vast events. We are picturing a given local recurrently connected network as being bombarded simultaneously by multiple highly ordered spatiotemporal patterns of activity called *sequential configurations* across millions of receiving synapses. Any one local network consists of about 30,000 neurons; each of these neurons is activated by from 10,000 to 100,000 input synapses. The local receiving networks are pictured as thrown into one or another of their own repertoire of coordinated firing patterns depending on the temporal and spatial intermeshing of these structured excitatory and inhibitory inputs and their relations to the patterns of relative synaptic strengths of the recurrent synapses of the local network.

We view this interpretation of activity at confluent synaptic junctional matrices as one of the significant predictions of this theory and one of the cornerstones of the theoretical mechanics of composite networks developed in this section.

A Simple Computational Model of Cortical Nets as Interacting Oscillators

The following sections describe a computational model and simulations of this theory. A listing of the simulation program, smmo02, is available from the author.

State Variable and Vector Descriptions of MultiModal Oscillations

The primary dynamic variables of this theory are not neuronal generator potentials, nor spike trains, but rather the higher order concept of sequential configuration or "mode." Therefore, the conceptual and mathematical descriptions of its behavior and mechanics are best carried out in terms of vector state variables, each of which contains information regarding the array of modes available to the individual local networks of the system. Specifically, consider the state variable representation of excitatory and inhibitory influences shown in Figure 19.3.

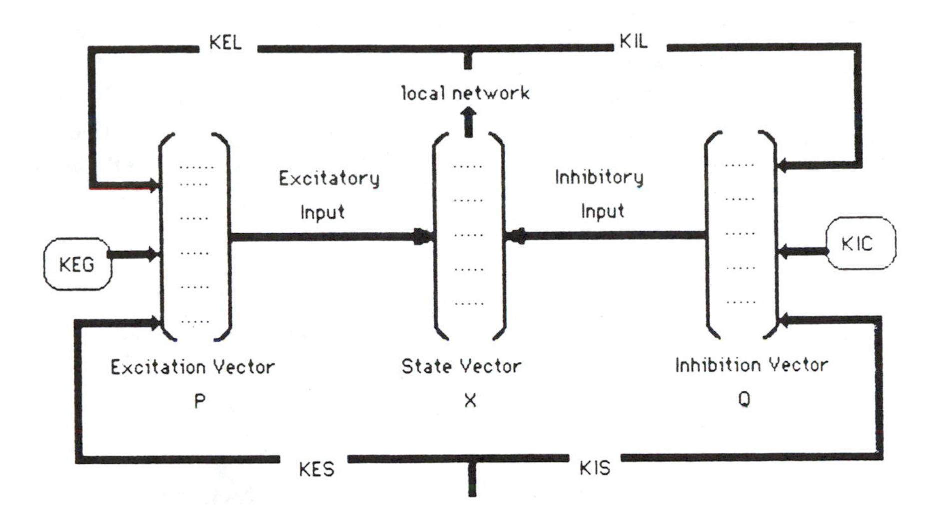

FIGURE 19.3. State vector and activation vectors to house mode-specific dynamics for a given local network.

We will imagine that a given local network makes a decision as to what mode of activity to go into on the basis of an internal state variable vector, X_j. The subscript j goes from 1 to M to represent all the modes available to the network. The values of X_j for each jth mode measure the driving influence for each jth mode in this local network at this particular time.

The influences which determine the X_j values are represented in two additional vectors, P_j and Q_j; these represent the net excitatory and inhibitory influences, respectively, directed to each of the j modes. That is, P_{j^*} represents the excitatory drive in support of the j^*th mode, and Q_{j^*} represents the inhibition supplied in support of the j^*th mode.

We imagine that these influences are carried into the P and Q vectors from the four activating systems as indicated in Figures 19.2 and 19.3. When a given local network is in mode j^* that network sends supportive or excitatory influences for that mode directly onto the P_{j^*} components of the excitatory P vector for every local network it influences. These include an increment of KEL recurrently back onto itself, and increments of KEG conducted globally to all other local networks in the net. Similarly, this same sending network projects inhibitory influences in increments of KIC in support of the j^*th mode in of the Q vectors in all the local networks to which it projects inhibition. Here we take this to be its four vertical and horizontal neighbors.

Similarly, excitatory and inhibitory influences are distributed to P and Q from the extrinsic stimulation, excitatory influences are distributed to P

from all networks in the system, and inhibitory influences are distributed to Q from all contiguous networks.

From the standpoint of a given receiving local network, P takes excitatory vector increments of KES from extrinsic stimulation, KEL from local recurrent connections, and KEG from global connections. Q takes inhibitory vector increments of KIS from extrinsic stimulation, KIL from local recurrent connections, and KIC from connections from contiguous networks.

The six parameters, KES, KEL, KEG, KIS, KIL, and KIC determine the relative weighting factors of these six extrinsic, recurrent, and lateral influences.

Mechanics of Mode Selection

The excitatory and inhibitory influences directed toward each mode for each local network are thus represented in the P and Q vectors. These influences are directed onto the X state variable vector for each of its j modes as indicated in Figure 19.4 and Eq. (19.1).

$$X_j = \sum_{i=1}^{M} \epsilon(i,j)P(i)/(1 + \mathrm{KIE}^* \sum_{i=1}^{M}(1 - \epsilon(i,j))Q(i)) \qquad (19.1)$$

The local network is determined to be quiescent, or in disorganized activity, or in a particular dynamic mode depending on this input activation. The values of X_{j^*} for every j^* are driven upward directly by values of its P input, and downward for the values of all the Q input. The factor, $\epsilon(i,j)$ represents

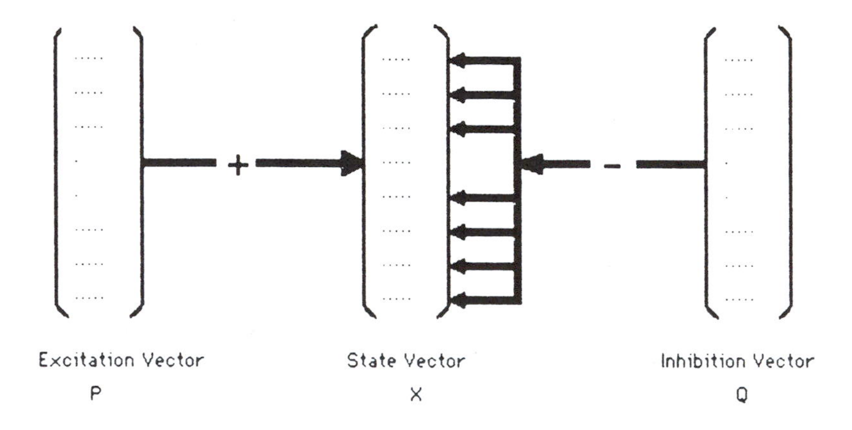

FIGURE 19.4. Manner of activation of state vector X by activation vectors P and Q.

the compatibility of the i and j modes. That is, if mode i^* has no compatibility with mode j^*, then $\epsilon(i^*, j^*) = 0$, and from Eq. (19.1) we see that the excitation in support of i^*, P_{i^*} gives 0 drive for X_{j^*}, while the inhibition in support of i^*, Q_{i^*}, gives maximum inhibition, Q_{i^*}, for X_{j^*}. For maximally compatible modes, $\epsilon(i^*, j^*) = 1$, and these effects are just reversed. Intermediate values of $\epsilon(i^*, j^*)$ reflect intermediate levels of compatibility and provide intermediate graded levels of supporting excitation and inhibition from i^* to j^*. That is, the values of P and Q excite and inhibit the X value of all modes according to the compatibility matrix ϵ. KIE is a parameter that defines the relative weighting of inhibition relative to excitation.

By means of Eq. (19.1), relative driving influences are obtained for every one of the M modes for each local network. These values are then used to determine the mode into which a given network goes as follows. Each of the values of X_j is compared to its corresponding value in a "threshold" vector, θ_j. The number of modes that have driving influences, X_j greater than threshold, θ_j, is labeled n_f. If n_f is 0, that is, no modes have drive exceeding their thresholds, then the output mode is 0; that is, the local network is quiescent. If n_f is equal to 1, that is, only one mode had drive exceeding its threshold, then the network mode is that mode. If n_f is greater than 1 but less than a parameter, THD, then the output mode is that suprathreshold mode which has the highest value of X_j. If n_f is greater than THD, then the output mode is 2, disorganized activity. THD is thus a threshold for disorganized activity. A parameter, KID, is used to weight the inhibitory influence of disorganized as compared to organized activity.

Computer Simulation

The results that follow are obtained from a computer program smmo02 constructed to simulate the interactions determined by these mechanisms. The composite system is given an intial distribution of mode values across its constituent local networks and is activated with a user-controlled extrinsic bombardment of excitatory input. This input may be either random or patterned and consists of the specification of a particular input mode for each network at each time. The local, contiguous, and global interactions among the local networks described previously then determine the progression of the overall distribution of mode activity for the entire composite network. In the simulations described here, we used mostly random mode stimulation to study the overall properties of this theoretical system and the influence of its constituent parameters.

```
KES,KEL,KEG,KIS,KIL,KIC,KID,KIE,DTH 10.00  8.00  0.03  4.00  1.00  0.20  1.10  4.00
MCM    1.000 0.000 0.000 0.000 0.000 0.000 0.000 0.000 0.000
MCM    0.000 1.000 0.000 0.000 0.000 0.000 0.000 0.000 0.000
MCM    0.000 0.000 1.000 0.000 0.000 0.000 0.000 0.000 0.000
MCM    0.000 0.000 0.000 1.000 0.000 0.000 0.000 0.000 0.000
MCM    0.000 0.000 0.000 0.000 1.000 0.000 0.000 0.000 0.000
MCM    0.000 0.000 0.000 0.000 0.000 1.000 0.000 0.000 0.000
MCM    0.000 0.000 0.000 0.000 0.000 0.000 1.000 0.000 0.000
MCM    0.000 0.000 0.000 0.000 0.000 0.000 0.000 1.000 0.000
MCM    0.000 0.000 0.000 0.000 0.000 0.000 0.000 0.000 1.000

N,NMC,NMR,NC,MM,LTSTOP:   900   30   30    4    9   12
THs:  2.00  2.00  2.00  2.00  2.00  2.00  2.00  2.00  2.00
PD,LSTRTD,LSTPD,INSEDs,NSPFS: 0.5000    0   999  5592  1227  1312    2
MSTM,LSTIM,INTERV,LSTP,NSTIMs:   3    1    1  999    9    9
MSTM,LSTIM,INTERV,LSTP,NSTIMs:   6    1    1  999    9    9

T,NM=   1   569    0   60   41   44   51   43   43   49
T,NM=   2   427    0   71   58   66   81   69   59   69
T,NM=   3   359   13   77   63   78   98   71   61   80
T,NM=   4   353    3   89   65   71  104   75   59   81
T,NM=   5   374    4   78   59   70   98   83   54   80
T,NM=   6   350    6   95   55   68  104   82   56   84
T,NM=   7   348    1   96   55   81  108   73   56   82
T,NM=   8   325    2  112   58   72  125   70   51   85
T,NM=   9   327    3  112   53   82  129   67   45   82
T,NM=  10   324    3  115   55   77  119   68   54   85
T,NM=  11   342    1  120   47   78  113   62   55   82
T,NM=  12   337    8  124   52   78  108   63   48   82
```

FIGURE 19.5. Spatial clustering of incompatible modes.

Results are presented first for the case of uniformly incompatible modes (the matrix $\epsilon(i, j)$ taken as a diagonal matrix). Second the basic characteristics of compatible traces are described. All cases utilized a system of 900 interacting networks each capable of resonating in seven distinct dynamic modes (labeled 3 through 9) or in disorganized activity (labeled 2). Quiescence is labeled by the number 1.

Behavior of the System with Uniformly Incompatible Modes

The primary operational behavior of this system is to progressively modify its spatial distribution of active modes across the individual local networks to attain globally stable equilibrium patterns. It does this by virtue of its local recurrent and systemic lateral internal activating systems. Therefore understanding of the model behavior depends on understanding what kinds

```
T,NM=  12   337    8  124   52   78  108   63   48   82

773851513163187159187115531115
168711991311611275535938513551
369351916421116918181137593357
114816914131131996615111644119
978541913132111116611561519917
613617169193769961875116181136
116187175135119932149533611811
195973411946133191451933755814
461133413111317191114111318171
613341113641791613171161166119
398163114814113613876166117511
115771819989193411611613746169
155573119167333113133788761613
951519111916273331331669511313 1
611574168169333315664911541599
361461955116533346116831115378
831173835511616141616111993114
184198581113311957124833166191
811791559141196514431318531131
191491177135163311166173611616
417116879133661811351317314731
171996111636148551611361419171
611698316668941614851467779865
891111771115166687836117149711
895333314115166677155971633199
133391591411166656115775119111
791196115513331111319165811414
611141711179165934116516315593
133661199813181141351641911114
341115732673183233313161365554
```

FIGURE 19.5. Continued.

of distributions occur, and how these relate to the main activation systems and their parameters.

The typical output behavior of the model is illustrated in Figure 19.5. This example shows a commonly occuring case where the equilibrium distribution consists of distinct spatially grouped clusters of distinct modes. Another common equilibrium configuration is where one or another individual modes pervades all constituent networks.

Influences of the Parameters

The broad features of the dynamic behavior of this model can be interpreted in terms of two main systemic influences and two local influences. The first systemic influence is that the global excitatory activations tend to produce a homogeneous distribution wherein a single dominant mode occupies all the constituent local networks. This process is dominated by that mode (or modes) initially represented most numerously among the local networks. Therefore, this influence can be characterized as a global democratic action, much like a voting process across the entire system. This influence produces effects ubiquitously throughout the system, in the manner of "action at a distance." The strength of this influence is directly represented by the parameter KEG.

The second systemic influence is the tendency of the contiguously distributed inhibition to encourage the production of regional clusters of local networks all exhibiting the same mode. This effect occurs as a result of the convergence of the contiguous inhibitory influences within a given neighboring region. The effect favors those modes active in the region relative to modes active distally. It can be interpreted as a tendency for a region of contiguous units to produce and maintain its own individuality. It does this in favor of modes represented simultaneously among numerous of the contiguous units. This system produces effects locally contiguous to the instantaneous boundary of the region. The strength of this influence is represented by the parameter KIC.

The first local influence is tendency of the local recurrent activations, both excitatory and inhibitory, to maintain the active modes in each of the local networks. This system thus acts to maintain any existing distribution of modes; it may be interpreted as a tenacity for maintaining the status quo. The strength of this influence is represented by the parameters KEL (tending to actively favor the regeneration of existing modes) and KIL tending to actively discourage the rise of competing modes).

The second local force is the extrinsic input system which may drive individual local networks toward any particular mode currently in its repertoire. This system is thus an independent extrinsic force, imposed from outside onto the system. The strength of this influence is represented by the supportive and inhibiting parameters, KES and KIS, respectively.

The behavior of the system is best conceptualized broadly in terms of its systemic operations. Thus, for a given typical level of local tenacity influences, and neglecting overpowering extrinsic stimulation, the main dynamic features of the system can be displayed in the parameter space shown in Figure 19.6. The strength of the global democratic voting system is shown on the horizontal axis, and the strength of the regional individuality system is on the vertical axis.

Five characteristic domains of behavior are exhibited by the model indicated in Figure 19.6. The first corresponds to the region where both

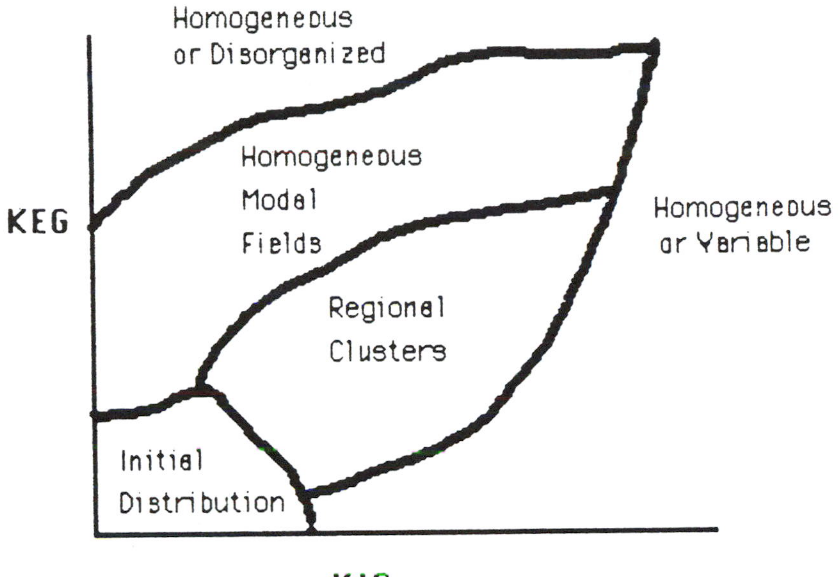

FIGURE 19.6. Parameter space map of final configurations.

systemic influences (KEG and KIC) are small. Here local network activations dominate. The tendency is for no change in distribution of modal activity.

If regional inhibitory influences (KIC) are kept small, but the global excitatory influences (KEG) are increased, the tendency for the convergence to homogeneous distribution of a dominant mode appears. If this global excitatory influence is increased to very high levels, so that many competing modes are driven above their thresholds in many local networks, then the model produces homogeneous fields or disorganized activity.

For any level of global excitatory activity, it is possible to increase the contiguous inhibition high enough that uniform quiescence, uniformal unimodal distributions, or other variations occur, depending on the particular initial and stimulating conditions.

Contiguous Inhibition, Clustering, and Boundary Effects

For more moderate values of both systemic influences, an interesting domain is defined where regional clusters occur. The mechanism of origin of these clusters is shown in Figure 19.7. This figure shows an illustrative example wherein a region of nine contiguous local networks initially exhibit

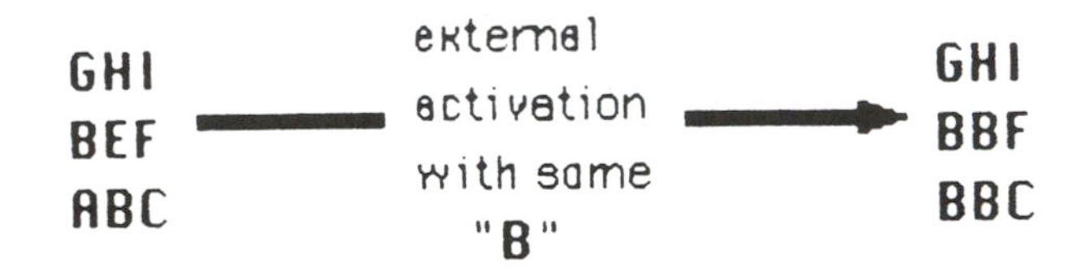

FIGURE 19.7. Cluster definition by contiguous inhibition.

eight distinct modes and are surrounded by quiescent networks. If there is some level of extrinsic or global sporadic excitatory bombardment of the various modes, there is an enhanced tendency for the modes *A* and *E* to be replaced by *B*'s as shown in Figure 19.7 because the convergent inhibition is driven at lower levels for *B* than for the other modes. Depending on the particular distributions defined for "contiguous," such clusterings may continue to spread step by step from their instantaneous boundaries.

Interactions of competing clusters and the corresponding deterioration of clusters also tend to be effected primarily at the boundaries of the clusters. Suppose that one or more clusters are active at a given time in a given spatial distribution, and one of these in mode *B* as shown in Figure 19.8. The local networks internal to the cluster—that is, those not on its boundary—are receiving multiple convergence of contiguous inhibition that selectively supports *B* by strongly inhibiting all other modes. The local networks of the cluster on its boundary, however, are supported to lesser degrees of this convergence. These same networks correspondingly receive some contiguous inhibition favoring those modes externally contiguous to the region. They are therefore more vulnerable to competing influences than the internal networks.

The fundamental points regarding these local domains, then, is that, if the balance of KIC to KEG is sufficiently favorable, there are strong pervasive tendencies for the emergence of regionally homogeneous clusters of activity to emerge, and these can be very tenacious within the overall distribution of

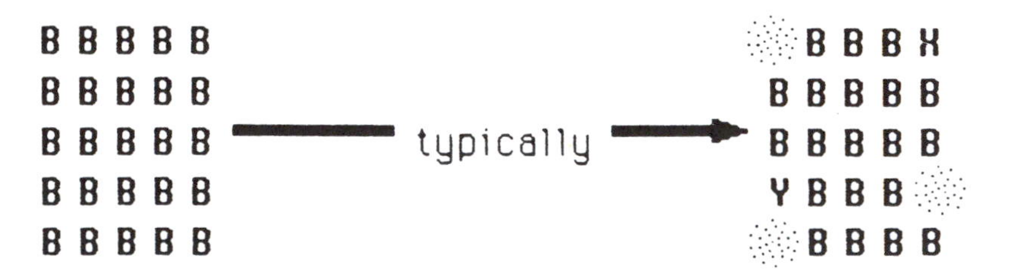

FIGURE 19.8. Erosion of clusters at boundaries by more weakly resisted external global and contiguous influences.

activity in the system. These regions tend to grow and decay around their boundaries, in response to extrinsic and global excitations that may be delivered sporadically and globally.

Behavior of the System with Compatible Modes

When various degrees of compatibility among modes are used in the model, interesting varieties of behavior occur, including higher-order grouping of mutually compatible modes and intermode competition within these higher order groupings. Figure 19.9, for example, shows a case where two higher order compatible groupings, 3–4–5 and 6–7–8, attain a shared quasi-stable equilibrium distribution.

Equilibrium Configurations as Dependent on Degree of Compatibility

More generally, the variety of behavior obtained for this situation is shown in the parameter space of degree of compatibility, ϵ, versus level of random background bombardment as given in Figure 19.10. At very low levels of these parameters (upper left corner), activity is distributed more or less uniformly across the individual modes, reflecting the input more or less directly with minimal influences from interactions. At intermediate levels of these parameters, equilibrium configurations are attained wherein a single mode generally dominates the entire field. At higher levels of these parameters (lower right corner), one or another of the higher order groupings often occurs. The example shown in Figure 19.9 is a rather unusual stable instance of situations, which more often pass transiently when activated with only random stimulation.

Patterned Stimulation

When a network is in a state driven by random stimulation such as those mapped in Figure 19.10, very small additional but structured input can throw the system into the higher order compatible groupings such as shown in Figure 19.9. For example, stimulating nine cells (1%) with 7's against a random background that is biased toward 3's (the position $p = .1$, and $\epsilon = .5$ in Figure 19.10) produces a pervasive dominant standoff of the 3–4–5 and 6–7–8 groupings similar to that shown in Figure 19.9.

However, when stronger levels of stimulation are provided in support of one or another of the higher groupings, so that this grouping tends to

```
KES,KEL,KEG,KIS,KIL,KIC,KID,KIE,DTH 10.00  4.00  0.05  4.00  1.00  0.20  1.10  4.00
MCM   1.000 0.000 0.000 0.000 0.000 0.000 0.000 0.000 0.000
MCM   0.000 1.000 0.000 0.000 0.000 0.000 0.000 0.000 0.000
MCM   0.000 0.000 1.000 0.500 0.500 0.000 0.000 0.000 0.000
MCM   0.000 0.000 0.500 1.000 0.500 0.000 0.000 0.000 0.000
MCM   0.000 0.000 0.500 0.500 1.000 0.000 0.000 0.000 0.000
MCM   0.000 0.000 0.000 0.000 0.000 1.000 0.500 0.500 0.000
MCM   0.000 0.000 0.000 0.000 0.000 0.500 1.000 0.500 0.000
MCM   0.000 0.000 0.000 0.000 0.000 0.500 0.500 1.000 0.000
MCM   0.000 0.000 0.000 0.000 0.000 0.000 0.000 0.000 1.000

N,NMC,NMR,NC,MM,LTSTOP:  900   30   30   4   9   15
THs:  2.00  2.00  2.00  2.00  2.00  2.00  2.00  2.00  2.00
PD,LSTRTD,LSTPD,INSEDs,NSPFS: 0.2000   0  999  5592  1227  1312   0

T,NM=   1  767    0   21   18   21   16   20   16   21
T,NM=   2  671    0   32   32   34   31   35   32   33
T,NM=   3  247  173   71   69   77   62   73   84   44
T,NM=   4   70   94  108   98  133   86  115  166   30
T,NM=   5   42  124  114   94  141   94  121  165    5
T,NM=   6   19  137  128   86  141   95  128  163    3
T,NM=   7   21  120  133   88  142  104  128  163    1
T,NM=   8   16  114  136   95  133  109  131  165    1
T,NM=   9   20  115  130   96  133  113  131  162    0
T,NM=  10   19  114  128   98  132  111  135  163    0
T,NM=  11   16  115  137   99  134  113  129  155    2
T,NM=  12   22  110  137  102  130  116  128  155    0
T,NM=  13   25  112  141   98  126  114  131  153    0
T,NM=  14   19  117  133  102  126  113  138  150    2
T,NM=  15   27  108  138  105  124  113  136  148    1
```

FIGURE 19.9. Selective compatability among modes: equilibrium balance between 3–4–5 and 6–7–8 systems.

dominate, usually one or another of the constituent modes of this grouping rather than the grouping itself, tends to dominate the entire field, unless the p and ϵ values are quite high. In these instances, the entire higher order groupings are nonetheless implicitly potentiated by the dominant activity of one of their members because of the intrinsic compatibility of the activity patterns.

Relation of Theoretical Local Networks to Cortical Anatomy

It has not proved possible to identify structurally distinct local networks in the cortex. Anatomically, the cortical networks appear homogeneously interpenetrating. However, functional columnar modules with a diameter about equal to the dendritic spread of single pyramidal cells (ca. 700 μ) have

```
T,NM=  4   70   94  108   98  133   86  115  166   30

347565477225888237223741644433
445552491441578815453344774853
885511266655554841355347777655
883836246621724918855336774653
883332333719128338245133388618
441516331771541333243339382428
948786675777331319555825554453
198243331266372817255182522227
448127381772877868841817988257
518877888664477882846717828556
566678888722298818786678887285
865638889988242289116683817983
886691478326453383776888872677
886615578316276191788688853777
888824538192277855191778423348
382911555663756845214788277348
889726755163327755888117591443
785288483866367716685547776672
825166988277786614877547122777
149482423497383188555544866948
377554453421588888865544562626
577166325528867778624444551665
211925832555847782334917486765
377265544547139882335477445335
678223261556621682515441885224
778833347826655773866788889782
438223317388235213952728855537
363553332884253344551453355519
333555558888655444554455445378
238885522882234343254256665544
```

FIGURE 19.9. Continued.

been definable in primary sensory areas on electrophysiological grounds. Figure 19.11 shows an estimate of the cellular and connectivity characteristics of such a cortical module, which is taken here as representative of the local networks of the present theory.

Plausibility of the Multimodal Oscillator Theory for Composite Cortical Networks

Given the experimental support of the modular interpretation of organization in cortical sensory receiving areas, this theory is a plausible view of the operations of composite cortical networks. However, it is far from an established or finished view. Its primary value is to help push

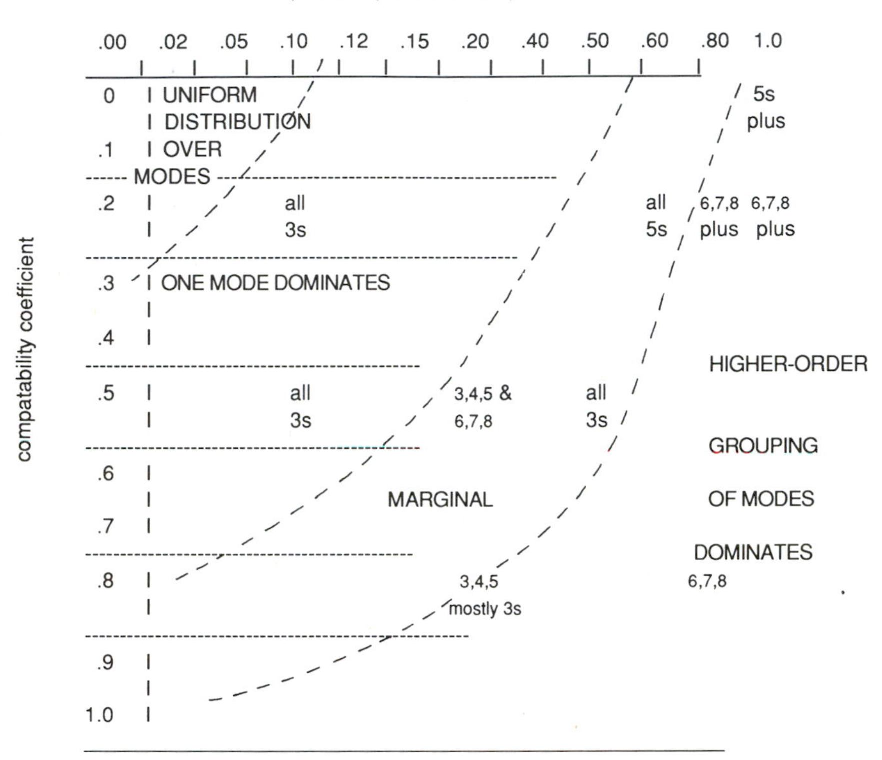

FIGURE 19.10. Map of final configurations by degree of compatability and level of random bombardment.

forward new lines of questioning and experimentation over this very difficult area.

The essential functional view of the theory is that cortical cognitive representations are made in terms of interrelated mutually supportive dynamic patterns among variable sets of local neural networks distributed diffusely through wide regions of cortex. This view suggests that experiments should be addressed to searching for such coordinated firing patterns among composite cortical networks with multiple microelectrodes.

Further, the composite patterns are seen as produced by the essential structural ingredients of the theory that are sequential configurations as coordinated unitary dynamic signals in local networks and synaptic junctions as carriers of such signals and determiners of their interactions. This view suggests the simultaneous use of several banks of microelectrodes to search for cortical patterns. Correlations from one bank to another would

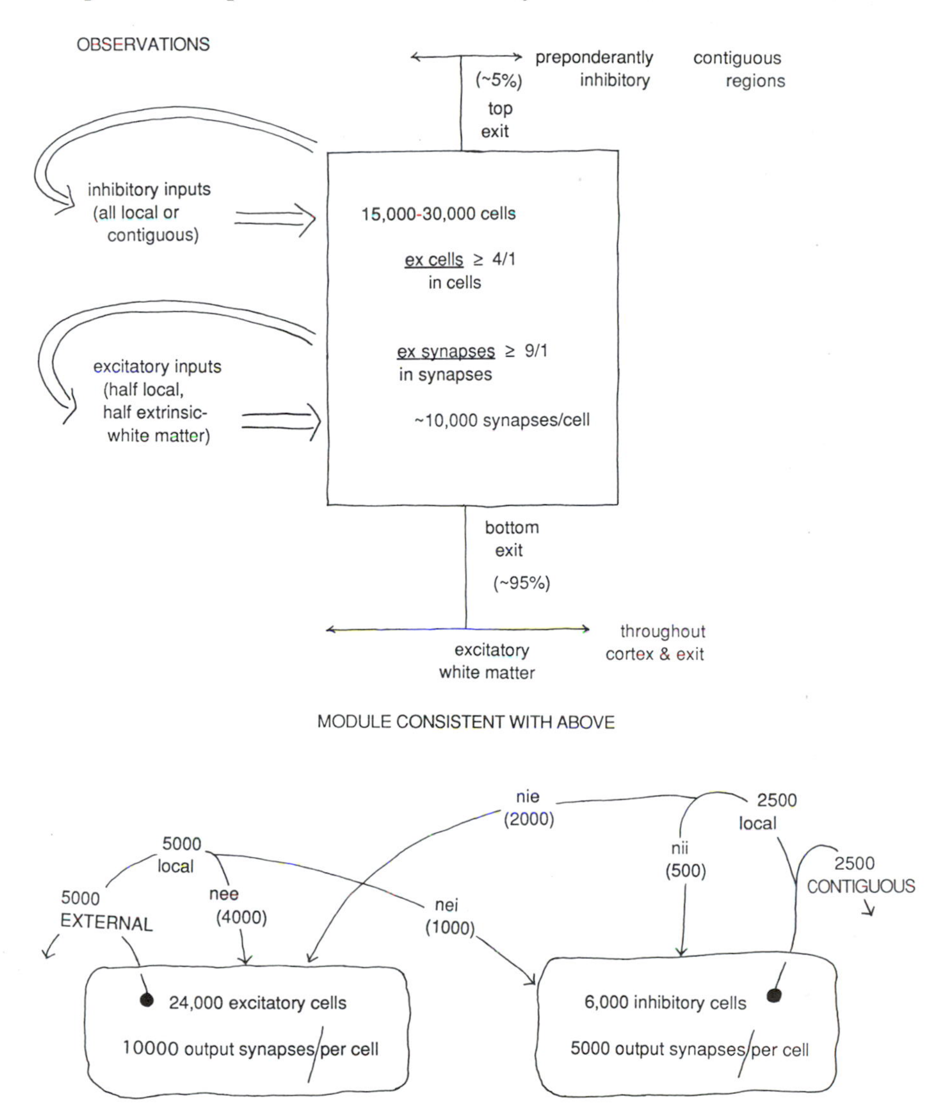

FIGURE 19.11. Estimates of connections for a local cortical network.

indicate coordinated firing between local networks; correlations within individual banks would reflect the sequential configurations within the local networks.

The next chapter discusses the possible relationships between the neuroelectrical patterns of this theory and cognitive representations, and provides a general overview of experimental approaches to cortical organization.

Chapter 20 Contextual Overview of Organization in Cortical Networks

This chapter discusses experimental approaches to cortical function, and principles, and then theories of cerebral organization, to provide a context for a speculative interpretation of the local and composite firing patterns considered in the previous chapter. These four topics, respectively, make up the four sections of the chapter. The next chapter includes a brief discussion of the systemic organization of nervous systems in broader terms, including the cortex.

The Experimental Context of Theories of Cortical Function

Reliable knowledge of brain operations have depended almost entirely on experimental and clinical observations. So far the complexity of the system has prevented the emergence of reliable broad theories. Only now are we approaching the point where sufficiently rich and varied amounts of observations have been accumulated to allow the formulation of tentative but meaningful broad neurobiological-based theories. In this context it is instructive to consider the experimental approaches that have and continue to both produce and constrain our understanding.

It is primary here that cognition is almost uniquely a human quality, whereas the fine-grained experimentation required to investigate it may be performed, with only very few exceptions, only on subhuman animals. This is a primary limitation and one of a number of ingredients that points out the need for penetrating neurobiological theory at this level.

Studies on Human Cerebral Cortex

Noninvasive Studies. Noninvasive and generally rather global observations can be and have been extensively performed on human subjects over much of this century. The main classes of observations include electroencephalography (EEG), the positron emission topography (PET), regional cerebral blood flow (rCBF), magnetic resonance imaging (MRI), and many clinical observations associated with brain-damaged individuals or brain surgery, some of these latter involving some small amounts of direct electrical stimulation or recording. Recently, some smaller amount of this type of information has also become available from magnetoencephalography (MEG).

EEGs record combined signals reflecting mostly large synaptic current loops simultaneously from very large numbers of synapses and neurons. They can not discriminate fine-grained patterns involved in information processing. They can and do show conditions of synchronous activity in large numbers of cells, in normal supposedly "idling" states or in certain clinical syndromes. They can also indicate in a general way the spatial distribution of activity across the cortex. They cannot discriminate fine-grained features of information processing but simply indicate by low-voltage, high-frequency signals that information processing of some type seems to be occurring. They can be used to show, however, the aggregate of activity in a given area in response to sensory stimulation. This type of EEG recording is called an *evoked response*. Such a recording shows several individual waves which are assumed to reflect aggregate behavior of different neuron and fiber populations in the region of the recording.

The PET and rCBF activity both measure the spatial distribution of nutritional or metabolic activity in the brain. When a given local region of the brain is particularly active that region utilizes more energy and therefore requires higher blood flow than normal. The rCBF measures this directly, and the PET measures it indirectly by measuring radioactive isotopes ingested by the subject. These techniques are very useful in showing what relatively local regions of the brain are highly active when subjects do particular activities such as read, remember, imagine a tune, do mathematics, think of a family member, and so on. A primary finding here is that most of these and similar activities involve selectively a number of local regions across the cortex.

MRI is based on sensing the radiation waves released by rotating tissue molecules in response to applied static magnetic fields. It is used primarily to distinguish different body tissues on the basis of their different chemical composition. It is particularly useful in clinical diagnoses.

Invasive Studies. Sometimes exploratory electrical stimulation and recording in brain regions can be performed during the course of surgery. This occurred particularly in the past before surgery for things like epilepsy and some forms of schizophrenia was replaced by drug treatment. These few cases have generally tended to confirm ideas about localization of function derived from animal studies and in some cases have provided very valuable additional information. The salient example of this type of information was provided by Penfield operating on patients with temporal lobe epilepsy in the 1930s.

Penfield believed that the responses of a large number of his patients to electrical stimulation in the parahippocampal regions of the temporal lobe showed the existence of a personal life history memory bank either stored in or accessed by this region. Everything an individual has attended to seems to be stored in this episodic, temporally arranged memory bank. These memories when stimulated carry an experiential flavor quite distinct from the more objective, structured memories of the type associated with cognition and assumed to be associated with cerebral cortex. Penfield's work has been criticized, but the basic concept seems to be solidly established by his efforts.

Another classical example of information obtained regarding the localization of function associated with surgical procedures arises from a small number of unfortunate subjects whose hippocampi were disconnected from surrounding tissue on both sides of the brain. These subjects were rendered incapable of further learning. They were unable to lay down new memories into either their episodic or their structured memory banks.

A second source of clinical information about brain organization derives from studies of brain-damaged subjects. After a century or so of relatively progressively systematic and widespread converging of such reports in the clinical literature, many common features and generalizations emerge as well as a large number of apparent contradictions and uncertainties. A representative and particularly comprehensive collection of this class of observations has been reported and theoretically integrated by the Russian neuropsychologist Luria. Many of Luria's findings are based on Russian soldiers experiencing brain damage in the two world wars.

Luria's inferences from these observations include the generalization that the human brain is functionally organized in terms of the operations and interactions of three main functional systems: a limbic-based system for regulation of tone and state; a posterior cortical system for dealing with information; and a frontal cortical system for programming and regulation. Luria further spells out theoretical overviews of the properties and localizations of function in these various regions in several books. He also describes the brain's operations in language in detail. We will return to his views in Chapter 21 of this book.

A representative classical example of this type of information is supplied by the case of Phinias Gage. The frontal lobes of Gage were destroyed by the penetration of a crow bar. He deteriorated from a socially responsible supervisor and family man to an uncontrolled, pleasure-seeking lout. The case illustrates the importance of the frontal lobes in the large-scale control of one's behavior and attitudes with respect to long-range plans and its place in the various schemes of things in the outside world.

Studies on Lower Animals

Studies on lower animals allow more systematic and deliberate invasive experimental techniques for exploring the qualities and localizations of various functions. One class of animal studies has been based on relating behavioral changes to the localizations of experimentally produced lesions or rather massive electrical stimulations. This class of study dominated experimental biopsychology throughout this century until about the 1960s and continues today. These studies showed the essential functional properties of many significant regions of the brain. The involvement of the amygdala in aggression, the hypothalmus in feeding and drinking, the brain stem reticular formation in arousal, the septum and medial forebrain bundle in reward, and so on.

A particularly significant body of work of this type was performed over a 50-year period by Lashley. Lashley studied the ability of rats to retain learned visually guided behavior in the face of lesions placed systematically through those parts of the cerebral cortex wherein he imagined the rat to have localized the associated memories. He found that the learned behavior degraded in proportion to the amount of tissue damaged in the critical areas. The central implication of his work is that memories seem to be distributed throughout regions, rather than precisely localized.

Further experimentation in this general category involves EEG-like recordings with larger (macro)electrodes in various brain regions associated with particular learning or behavioral paradigms. E. Roy John, for example, provided a very penetrating and influential discussion of the fundamental nature and fabric of neuroelectric signalling with respect to learning and cognition based on such studies in cats. His work called into question many common assumptions regarding the representation of memories in neural networks. Particularly, John emphasized the stochastic nature of memory representation. He also claimed a large degree of apparent independence of functional units to specific anatomy.

Considerable confusion is associated with much of this literature, however, because of the relatively large scale of the techniques involved and the fine-grained and distributed nature of the representation of information in neural networks of the brain. Therefore, many of these stimulations and lesions influenced multiple, and sometimes opposing, interpenetrating neural networks, thus producing multivariate results. Results from different laboratories often contradicted each other.

Microelectrode Studies. Microelectrode studies became relatively common in brain research only in about the 1960s. These allow the recording or stimulating of individual neurons and thus a much finer level of control over experimentation. The fundamental problem here, however, is the sheer quantity of nerve cells and the even larger quantity of synaptic interconnections involved in brain activity. Microstimulation of one neuron, for example, does not usually produce a noticeable effect in any typical brain region.

Nonetheless, microelectrode recordings from single neurons have and continue to corroborate and refine many features of localization of function in the nervous system and various qualities of its representation of information. Basic qualities observed repeatedly involve the stochastic nature of single neuron activity and the pervasiveness of apparently random activity.

Microelectrodes have been and continue to be used extensively to delineate the detailed neurophysiological mechanisms underlying the observed electrical signals they record. Thus, the mechanisms of production of synaptic potentials and action potentials were uncovered by these means in the 1950s and 1960s, respectively. These resulted in Nobel Prizes for

Hodgkin and Huxley and for Eccles. Other studies showed the neural phenomena of accommodation and adaptation.

The essential functional units involved in the fine-grained representation and processing of information in the brain, however, are networks containing usually large numbers of richly interconnected neurons. These processes are not revealed either by the older, more massive techniques nor by individual single microelectrodes.

Over the last 30 years a number of investigators have developed the capacity to record and analyze the simultaneous activity of tens of neurons from a given local network. By analyzing the correlations in activity among such units inferences can be drawn regarding the nature of the functional pattern signalled in the net. Verzeano and Gerstein pioneered in these efforts. Perkel, Moore, and Gerstein developed a number of mathematical and visual means of displaying and interpreting multiunit relationships. Currently a dozen or so laboratories around the world are more or less routinely using banks of up to about 30 microelectrodes to study coordinated firing patterns in local networks. Current extensions of the technology include devices that record simultaneously from up to 30 microlectrodes and at the same time measure the amounts of several types of transmitters activating the neurons at a similar number of sites—all contained within a three-dimensional spatial region some 500 microns on a side.

These studies are providing increasingly fine-grained information on the nature of the representation of information and neurophysiological mechanisms of interaction in local networks. Generally, so far, the work shows that significant numbers of pairs of neurons exhibit correlated firing in apparently normal activity, these correlations often change according to behavioral mode, many pairs of neurons are uncorrelated in firing. We should anticipate great strides in understanding the detailed neurophysiological mechanics of neural networks over the next decade with the help of these kinds of observations. A caution is that most significant neural networks involve thousands and tens of thousands of neurons, so that these devices are typically recording some 1 percent of the neurons of a network. Clearly, these data need to be interpreted within the framework of a companion theory for the operation of the net. The interaction between the theory and these penetrating but partial data will likely be the most propitious vehicle for advancement.

Another, but less developed, technique for observing activities of numbers of neurons simultaneously is that of voltage sensitive dyes. Certain dyes become visible in conjunction with local electrical fluctuations, such as those generated by neurons. It is possible to view and record the unitary electrical signals of very large numbers of neurons in neural networks by this technique. The procedure works well only for two-dimensional networks. An in vitro preparation such as an isolated slice of hippocampus, for example, has been used advantageously for this purpose. The technique is elaborate and difficult and is not expected to be widely used for some time.

Principles of Organization in Cortical Networks

Collectively, these various paths of experimental work have led to a number of principles regarding the systemic functional organization of cortical networks. In addition, recent studies in the areas of molecular neurobiology, neurochemistry, and development of the nervous system have also contributed to this picture.

First, memory traces seem to be diffusely spread through regional composite cortical networks. Perhaps it is simply that particular individual dynamic patterns occupy, when active, a relatively extensive continuous region; but, perhaps equally likely, distinct functional networks (each of which may signal any of a number of distinct patterns or modes of activity) anatomically overlap and interpenetrate as they do in the hypothalamus and brain stem.

Second, multiple representation and perhaps redundancy seem prevalent. There is great divergence of representation centrally. The peripheral auditory tract, for example, projects some 30,000 fibers to the central nervous system, but this information is distributed among some 100 million neurons in primary auditory cortex. Further, most sensory channels project significantly to several distinct cortical and subcortical regions.

Third, any higher form of activity of the nervous system involves the coordination of a significant number of different regions.

Fourth, these three points suggest a higher level "system" organization in addition to a "mechanism" organization. Such a system is to some variable degree protected against the failure of individual components; if a given component region fails to function properly, other regions may partially compensate for the lost or diminished function, or the system may perform higher level tasks more or less adequately without it. This is akin to the point of view developed in the field of "general systems theory" in the 1970s by Lazlow and von Bertalanfy.

Fifth, neuroelectric signalling is stochastic, or noisy. Any single neuron participating in a given pattern may or may not participate in a given single realization of the pattern. Many neurons appear to fire randomly, probably prompted by a number of separate and independent factors.

Sixth, neural networks, not single neurons, are the fundamental unit of organization of the nervous system. This is underscored by the findings that in development synapses and neurons are significantly overproduced and only those actively used during development survive; unused synaptic terminals degenerate and unused cells die. Also, as indicated previously, neurons change firing relationships with other neurons in different behavioral modes, even though their individual spike trains often do not exhibit noticeable changes.

Seventh, the cortical operations and organization of information are dependent on highly idiosyncratic individual structuring. Since neural networks develop their very structure in conjunction with the unique experience in each individual, the structure of each individual's brain is unique. Cognitive operations are carried out within the framework of such idiosyncratic structures of one's cortical circuitry. Persons interpret sensory input according to these prior and often idiosyncratic internal categories.

Eighth, brain networks, particularly cortical networks, seem malleable at least to some degree, even in adulthood, by virtue of the malleability of synaptic receiving spines on the dendrites of pyramidal cells. Such changes are associated with the action of calcium ions that participate in and link electrical and morphological dimensions. The changes occur over periods on the order of 30 minutes, which is about the same time period required to embed new memory traces. Recent work has suggested that particular NMDA receptors may serve such long-term plastic effects.

Ninth, it seems that the nervous system is organized primarily according to function and that anatomy is secondary. This has been emphasized by Luria on the basis of properties of recovery of function in brain-damaged soldiers. It has been illustrated recently by experiments showing plastic changes of somatosensory representation of hands following removal of a finger in monkeys. Related to this, is the less believable but suggestive and provocative theory of John that significant patterns retain spatiotemporal coherence even as they move across distinct anatomical regions.

Tenth, network firing patterns and operations are fundamentally dependent on basic neurochemical states relating to synaptic transmitters, nutritional and metabolic states, fatigue levels, and so on. Disruption of various transmitter systems is currently believed to underlie many central pathological syndromes, including for example, schizophrenia, depression, Parkinson's disease, and Alzheimer's disease.

Theories of Organization in Composite Cortical Networks

Significant broad theories for the overall functional organization of the nervous system with various implications for cortical organization have been presented by Jackson, MacLean, and Luria. These broad theories will be considered in Chapter 21 of this book. The current part deals with more specific theorizing applicable to regional areas within the cerebral cortex.

Neuroanatomical Nontheories

The great neuroanatomists (e.g., Bell, Golgi, Cajal) have been remarkably reticent with regard to theoretical conjectures on the networks they so lovingly described. They most certainly were fundamentally respectful of the overwhelming richness and complexity of these circuits. The attitude is well expressed by Sir Charles Bell who noted that "those who see the least know the most, and those who see the most know the least."

Neurophysiological Theories

Prior to this century the nervous system was viewed fundamentally in terms of its transmission of signals from one place to another. The concept of "excitatory states" in central networks of the spinal motor system suggested by Sherrington in the first decades of this century was likely among the first theoretical speculations on intrinsic network operations by experimental neurophysiologists. Kubie, in the 1930s, discussed "circularities" of activities in central neural networks.

Only in the time of Eccles and the advent of reliable single unit recordings could experimental neurophysiologists begin to envision more clearly the possible shapes and forms of such circularities and excitatory states. Throughout his illustrious career Eccles espoused numerous theoretical conjectures regarding the higher order relations of neuroelectric activity to mind, consciousness, the soul, and various attributes of these. These conjectures frequently included specific visions of functional organization and dynamic patterns of activity in neural networks. Eccles, for example, discussed wavefronts moving through cortical tissue and contributed theoretical ideas regarding the module hypothesis of cortical organization discussed later. He further was influential in directing a great deal of communal effort onto the study of the circuitry of the cerebellum in the 1970s. Eccles's lasting contributions, however, are in the area of the mechanisms of synaptic activations of neuronal membrane potentials. His career forms a bridge from the premicroelectrode era through its first illuminations of single-unit signalling.

Neuropsychological Theories

Several significant theories of cortical organization from the area of neuropsychology appeared during this same time period. Lashley's lesion studies led him to the idea of diffuseness of representation of the memory trace in composite cortical regions as indicated earlier. Lashley stated that such regions exhibited "equipontentiality" (any part of the region is equally effective in storing a trace) and "mass action" (the more tissue, the better is the response) with regard to their ability to store memory traces. These properties are consistent with a holographic view of cortical memory storage as a particular type of diffuse representation, and a few theorists including Pribram advocated this latter interpretation.

The Grand Trace System. Less well-known however is Lashley's grand trace theory of cerebral organization, which he saw published in 1950, near the end of his long career. In this theory, Lashley imagined that the billions of neurons in the cortex are organized into a large number of functional systems. Each system houses a large number of individual memory traces. The systems consist of different arrangements of the same massive pool of neurons. The same neurons may participate in different permutations in many systems.

Any such system may be thrown into a state of tonic activity, in which case it dominates the field. It maintains its dominance by selective spatiotemporal inhibition of competing systems and by supportive excitations of traces within itself.

Lashley further speculated that one may move mentally quite easily among the traces within a given system, but moves from traces in separate systems require the shifting into dominance of the second system. He suggested that fixation into memory is generally possible only when the new material forms part of such a dominant system. He further speculated that the content of experience might relate specifically to firing neurons, whereas the direction of attention might be determined by the distributions of graded generator potentials throughout the entire system.

Cell Assemblies. The second major neuropsychological theory was presented in 1949 by Hebb. Hebb's central idea was that neurons in cerebral cortex formed themselves into functional groups by selectively

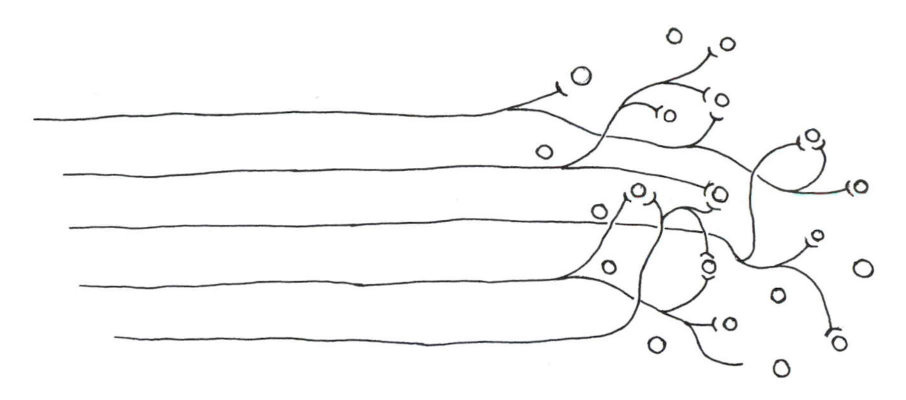

FIGURE 20.1. Representative synaptic junction to illustrate the synaptic learning theory of Hebb.

changing their synaptic strengths in accordance with the activity pattern being learned. This idea remains today the cornerstone of theorizing about learning and memory.

Picture a receiving population of neurons on the right and an activating population of fibers on the left as shown in Figure 20.1. Suppose that any one fiber projects synaptic terminals to a large number of receiving cells and that any one receiving cell is activated by a large number of distinct sending fibers. Imagine now the firing at a particular time of a specific subset of sending cells and the corresponding responsive firing of a specific subset of receiving cells. The essential ingredient here is that the entire populations and their junctional synaptic linkage are important.

Hebb hypothesized that these two sets of active neurons could become linked and thereby represent the embedding of a memory trace in the following particular way. Suppose that all those individual synapses connected between the firing senders and the firing receivers were increased in strength as a result of the physiological coincidence of firing in the pre- and postsynaptic cells. Note that the linkage is important: not all the input synapses of the receiving cells are increased in strength and not all the output synapses of the sending fibers are increased in strength, only those that link the active senders with the active receivers.

The result of this is that the next time this set of senders is active, its influence on the receivers will be stronger. It will be more likely to override competitors and bring about the response it initially triggered. Further, note the following. Since the activation of the initial input pattern was enough to trigger the response originally and the synapses were subsequently increased in strength, it follows that, under the same conditions in the future, the response may be triggered by subset of the initial pattern. If all the synapses were doubled in strength, for example, then any subset of half the original input would suffice. By this means, for example, if the original pattern represented a cat; one might trigger the entire response "cat" to some portion of the original stimulus, say a cat's head.

Hebb's ideas have pervasively influenced subsequent theories and modeling of neural networks. A vast proportion of this work has been focused on the processes of learning implied by this mechanism he suggested.

Statistical Configurations. A third influential neuropsychological theory of representation of memory traces is the idea of statistical configurations presented by John in 1972. The main value of John's thinking is to underscore many of the principles of organization enumerated previously, particularly the ideas that memory traces are stochastic, diffuse, redundant, and related primarily to function rather than anatomy. "Statistical versus switchboard" is the banner of this conceptualization. John's ideas that spatiotemporal patterns may retain their coherence while moving freely across anatomy has repelled many neuroscientists and may be overdrawn. Nonetheless the essential texture of his entire view should not be ignored.

Contemporary Streams

It is only in the present times is the theorizing of neural networks approaching its maturity. With respect to cortical circuits, one may identify three main contributing streams: the modular view of cortical organization, multiple microelectrode recording, and mathematical and computer modeling of neural networks.

The Modular View. The modular view of cortical organization is that cortical tissue is arranged into localized functional groups of about 30,000 neurons. These modules have been defined most clearly in the classic sensory receiving areas of posterior cortex. Their occurrence in association, motor, and frontal areas is less clear. In the visual cortex these modules are called *columns*. All the cells in a given column coordinate to signal certain elemental features of the visual field. For example, a column may signal the angle of a line at a particular point in the visual field. Modules in auditory and somatosensory cortex are called *barrels*. Barrels signal particular frequencies of sound in auditory cortex and elemental units of somatosensory stimulation in somatosensory cortex.

Cortical modules were first described by Mountcastle in 1957. Hubel and Wiesel discovered them in visual cortex, and Rose described them in auditory cortex. Eccles and Szentagothai theorized about their generality in the 1970s.

Cortical modules are assumed to work together to represent higher order generalizations based on their individual properties. How this is done is not generally clear. One might imagine that activity over composite regions might simultaneously be representing higher order features simultaneously as their constituent elements may be representing more elemental features. How and whether the patterns in constituent modules might vary depending on such integration over composite networks is not clear.

Further, the integration of elements into higher order features might be expected to occur as composite networks projected serially to other composite networks, for example, from lateral geniculate nucleus to visual areas, or visual area I to visual area II. Some modules visual area II are sensitive to angles of boundaries independent of where in the visual field they occur. It is possible to interpret this as a generalization based on activation from the representation in Visual area I, where angles of boundaries are represented for particular locations in the visual field. However, the situation is clouded by the current thinking that parallel projections to various cortical areas are as significant as serial projections between them. A number of recent studies have seen synchronous oscillations in cortical networks as a means of linking together neural representations of various features of single objects.

Multiple Microelectrodes. As indicated earlier, multiple microelectrode studies are currently promising to fill in considerable fine-grained detail with respect to this level of representation and processing.

Computational Neural Networks. Models and computer simulations of general features of these theories (most notably the cell assemblies of Hebb) began in the 1950s and continued at a low level for some 30 years. Recently an explosive increase of activity has occurred in the general area known as *computational neural nets*. These studies have been instructive but limited. The early studies were limited by weak neuroscientific knowledge and weak computational capabilities. The more recent studies have focused centrally on functional characteristics and have not made contact with neurophysiological foundations. Often constructed within the context of devices for artificial intelligence, the relation of model operations to neural mechanisms are weak, and the relation of model behavior to observable properties of physiological neural networks are not explored. Moreover, these recent studies have focused almost exclusively on spatial distinctions among patterns, whereas experimental evidence from multielectrodes has shown repeatedly the central importance of temporal features in firing patterns.

A View of Intermediate-level Functioning by Composite Cortical Networks

The composite networks considered in the last chapter likely operate at an intermediate level within psychological organization generally and cognitive operation more specifically. Specifically, composite cortical regions are seen there as composed of collections of local networks or modules, of the same general size as the modules of sensory receiving areas, and considered in the previous part of this book.

Let us now further imagine that the modes of activity in these individual local circuits signal some rather rudimentary cognitive elements. In visual cortex, for example, these rudimentary elements appear to be angles of boundaries or lines in the visual field. In association cortex, the elements represented in the multimodal local circuits are likely more abstract, but still of a somewhat rudimentary nature.

We have seen that the activity of the composite regions consists of equilibrium configurations attained by complex neurophysiological lateral interactions among the modes signalled by the individual local circuits. These equilibrium configurations then can be thought to represent some higher level cognitive elements. In the visual cortex, for example, the systemic collection of the elemental lines signalled by individual circuits can be envisioned to define and represent the morphological outline of a psychological entity in the visual field or some component of such an entity.

Similarly, the equilibrium configurations in the systemic collections of local circuits in association cortex are imagined to represent some higher order psychological units based on the rudimentary elements signalled by the constituent local circuits. What these might be is very uncertain. For illustration, however, one can suggest that entities relating to generalizations of sensory objects, such as concepts of rat, cat, dog, mammal, and the like, are signalled in entirety or in part by such configurations. Fundamentally, however, the theory developed here for the neurobiological foundations is noncommittal as to exactly what intermediate-level cognitive concepts are represented.

In this context it is necessary to point out that any region of cortex operates in conjunction with much of the rest of the brain in mediating virtually any higher level psychological functions. Therefore, it is not accurate to think of higher order functions as mediated by any particular composite cortical area. For example, purposeful deliberations or actions must involve links between the posterior and frontal cortical regions; rewarding or appetitive experiences and behaviors must involve interactions of cortical regions with limbic, hypothalamic, and brain stem regions; states bringing on associations with significant prior personal episodes likely involve interactions with hippocampal and temporal lobe regions.

Further, higher level cognition very likely involves pervasive interactions with the language areas. Therefore, interactions with both posterior auditory and frontal speech areas are implicated in much verbal-related cognition. In this context, it is not clear to what extent the entities dealt with in the present theory of composite cortical nets might or might not equate to words. Thus, a given word might likely involve connections with both the frontal and posterior speech areas mentioned previously, and also connections to visual areas for housing the images of the written word and the object it may represent, connections to any emotional or appetitive regions associated with the word, and perhaps some integration within a composite circuit of the type considered here.

Some background for some of these considerations is given in Chapter 21. The point for current purposes is that, first, the precise nature of the psychological elements implied by the theory are not clear at this time, and second, they are almost certainly intermediate in level since higher level concepts most likely involve patterns of activity distributed widely over numerous brain regions.

The Representation of Cognition by the Multi-Modal Oscillator Theory

The linkage of the multimodal oscillator theory of Chapter 19 to the psychological dimension, although highly speculative, is both intriguing and instructive. Particular sequential configurations in local networks would correspond to particular lower level psychological elements or basic building blocks from which psychological elements are constructed. Particular overall distributions of sequential configurations in the composite nets would be the neurobiological manifestation of intermediate-level psychological ingredients such as just discussed. Similar or significantly overlapping configurations would represent similar and related psychological ingredients.

The critically significant link here is that much of the mechanics of cognition would be fundamentally determined by these systemic lateral interactions mediated by synaptic junctional matrices among the constituent local networks. These cognitive mechanics would be embedded within the detailed neurophysiological interactions of the various sequential configuration patterns active in the individual confluent junctions, as the projections of these patterns laterally among the modules are harmonized and integrated within the fabric of the entire composite network. One begins to see that a neurobiologically based mechanics of cognition might come within sight if not yet within reach.

A normal equilibrium cognitive state would reflect the attainment of a systemic equilibrium distribution of sequential configurations in the composite network. Such an equilibrium state would be expected to consist of a pervasive occurrence of similar or compatible sequential configurations. The means by which particular sequential configurations become "compatible" in this sense of interacting excitation and inhibition in confluent synaptic junctions would be a fundamental component in spelling out the mechanics of cortical distributions and, therefore, the mechanics of cognition. Global or microlevel malfunctions at synaptic junctions would produce, respectively, global or highly discrete malfunctions of cognition.

Closing Comments

There are numerous difficulties in investigating this area. The first is the strategic problem related to performing penetrating controlled experiments only on subhuman animals when much of the real interest is in uniquely human capabilities. A second main difficulty is with the experimental technology. We have a vast, rich, complex, and complicated system, the detailed workings of which can be glimpsed at present, even under the most favorable and fortuitous conditions, only through small windows involving less than one or two dozen neurons.

In this context it is clear that there is a need for insightful and highly specified theory. The conceptual distinction between a mathematical or computational model, on the one hand, and a penetrating conceptual theory, on the other needs to be emphasized. A model in this sense can only provide an interface between theory and experiment. What is needed is the higher order classical approach of "theory guides and experiment decides." The challenge here to theorists and modelers is clearly great—as great as it has been to the many great experimental brain scientists active over the last century.

Chapter 21 Broad Principles of Systemic Operations in Nervous Systems

This chapter considers theoretical principles governing the engineering operations of nervous systems at their large-scale systemic levels of organization. Many of the concepts regarding these operations are very speculative. This is in large part because the principles of systemic organization themselves are for the most part far from clear. The guiding rationale for the chapter is that biologically based principles of systemic operations such as are considered here are a vital component of the fabric of large-scale systemic organization, and should play a part in the progressive unfolding of our understanding of this organization.

Systemic organization is approached under the four headings of invertebrate and autonomic systems, brain stem (reptilian) systems, limbic (paleomammalian) systems, and neocortical (human) systems. This structuring is due primarily to Paul MacLean. Relevant literature citations are given in the appendix.

Invertebrate and Autonomic Systems

The most successful invertebrate structural plan is the paired-ganglia system of annelids, molluscs, and arthropods.*

Structure

The paired-ganglia system consists of a dozen or so ganglia of neurons arranged in two parallel symmetric longitudinal tracts. Typically each ganglia contains from a thousand to tens of thousands of neurons. The neurons are densely interconnected longitudinally up and down each tract, laterally directly across the tracts, and internally within each ganglion. Sometimes the forwardmost pair of ganglia tend to partially merge, forming a primitive type of cerebrum. The overall structure has the general form of a ladder.

Functions

The primary functions of these systems are rudimentary life support, protective reflexive action, and primitive locomotion. These include such basic functions as cardiac contraction, food ingestion, reproduction, and walking, swimming, or flying.

Neurophysiological signalling is significantly different in texture from that of typical vertebrate nervous systems. There are large numbers of spontaneously active neurons. Many of these are pacemakers, firing regularly at fixed intervals. A common mechanism for this is an excessively high resting conductance for sodium, so that the resting potential is higher than the threshold. There is ubiquitous fatigue, accommodation, and postinhibitory rebound. Individual neurons often tend to fire in bursts. In addition to normal chemically transmitting synapses there are specialized

* Broadly speaking, one may identify four main classes of invertebrate nervous sytems: (1) the primitive continuous, reticulum-like nerve-nets of coelenterates and Ctenophores. (these nets are radially symmetric and exhibit two-way conduction; their behavioral repertoire is primitive as illustrated by the passive sensorimotor withdrawal reflexes of jellyfish); (2) the nerve-cord systems of flatworms and roundworms (these systems are longitudinally symmetric and consist of distinct individual neurons; they operate largely in terms of longititudinally conducted sensorimotor reflexes; there are minimal groupings of neurons in to pools or ganglia and hence minimal network integration operations); (3) the paired-ganglia systems of annelids, molluscs, and arthropods; (4) the specialized fast reflex systems of echinoderms (these systems mediate fast reflexive actions such as withdrawal swimming in starfish or retractive closure in sea anemones).

electrical junctions and considerable electrotonic coupling not involving specialized junctions. Some synapses produce both excitatory and inhibitory effects. Patterns are largely sculptured by inhibition. Extensive endogenous activity and sensory drive produces widespread fields of excitatory reverberations against which inhibition sculpts a pattern. In this context, disinhibition is a widely used mechanism.

Systemic Engineering

The broad-scale operations of these systems are those of sensorimotor reflexive action, central rhythm generation, and systemic feedback control around chemically defined biological "set points." These operational principles can be seen as the systemic engineering of the system.

Sensorimotor reflexive action can be easily understood as a rudimentary longitudinal connection from extrinsic stimulation to desired effector response. The extrinsic stimulation can arise either in the external or internal environment. Sensorimotor reflexive action is the essential basis of the responsiveness of these systems to environmental needs (food, mating, and so on) and threats and to internal conditions calling for rudimentary life support reactions.

Central rhythm generators are pools of interconnected neurons that generate a simple rhythmic recurrence of coupled bursty firing. In central rhythm generation, the primary driving initiation can come from an externally initiated external drive eventually fed forward to bombard the generator pool or from within nervous system itself from endogenously active cells. The rhythm typically is defined primarily by recurrent inhibition (or mutual inhibition between paired pools) and maintained or abetted by recurrent excitation. Fatigue, accommodation, and postinhibitory rebound help define the rhythm.

Central rhythm generation is the essential driving mechanism for many rudimentary life support functions, and for all the basic locomotor operations (swimming, walking, flying). A single central rhythm generator is often used to drive several additional ganglia into rhythmic activity in various relative phase settings. A common example is the alternation of drive for two classes of muscles (each on both sides) to produce the rhythmic alternations of locomotion.

Systemic feedback control toward maintenance of desired "set point" levels of key biological factors such as nutrient concentrations is the third systemic engineering principle of these systems. It is primarily definitive of the life support operations. It is indeed the principle pointed out in the early part of this century by Claude Bernard as the overall governing principle of the vertebrate autonomic system. Bernard identified this drive as the drive to maintain a constant "milieu interne".

Vertebrate Autonomic Systems

The vertebrate autonomic system is fundamentally very similar in structure, function, and systemic engineering to the invertebrate system just discussed. It consists of a few million neurons arranged in interconnected pools or ganglia each containing tens of thousands. Its main functions are rudimentary life support: cardiovascular, respiratory, digestive, sexual and reproductive, various secretions like crying, perspiration, and so on. Its neurophysiology is in many ways similar to that of the invertebrate systems: endogenous active neurons, accommodation, fatigue, multiple types of unusual synaptic actions. Its primary principles of systemic engineering are also sensorimotor reflexive action (driven here entirely by the internal environment), central rhythm generation (cardiac and respiratory driving), and systemic feedback control (the milieu interne).

Brain Stem (Reptilian) Systems: Instinctive, Appetitive, and Higher order Sensorimotor Behavior

We can define the reptilian system in which the brain stem houses the highest levels of central control as prototypical of the highest level of development in lower vertebrates. There are good functional and anatomical grounds for considering this system as a stable prototypical one as will become clear in this chapter.

Structure

The governing nervous structures of this brain stem system consist of (1) the autonomic system for rudimentary life support discussed previously; (2) the spinal muscle systems for approach–avoidance reflexes and various coordinated multimuscle patterns associated with locomotion and other basic coordinated movements; (3) brain stem centers for higher order sensorimotor reflexive action; (4) the basal ganglia, paleocerebellum, and very primitive thalamocortical tissue for higher order motor control; (5) the hypothalamus for control of appetitive behavior through its actions on the autonomic and hormonal systems; and (6) the pervasively distributed interconnected and differentiated networks of the reticular core of the brain stem for the representation of fixed action patterns of instinctive behavior, arousal, and biorhythmic control.

Functions

The functions of these lower vertebrate systems can be summarized as higher order sensorimotor reflexes, arousal appetitive behavior, instinctive behavior, and biorhythmic control.

Systemic Engineering

Sensorimotor Reflex, Central Rhythmic Generation, and Systemic Feedback Control. The first level of systemic engineering in lower vertebrates can be identified as that of the primitive sensorimotor reflexive action, central rhythmic generation, and systemic feedback control that govern the rudimentary life support operations of the autonomic system.

Switchboard in Support of Higher Sensorimotor Reflexive Action. A second level of essential systemic engineering can be defined as that of a switchboard in the extremely obscure and highly interconnected networks of the central core of the brain stem (reticular formation) and the spinal cord (largely substania gelatinosa). These richly interconnected obscure regions are taken here as the housing of large numbers of functionally discrete networks allocated by their particular fine-grained input and output connections to particular specific actions. Although functionally discrete, these networks are taken as anatomically diffuse and extensively interpenetrating. Acting as a switchboard for these largely hypothetical discrete functional networks is seen as serving both several levels of sensorimotor reflexive action, and thereby dominating the motor system of lower vertebrates, and a large part of the instinctive behavior of vertebrates.

This construction of switchboard connections in the core of the central nervous system of vertebrates allows the definition of a three-tiered hierarchy of organization of sensorimotor operations at the heart of the vertebrate motor system. The first tier is the *elemental sensorimotor reflexive action of the muscle systems* themselves involving the classic pathways through two and three serial synapses in the local horns of the spinal cord. By direct feedback from spindle receptors into the spinal cord, muscle action is intrinsically feedback controlled—a remarkable buffered engineering design. Much of the activating input directed from above to spinal motoneurons acts on gamma motoneurons, which do not activate muscles directly but rather change the responsive set points of muscle spindles to secondarily induce muscle action. Even those muscles directly activated by higher input are buffered in the response range by sensory signals from the muscles sent back to the spinal cord. All of this is quite beyond simple push–pull action.

The second tier is the reflexive *switchboard within the central gray core of the spinal cord*. These networks are known to house the essential differential muscle coordination patterns associated with vertebrate locomotion. These systems can operate independent of higher order directives. They most likely house many other basic patterns of coordinated muscle combinations as well. Activating input from without or above to the spinal cord in the lower vertebrate motor system progresses within the obscure densely interconnected central gray core. These actions are taken here as effecting selective switchboard connecting between input directives and output muscle combinations.

The third tier of reflexive sensorimotor action is the highly important *switchboard of multisensory input through the optic tectum* (called the *superior colliculus* in mammals). This switchboard mediates the primary behavioral response of lower vertebrates to sensory input. It seems to be the highest function of sensory input in these systems. Somatosensation, audition, and vision all send their rich central projections primarily to the optic tectum. (Taste and smell send theirs primarily to the amygdala and hypothalamus for more narrowly defined biological response.) The optic tectum projects extensively to the central gray matter of the spinal cord through the vestibular nuclei. The intelligence guiding the correct particular motor combinations to be activated in response to particular objects and situations in the external environment is embedded in the switchboard junctional matrices of these reflexive sensorimotor reflexive pathways.

Capstone control of the motor system of lower vertebrates is supplied by the central reticular formation of the brain stem operating in conjunction with functional nuclei within it, the basal ganglia, and paleocerebellum. The operations of these networks are obscure. We can only speculate that likely more switchboard action and perhaps some more developed but still largely primitive representations are used.

A particular, fundamentally important sensorimotor reflex is that of *behavioral arousal*. It brings about a much more widespread, massive, and unitary response to a wide variety of disparate stimulation than do the more selective responses just considered. Significant responses are felt throughout the autonomic system, the motor system, the sensory channels, and, in higher vertebrates, the higher limbic and neocortical areas. It is governed by a wide, diffuse region of the brain stem reticular formation.

Switchboarding in Support of the Fixed Action Patterns of Instinctive Behavior. The definition of discrete functional nets in the diffuse central gray core also provides the basis for defining a third level of systemic engineering in vertebrates: switchboard action as the prime mediator of instinctive behavior. Lorentz and Tinbergen have shown that instinctive behaviors can be decomposed into clear sequences of discrete sensorimotor reflexive actions called *fixed action patterns*. It seems likely that the neural substrates for these fixed action patterns are particular individual neuronal networks extending diffusely through the central core of the brain stem and spinal cord. The triggering of these networks during various critical periods could be primed according to some biochemical factor affecting the sensitivity of the net. Their operations would be governed by the principles of sensorimotor switchboarding.

Systemic Control of Appetitive Behavior. The term appetitive can be applied to a wide range of behavior associated with various kinds of biological deficiency, arousal and goal-seeking, and ultimate consummation and reward. Such behavior can be nicely represented by a systems control model (cf. NBM, page 101, and the Lindsay and Norman citation there). In this view arousal, goal-seeking behavior, priming of target images, and state goal comparisons are all triggered by the biologically sensed mismatch between the desired and existing level of some chemical signal. When comparison of sensory fields with target images and context reveals both an incentive stimulus and an action contingency situation, both reward and consummatory behavior are triggered. Consummatory behavior, in turn, changes the originally sensed chemical state so as to remove the chemical mismatch and turn off the aroused, goal-seeking behavior.

This model is a very nice example of a highly discriminating engineering control system. The basic engineering design is quite simple and not fundamentally different from that of an autonomic control system, or indeed of a residential central heating system. The primary distinction is in the richness of variation available to the system in both its sensory and motor processes. These latter are in turn seen as clearly subordinate functions within the purposes of this essential central control system. The main engineering operations are comparisons (of biochemical states, target and sensory images, and so forth) and acting as a switchboarding for particular behavior selection and target images according to the particular drive state triggered. Most of this can be seen to be operative in terms of state sensing drawn from or similar to that of autonomic function and switchboard action similar to that used in sensorimotor reflexive action and instinctive behavior. The higher reaches of this kind of behavior, however, require a rich universe of internal representation available only to higher mammals.

Biorhythmic Control Biorhythms are mediated primarily by discrete functional networks defined by means of connections and chemical transmitters within the diffuse reticular core of the brain stem and its rostral extensions. They are essentially complex timing circuits whose periods are determined by interacting elemental physiological processes. In principle, although not in biological mechanism, they are akin to the central rhythm generators of invertebrates and the autonomic system.

Limbic (Paleomammalian) Systems: Affective and Representational Capacities

The structures and functions of the limbic system introduce in the earliest mammals significant central capacities beyond those of the brain stem reptilian systems. The limbic system capacities discussed here function as well in higher mammals including humans.

Structures

The word *limbic* means bordering. These structures are seen as bordering the top and rostral end of the brain stem and separating it from the higher subcortical and cortical tissue above it. The structures include (1) a group of anterior structures—the medial forebrain bundle, septum, preoptic area, lateral hypothalamus, nucleus acumbens, and surrounding tissue; and (2) a group of dorsal structures—the hippocampus, the dendate gyrus, the subiculum, and entorhinal cortex, and other surrounding regions that collectively define the paleocortex and mesocortex. A third group of structures associated with the neocortex—the neocortex, the thalamus, and the cerebellum and surrounding tissue—also begin to develop significantly in mammals, but are not markedly developed in earliest mammals. This last group is not considered a part of the limbic system and is discussed in the next section.

Functions

The anterior group of structures deals with affective features of experience and behavior: reward and aversion, and visceral states. The more ventral group deals with highly abstracted representations derived from highly processed sensory input and, particularly in higher mammals, from previous experience. This representational input is directed into these structures from the highest sensory and cortical levels of development, from the association cortex in humans and more primitive thalamic and cortical regions in lower mammals. These representations provide the foundation for primitive (or paleo) cognition.

Systemic Engineering

Learning is also introduced on a large scale in these limbic representations. This capacity for representation and learning such representations is a profound evolutionary step that markedly separates mammalian and reptilian capacities.

Further, the affective functions, particularly with its reward and aversion systems introduces the highly significant valencing of representations (as positive or negative and in varying degrees) and the experiences associated with them. These affective functions are associated with particular groups of locally interconnected neural networks; in addition, likely to reinforce their vitally important controlling functions, these networks become chemically differentiated as well, particularly in association with monaminergic neurostransmitters. Dopamine, for example, becomes a primary mediator of reward.

A specific theory for the coordinated systemic embedding of representational memory traces in these limbic structures has been presented in Chapter 18. In brief, this theory advocates the following. First, candidate abstracted representations reflective of the qualities of the external world are recurrently reverberated through a closed loop including the entorhinal cortex, the dendate gyrus, the CA3 and CA1 hippocampal regions, and the subiculum. (These representations are embodied in coordinated network firing patterns; for example, sequential configurations. They are seen as operated on by the synaptic junctional matrices of the loop by the mechanisms introduced in Chapter 17, including switchboard action, information sculpting by inhibition, and local recurrent mode selections.) Second, particular representations from these are chosen to be permanently embedded in the feedforward and locally recurrent synapses of these constituent networks when input reflecting "significance" is coincidentally supplied to the networks of this system by the medial septum. This significance is defined in terms of inputs into the septum from the reward and aversion systems, the brain stem arousal system, and a "newness" detection mechanism within the representational loop itself. This theory sees these processes as the source from which mammals (including humans) begin to build their internal cognitive representation of the external world. The theory is discussed more fully in Chapter 18.

A second theoretical projection suggested here is that the limbic and neocortical structures of mammals early on developed the use of recurrent "representational loops" in at least three centrally important places in cognition. The first is the hippocampal loop identified previously. The second and third are respectively the several precise, topographically organized thalamocortical loops involved in all three major sensory modes and the recurrent loop interconnecting the neocerebellum, the motor thalamus, the motor neocortex and the inferior olive.

All three of these loops are seen in this theory as serving the fundamental new development of representation. They are representing particular representational traces for execution and processing in their respective systems in terms of coordinated network firing patterns such as sequential configurations. Thus, at this earliest level of mammalian development, one interprets the highest levels of both sensory and motor systems as taking on a significant finely filligreed representational dimension and beginning to clearly transcend the immediate sensorimotor nature of the systemic engineering of their lower reptilian structures and processes. The introduction of abstracted cognitive representation and its dependence on an effectively functioning affective valencing system becomes defined at essentially the same level of development.

The relation of these theoretical ideas to earlier theories of Paul MacLean and others are discussed briefly at the end of this chapter.

Neocortical (Human) Systems: Highly Elaborated Representationalism

The human system outlined here develops more or less continuously from the limbic affective and paleo- and mesocortical regions of the earliest mammals.

Structures

The overriding development of the nervous system in higher mammals is a tremendous growth in neocortical tissue and tissue related to it. Vast regions of richly populated and densely interconnected tissue evolve in both the posterior and anterior cortical regions. These are highly and richly interconnected with each other, and virtually all important brain stem, limbic, and other subcortical regions.

Functions

The functions associated with these regions may be interpreted as continued extensions of the affective and representational functions of the limbic structures.

The anterior regions are associated more with governing overall systemic activity and behavior and motor output over both very short and very long time periods. The anterior cortex is intimately involved with large-scale integration of personality, including both long-term life organizations in terms of one's motivations, ambitions, and degrees of intermeshing with the social context; and the web of controls on one's behavior in all sorts of social situations. It also controls sequencing of behavior patterns in hierarchical arrangements from the most mundane of daily tasks to the largest scale life plans.

The posterior regions are associated primarily with representing highly abstracted information more closely related to the sensory input representational fields and to internal representations of qualities of the external world generally.

In both anterior and posterior regions there are the significant developments of (1) large regions of uncommitted tissue (that is regions that seem not to be dedicated to any particular sensory or motor operations, but rather are open to one's individual experience and learning, and (2) progressive left–right asymmetry, becoming more marked in these uncommitted areas that seem to be the cortical regions most internal or most removed from connections to the sensory and motor interfaces with the external world. This left–right asymmetry is unique to humans. It seems to provide the foundation for a higher dimension of reflective cognition. This includes the phenomenon of self-awareness, which appears to be restricted to humans, chimpanzees, and orangutans.

Systemic Engineering

We can make the following broad speculations regarding the systemic engineering organization of the human neocortical regions.

1. The central functional ingredient of this tremendous cortical

development is representation. This includes a potentially very rich and, by human standards, unlimited internal model of the external world and its qualities. It also includes the capacity for the creation of unlimited groups of representations of abstractions of any imaginable kind and wide varieties of interactions among these representations.
2. The representation is served by learning.
3. The representation and learning occur and exist in close conjunction with affective processes of the limbic system. (a) The representational loop of the hippocampus discussed earlier and in Chapter 18 is a core component of the general human representational system; the operations of this system are seen here as fundamentally dependent on limbic affective input. (b) There is a greatly elaborated human will and volition often including a highly sophisticated and culture-dependent motivational complex that greatly transcends basic biological drives. Ultimately, these valenced motivational and value weightings likely relate back to widely integrated or filtered involvement of the limbic affective systems.
4. The overall organization of the human brain is fundamentally centrocenphalic in the manner suggested by Penfield. That is, fundamental values, motivations, and judgments refer back to limbic weighting systems but are seen as weighted around this central core in terms of highly elaborated representations of the various qualities of one's external world and perceived relations and positions within it.
5. All the broad functions of the brain stem reptilian level as defined earlier seem stable and operative in the neomammalian and human brain, although reduced in relative significance in accordance with the great development of internal representational features and learning. For example, both the motor and the higher order sensory systems have developed in mammals higher order control loops that appear distinct from and parallel to the lower order reptilian higher order control loops that still operate as coherent wholes in mammals. The new systems can be seen as operating in terms of finely filligreed representation and learning; the old systems, in terms of a diffuse sensorimotor switchboard.
6. The overall organization of the system can be seen as compatible in general terms with Luria's view of a central core for tone and state control in brain stem and limbic structures; a frontal system for overall regulation, planning, and sequencing; and a posterior system for representation and processing of information. The system should be seen as highly interrelated in operation, with any significant process

making use of interrelations among ingredients in many areas of all three domains. Particular large-scale cognitive memory traces should be seen as diffuse structures that interrelate local cortical circuits across widely separate cortical areas in both frontal and posterior regions.

7. The view outlined in this chapter is basically dependent on and consistent with Paul MacLean's theory of the triune brain, which recognizes stable levels of organization in the reptilian, paleomammalian, and neomammalian levels along the lines outlined.

However, two points of difference are worth noting. First, the orginal emphasis on the limbic as opposed to neocortical organization emphasized emotion as the property of combined linkage between affective and hippocampal components of the limbic system. (This was essentially Papez's contribution in the 1930s.) This emotion was contrasted with more objective cognitions attributed to the neocortex. The present formulation sees this distinction between emotion and objectivity as off the point and basically misleading. Fundamentally, affective input, valencing, and the implication of value weightings are seen here as vitally involved, although at a more generalized and broadly integrated level, in the most abstract neocortical representations, through the diffuse rostral connections of the limbic affective systems with the frontal neocortex and through the involvement of these affective systems in the laying down of all vital experiential memory traces in their connections through the septum to the hippocampal loop.

Second, and relatedly, MacLean was following Hughlings Jackson in the idea that certain stable levels of organization would occur in evolution that would be fundamentally not changed, but supplemented, by subsequent development. In oversimplified terms, the view was that sensorimotor behavior and the like was the reptilian level; emotion was the paleomammalian level; and representation was the neomammalian level. In MacLean's view the reptilian and paleomammalian levels were each at similar levels of stability beneath the neomammalian level. In the view developed in this chapter, the relation of the two lower levels to the neomammalian (human) level are not the same. The relation of the reptilian level is the same in both views—the reptilian level is fixed, stable, not developed in significant directions. However, in the view developed here, the neocortical developments are seen as highly dependent on and interactive with the limbic affective, representational, and combined learning processes.

Afterword

This book has attempted to formulate essential theoretical structures for four essential levels of neuroelectric signalling, developed correspondingly in its four main parts. These four levels are membrane and ionic mechanisms of neuroelectric signal generation in single neurons, firing rate sensitivities and transfer in single neurons, coordinated firing patterns (such as sequential configurations) as representations of meaning in local neural networks and fiber systems, and the mechanics of interaction of coordinated firing patterns by means of synaptic junctional matrices in composite neural networks.

Value, Usefulness, Completeness, and Established Validity of These Theoretical Structures and Implied Future Work

Let us consider now the degrees of value and usefulness, completeness, and established validity of these theoretical structures and the future work required for their continued cultivation and application. This book advocates that all four parts present and imply material of considerable potential value and impact for neuroscience. The theoretical development is substantially complete for the first three parts, but is only begun for the last part on composite networks. The material in Part I is much better established than that of the last three parts; Part IV needs continued theoretical cultivation; Parts II, III, and IV need application to particular biological networks and interaction with experimental work.

The theoretical formulation of the mechanics of neuroelectric signalling presented in Part I presents a universally structured and biologically rooted hierarchical mechanics that, when properly applied, can help show the relative significance of the many biological details and their relative places within overall network and systemic operations. On this basis, the formulation is potentially extremely valuable in its capacity to provide a common foundation for linking computational neural network studies with both theoretical and experimental neurobiological studies.

The theory in this section is substantially complete and of established validity. Future developments are to include within its hierarchy primarily the details of molecular gating operations as they become more clearly known. The bigger need with respect to the broader range of neuroelectric signalling is to intelligently incorporate realistic computer simulation models into ongoing experimental and theoretical studies of particular neural networks and particular types of neurons, to help integrate understanding of the higher order network operations and their degrees of dependence on these basic mechanisms.

The dynamic similarity theory developed in Part II is potentially valuable as a device to gain insight into the comparative structural design of different types of neurons in different network settings. The theory provides a way to succinctly summarize on a universal comparative basis the firing propensities of neurons to their particular input systems in terms of nondimensional characteristic numbers representing their anatomy and physiology. The theory is essentially complete. Its usefulness, however, remains to be validated. Future work needs to develop specific numerical predictions for various particular neurons, particularly for comparisons across types of neurons, and to compare these predictions with experimentation.

The theoretical concept of a higher order view of coordinated firing patterns in neural networks (as opposed to single neurons) as developed in Part III seems a very valuable and indeed necessary ingredient for interpreting the operations of neural networks, both local and composite. The theory is sufficiently complete at this level of development. However, what is needed now is a combined theoretical and experimental search for such coordinated patterns in specific networks. Multiple microelectrode experiments with much thought out direction and companion computer simulations are called for here.

The greatest payoff of the theoretical mechanics developed in this book as a whole lies before us in the area of composite neural networks. In a sense all the material of the first three parts of the book may be seen as tools to be used in the cultivation of this vast and largely uncharted area. This is the area of interface between neuroelectrical mechanism and operation and contextual psychological and behavioral function. This book advocates that the neuroelectric operations of this area in particular need to be conceived in a language of its own—a language representative of networks and populations and not of single neurons. The book particularly advocates that composite networks need to be studied and conceived in terms of the coordinated firing patterns (sequential configurations) of their constituent local networks and the interactions of these in their component confluent synaptic junctional matrices.

The theory for composite networks is introduced only in outline in this book and applied in broad terms to composite networks of the hippocampus and cerebral cortex. An immense amount of theoretical and combined theoretical–experimental work remain ahead. In particular, future theoretical work needs to clarify the theoretical interpretations and predictions of operations performed by synaptic junctional matrices introduced here. Computer simulations will be the indispensable tool for this development.

Further theoretical work needs to be performed regarding the properties of interacting sequential configurations carried in various composite network structures and acted upon by various synaptic junctional matrices. Purely theoretical studies can be instructive here and should be performed. However, much of this work should be undertaken in explicit companionship with experimentation on specific composite networks. The networks of the hippocampus and the cerebral cortex discussed in Chapters 14, 16, 18, 19, and 20 are good targets for this approach. The most fruitful experimental design would include several distinct banks of multiple microelectrodes. These would be seen as searching for signs of both individual coordinated patterns in local networks and interactions among these patterns across local networks. This level of work is seen as using all the theoretical ingredients developed in this book, including realistic biologically based computer simulations as developed in Part I, dynamic similarity analysis as developed in Part II, as well as the coordinated firing pattern and synaptic junction concepts introduced in Parts II and IV.

Higher Dimensions of Brain and Mind

The substantive thrust of this book is on the biologically rooted mechanics of neuroelectric signalling, seen essentially from a bottom-up perspective at the levels of membrane–ionic mechanism, neuronal firing rates, coordinated firing patterns in local networks, and synaptic junctional interactions in composite networks. The higher levels of network operations, however, necessarily take one into the higher dimensions of systemic organization, psychology, and behavior. The example networks of the hippocampus and cerebral cortex discussed in Chapters 18 through 20 have taken us necessarily into the areas of cognition, representation of the external world, experiential significance, and learning and memory.

In recognition of the necessary involvement of such higher contextual dimensions in studying composite networks, Chapter 21 has presented in outline a speculative view of some essential features of systemic functional organization of the nervous system and brain, with particular reference to its implications for systemic neuroelectric operations. Although these higher dimensions are beyond the substantive scope of this book, it is nonetheless only natural and obvious to conclude with some passing comments on the higher dimensions of brain and mind.

Higher Dimensions of Brain

In accordance with MacLean, Penfield, and Luria we have outlined the brain in Chapter 21 as essentially centrocenphalic in overall organization with a core in the brain stem that mediates basic instinctual, biological, and reflexive behavior; a core in limbic system that provides affective valencing and organismic significance to representations of experience; and a highly elaborated cortical system concerned primarily with representations and manipulations. Frontal regions are seen as dealing with overall holistic psychological organizations and operations; posterior regions are seen as dealing with more restricted cognition representations. Any significant process is seen as massively distributed through cortical and subcortical regions and as essentially stochastic in operation. The linkages of some features of these larger brain functions to composite network operations of the hippocampus and the cerebral cortex are touched upon at the higher limits of the bottom-up approach of this book in Chapters 18, 19, and 20.

Higher Dimensions of Mind

What about higher dimensions, beyond behavior, emotion, and cognition? What about, for example, Goodness, Truth, and Beauty, by which general concepts Socrates supposed to represent most of the higher resonant qualities of experience? It is instructive to consider the relation of brain activity to the external world. It seems most likely that our experience is directly in contact with our brain activity and that this brain activity is a reflection of the external world. As a result of years of forming and embedding internal representations of various features of that external world, one's brain has an internal structure that reflects the character and qualities of that world, biased by one's particular experience in it.

In this view, the external world itself, however, is perennially beyond our direct experience. We experience, indeed, "through a glass, darkly." There is a fundamental, largely impassable distinction between what we experience and what is out there beyond our brain. Kant formalized this distinction in terms of "phenomenal" and "noumenal" worlds. The distinction is incorporated in the fundamental laws of modern physics in the quantum mechanical formulation and its associated uncertainty principle. In this science, our images of the outside world are necessarily uncertain, but whether the outside world itself is uncertain is uncertain.

In this light one may speculate that the qualities of Goodness, Truth, and Beauty, which represent qualities both experiential and transcendent and which are widely represented in the individual experiences and collective culture of humankind (and perhaps other higher dimensions of experience as well), may reflect factors at the interface of our experience with the outer world that have a significant impact on our contextual situation within that external universe.

For example, the essential nature of Truth (and understanding) can be taken as the result of a match–mismatch operation between various related internal representations. Matching of representations corresponds to Truth; mismatching corresponds to anti-Truth ingredients. Matching can be graded; Truths can be graded. Representations can vary in degree of generality and inclusiveness; Truths may be large or smaller. This concept of Truth can be applied broadly to any set internal representations, ranging from one's accumulated embedded experiences and models of the outside world to immediate sensory input fields to abstract learned forms. It is static, basically a measure of the relationship of one's internal model with features of external reality. Clearly it helps us create a successful interface with the external world.

In the theoretical mechanics of this book, such match–mismatch operations of representations would be mediated by interacting coordinated firing patterns operating through synaptic junctional matrices. The concept of compatibility of distinct coordinated firing patterns would be intrinsically interwoven into the match–mismatch processes and contribute significantly to the qualities and properties of what we call Truth and understanding. Details could be hypothesized; computer simulations could be performed.

Second, one may take the essence of Beauty as an overall quality of collections of representations—likely related to overall order, arrangement, and harmony of the individual elements of the collection. Likely our sense of Beauty is secondary to order in the external world. We are sensitive to order and stability in the external world because our own continued existence is dependent on this. Any large-scale disorder of the external world (earthquakes, tidal waves, hurricanes) initiates deep-seated intrinsic alarm signals in us. In this view, Beauty can be taken as the positive quality of the absence of (potentially threatening) disorder—poetically as a deep response of gratitude for order and stability in the universe.

Again one may imagine a representation of this idea of Beauty in neuroelectric signalling over sets of representations across wide regions of neural networks. Again, details could be hypothesized; computer simulations could be performed.

With this interpretation of beauty, one might further suggest more broadly, that Truth is a subset of operations of the nervous system dealing largely with representations. It shows at least two main domains: verbal–analytical, which deals with sharp distinctions and comparisons among the various representations as discussed earlier, and global, akin to beauty, which deals with how the various representations fit together to make harmonious wholes. These two domains of Truth can be seen as primarily sympathetic with science and philosophy, respectively, although both involve both ingredients.

Finally, one can speculate that Goodness (approval, rejection, ought–ought not, conscience) is rooted fundamentally in the affective valence systems of the limbic system. In this view, various large, combined constellations of related representations associated with widely distributed cortical regions are seen as weighted and valenced by some unknown interactions with limbic reward–aversion influences. One can note the extremely pervasive significance of this deep hypothetical mechanism for forming judgments and organizing the overall cohesive fabric of our internal universe.

One may further note the relationship of this mechanism with other profoundly significant attraction or attraction–repulsion pairs in organizing our external world: gravitation in planetary and galactic structures; electromagnetic forces in molecular structures and processes; sexual polarity in human and animal societies.

Conscious Awareness

Conscious awareness is a fundamental enigma associated with brain activity that neurobiology will eventually need to comprehend. Conscious awareness (CA) defines itself as a primitive given for each of us. The first central fact is that what we consider our essential self, our CA, and our body seem to form an inextricably interwoven common identity. The second central fact is that CA seems to depend fundamentally on the state of the brain, to be intimately associated with its integrity of function and localized largely with it. Our CA seems often the servant of deeper currents of motivation and various forward motions of our nervous system and will within us, but often seems to exert at least some willful control over these currents and our directions of behavior by focused efforts that we interpret as free will.

Despite the central significance of this quality of our being, its fundamental basis remains strangely obscure. An essential point is that we do not know whether CA is linked to some particular physical structure, process, or other attribute or whether it is rather a fundamentally distinct dimension of existence in its own right.

If it is associated with some particular physical attribute, difficult and disturbing possibilities seem to exist regarding particularly the possible distributions of CA in the universe, and the specter of various artificially created CAs. For example, we might imagine three levels of possible physical correlates of CA. First, CA might be associated with particular brain networks or particular brain regions. A salient example of this type of hypothesis is the idea, generated in the 1960s, that the brain stem reticular formation, and particularly its ascending arousal system, is the anatomical substrate for CA. In this view CA could be seen as an attribute associated with this level of structural development in the same way that affective valencing, for example, can be associated with limbic mechanisms. The distribution of CA in the universe might be definable on the basis of the development of BSRF, but the likelihood of additional development in higher structures would remain a central likelihood. A more open view of this main possibility is that some as yet undefined collection of brain structures encourages the emergence of CA; that additional higher structures augment and brighten its character; and that one eventually will be able to predict the distribution of CA among animals by the extent to which they contain these particular or similar structures.

A second possibility is that CA reflects particular physiological and dynamic states, either neurochemical or neuroelectrical or both, rather than particular anatomical structures. In this event one could picture the locus of CA as mobile, moving through different brain regions in time, and highlighting different experiential content according to the regions through which it moved.

A third, more exotic general class of possibilities is that CA reflects some physical attribute of which the occurrence in the brain is simply one, perhaps particularly strong, realization. Possibilities here could include the following. (1) CA could be associated with basic physical duality like energy and matter, waves and particles. Here, CA could be a pervasive omnipresent feature of the universe. Its content and nature would likely be very foreign to us in other realizations. In such a scenario, one would presume that the manifestations of CA would be limited by the physical structures associated with it, in the same general manner that our CA is limited by our physical makeup. (2) CA could be associated with larger-scale organizational attributes of inorganic matter as well as with neurobiological matter. Here one would envision the spectra of conscious planetary systems, conscious galaxies, conscious mountain ranges, and so forth. (3) CA could be associated with organization of organic molecules. Molecular neurobiology, rather than neural network or systemic neurobiology, would hold the key to CA. Here one would envision CA distributed freely among living plants and animals. Again, any of these possibilities of a distinct physical foundation for CA raises disturbing possibilities somewhat akin to the genetic engineering concerns confronting advances in molecular biology.

The alternative seems to be that CA is a clear manifestation of an alternative dimension of reality; one of which we are a component as directly as we are a component of physical reality. In this instance, the traditional possibilities of dualism, parallelism, idealism, monism, and so on exist.

The point for this work is that this fundamental mystery remains an enigma with a significant impact on the core of higher levels of systemic organization of the brain and the mechanics of its composite neural networks, as on the larger questions concerning the ultimate nature of ourselves, the external universe, and our interrelation with it.

Appendix

Comments: A Note on Compartment Selection in Representing Dendritic Trees

When forming a compartmentalized representation of a neuron as introduced in Chapter 5 it is advantageous to choose compartments so that all branching points of the dendritic tree are at the effective electrical center of one of the compartments. This is indicated in Figures 5.1 and 10.1. Such an approach ensures that all compartments will share longitudinal current loops with at most only one other compartment at each of its ends. When this is not done, one must represent three intercompartmental current loops across bifurcations as shown in Chapter 21 of NBM. When branch points are centered in compartments as recommended here, one also may typically take each intercompartmental resistance as determined by only one cylindrical radius.

The parametric characterization of dendritic trees in Chapters 9 and 10 defines compartments so that bifurcation points of the dendritic tree occur at their effective centers. Therefore, the dynamic similarity characterization of Part II involves compartments that share longitudinal current loops with at most only one other compartment at each of its ends, and intercompartmental resistance is determined by only one cylindrical radius.

Selected References from the Relevant Literature

Part I

General Neurobiology

Shepherd, G. M. (1988). *Neurobiology*. 2nd ed. Oxford University Press.

Kandel, E. R., J. H. Schwartz, and T. M. Jessell, eds. (1991). *Principles of neural science*, 3rd ed. Elsevier.

Bradford, H. F. (1986). *Chemical neurobiology*. W. H. Freeman and Co.

General Neural Modeling

MacGregor, R. J., and E. R. Lewis. (1977). *Neural modeling*, Plenum Press.

MacGregor, R. J. (1987). *Neural and brain modeling*, Academic Press.

Chapter 3. Biophysics

Hille, B. (1984). *Ionic channels of excitable membranes*. Sinaver, Sunderland, MA.

Jack, J. J., B. D. Noble, and R. W. Tsien. (1975). *Electric current flow in excitable cells*, Oxford University Press.

MacGregor, R. J., and E. R. Lewis. (1977). *Neural modeling*, Plenum Press.

Chapter 4. On the membrane Capacitive Effect

Takashima, S. (1989). *Electrical properties of biopolymers and membranes*, Adam Hilger, particularly Chapter 10.

Electrical properties of biological polymers, water, and membranes. (1977). Volume 303 of the Annals of the New York Academy of Sciences, particularly articles by S. H. White, S. Takashima and R. Yantorno, H. Meves, and R. E. Taylor.

Chapter 5. Electrical Signals in Dendritic Trees

Rall, W. (1979). "Core conductor theory and cable properties of neurons," Chapter 3, *Handbook of Physiology: Nervous System*. American Physiological Society.

Davis, L., and R. Lorente de No, (1947). "Contribution to the mathematical theory of the electrotonus," *Studies Rockefeller Inst. Med. Res.* **131**, 442–496.

MacGregor, R. J. (1968). "A model for responses to activation by axodendritic synapses," *Biophysical J.* **8**, 305–318.

Chapter 6. Representing Generation of Action Potentials

Hodgkin, A. L., and A. F. Huxley. (1952). "A quantitative description of membrane current and its application to conduction and excitation in nerve," *J. Physiology* **117**, 500–544.

Traub, R. D., (1982). "Simulation of intrinsic bursting in CA3 hippocampal neurons," *Neuroscience* **7**, 1233–1242.

MacGregor, R. J., and R. M. Oliver. (1974). "A model for repetitive firing in neurons," *Biological Cybernetics* (formerly *Kybernetik*) **16**, 53–64.

Chapter 7. Computer Simulation of Neuroelectric Signalling

MacGregor, R. J. (1987). *Neural and brain modeling*. Academic Press.

Part II

Dynamic Similarity Theory

Schlichting, H. (1960). *Boundary layer theory*, 4th ed. McGraw-Hill.

Prandtl, L., and O. G. Tietjens. (1934). *Fundamentals of hydro- and aero-mechanics*, J. P. den Hartog, trans Dover.

Reynolds, O., (1883). *Collected Papers*.

MacGregor, R. J. (1988). "Theory of dynamic similarity in neuronal systems," *J. Neurophysiology* **60**, 751–768.

Probability Theory

Walpole, R. E., and R. H. Meyers. (1978). *Probability and statistics for engineers and scientists*, 2nd ed. Macmillan.

Part III

Coordinated Firing Patterns in Local Networks

MacGregor, R. J. (1991). "Sequential configuration model for firing patterns in local neural networks," *Biological Cybernetics* **65**, 339–349.

MacGregor, R. J., and G. L. Gerstein. (1991). "Cross-talk theory of memory capacity in neural networks," *Biological Cybernetics* **65**, 351–355.

MacGregor, R. J., and T. McMullen. (1978). "Computer simulation of diffusely-connected neuronal populations," *Biological Cybernetics* **28**, 121–127.

Harth, E. M., et al. (1970). "Brain functions and neural dynamics," *J. Theoretical Biology* **26**, 93–120.

Part IV

General Theory of Biological Neural Networks

Sherrington, C. S. (1906). *The Integrative Activity of the Nervous System*, Yale University Press.

Kubie, L. S. (1930). "A theoretical application to some neurological problems of the properties of excitation waves which move in closed circuits," *Brain* **53**, 166–177.

MacGregor, R. J., and T. McMullen. (1977). "Theory of monosynaptic transfer between neuron populations," *Behavioral Science* **22,** 207–217.

Shepherd, G. M., ed. (1960). *Synaptic organization of the brain*, 3rd ed. Oxford University Press.

See also entries for hippocampus and cerebral cortex.

Composite Networks in the Reticular Formation

MacGregor, R. J., (1972). "A model for reticular-like networks: ladder nets, recruitment fuses, and sustained responses," *Brain Research* **41**, 345–363.

Composite Networks of the Hippocampus

Ramon y Cajal, S., (1968). *The structure of ammon's horn*, Charles C Thomas, from the original of 1883 by L. M. Kraft.

Lorente de No, R. (1934). "Studies on the structure of the cerebral cortex, II," *J. Psychology and Neurology* **46**, 113–134.

Brown, T. H., and A. M. Zador. (1990). "Hippocampus," Chapter 10 in G. M. Shepherd, ed., *Synaptic organization of the brain*, 3rd ed. Oxford University Press.

Traub, R. D., and R. Miles, (1991). *Neuronal networks of the hippocampus*, Cambridge University Press.

Rolls, E. T., (1990). "Theoretical and neurophysiological analysis of the functions of the primate hippocampus in memory," Cold Spring Harbor Symposium on Quantitative Biology **55**, 995–1106.

O'Keefe, J., and Nadel, L., (1978). *The hippocampus as a cognitive map*, Oxford University Press.

Freedman, R., et al. (1987). "Neurobiological studies of sensory gating in schizophrenia," *Schiz. Bull.* **13**, 669–678.

Adler, L. E., et al. (1990). "Sensory physiology and catecholamines in schizophrenia and mania," *Psychiatric Research* **31,** 297–309.

Chapters 19 and 20. Composite Networks of the Cerebral Cortex

Lashley, K. S. (1960). "The cerebral organization of experience," in F. A. Beach, D. O. Hebb, C. T. Morgan, and H. W. Nissen, eds., *The neuropsychology of Lashley*, McGraw-Hill.

Lashley, K. S. (1950). "In search of the engram," *Symp. Soc. Exp. Biol.* **4,** 454–482.

Hebb, D. O. (1949). *Organization of Behavior*, McGraw-Hill.

John, E. R. (1972). "Switchboard versus statistical theories of learning and memory," *Science* **177**, 850–864.

Mountcastle, V. B. (1957). "Modality and topographic properties of single neurons of cat's somatic sensory cortex," *J. Neurophysiol.* **20,** 408–434.

Hubel, D. H., and T. N. Wiesel, (1962). "Receptive fields, binocular interaction, and functional architecture in the cat's visual cortex," *J. Physiol.* **160,** 106–154.

Eccles, J. C., (1981). "The modular operation of the cerebral neocortex considered as the material basis of mental events," *Neurosci.* **6,** 1839–1856.

Szentagothai, J. (1975). "The 'module-concept' in cerebral cortex architecture," *Brain Research* **95,** 475–496.

Sztengaothai, J., (1983). "The modular architectonic principle of neural centers," *Rev. Physiol. Biochem Pharmacol.* **98,** 11–61.

Szentagothai, J. (1978). "The neuron network of the cerebral cortex: a functional interpretation," *Proc. Roy. Soc., B* **201**, 219–248.

Brazier, M. A. B. and H. Petsche. (1978). *Architectonics of the cerebral cortex*. Raven Press.

Braitenberg, V. (1978). "Cell assemblies in the cerebral cortex," p. 171, in R. Heim and G. Palm, eds., *Theoretical approaches to complex systems*. Springer.

Gerstein, G. L., D. H. Perkel, and K. N. Subramanian. (1978). "Identification of functionally related neural assemblies," *Brain Research* **140**, 43–62.

Gerstein, G.L., P. Bedenbaugh, and A. Aertsen. (1989). "Neuronal assemblies," *IEEE Trans. BME* **BME-36**, 4–14.

Gerstein, G. L., M. Bloom, I. Espinosa, S. Evanczuk, and M. Turner. (1983). "Design of a laboratory for multi-neuron studies," *IEEE Trans SMC* **SMC-13**, 668–676.

Abeles, M., (1982). *Local cortical circuits.* Springer-Verlag.

Abeles, M., and M. Goldstein. (1977). "Multiple Spike Train Analysis," *Proc. IEEE* **65**, 762–773.

Vaadia, E., H. Bergman, and M. Abeles. (1989). "Neuronal activities related to higher brain functions—theoretical and experimental implications," *IEEE Trans, BME* **36,** 25–35.

Kruger, J., and M. Bach. (1981). "Simultaneous recording with 30 microelectrodes in monkey visual cortex," *Exp. Brn. Res.* **41**, 191–194.

Aertsen, A., and G. Palm, eds. (1986). *Brain Theory.* Springer.

Palm, G. (1981). "Towards a theory of cell assemblies," *Biological Cybernetics* **39**, 181–194.

Edelman, G. M., and G. N. Reeke, Jr. (1982). "Selective networks capable of representative transformations, limited generalizations, and associative memory," *Proc. Nat. Acd. Sci.*, USA **79,** 2091–2095.

Crick, F. H. C., and C. Asanuma. (1987). "Organization of the neocortex," in J. L. McClelland and D. E. Rumelhart, eds., *Parallel distributed processing: explorations in the microstructure of cognition, vol. 2, Applications*, MIT Press.

McClelland, J. L. and D. E. Rumelhart, eds., (1987). *Parallel distributed processing: explorations in the microstructure of cognition.* MIT Press.

Chapter 21. General Systemic Organization

MacLean, P. D. (1964). "Man and his animal brains," *Modern Medicine* **32,** 95–106.

Penfield, W. (1975). *The mystery of the mind.* Princeton University Press.

Luria, A. R. (1973). *The working brain.* Basic Books.

Milner, A. D., and M. D. Rugg, eds. (1992). *The neuropsychology of consciousness.* Academic Press.

Shepherd, G. M. (1988). *Neurobiology*, 2nd ed. Oxford University Press.

LISTS OF SYMBOLS

Part I–Given in Approximate Order of Occurrence

J_i	flux vector of ith ions: ions per area per time
q	unitary electronic charge
z_i	electronic valence of ith ions
μ_i	mobility of ith ions
c_i	concentration of ith ions

∇	dell vector operator
V	electrical potential
k	Boltzmann's constant
T	absolute temperature
P_i	molecular pump for ith ions
x, y, z	Cartesian coordinates
i, j, k	unit vectors in x, y, z directions
t	time
ρ	electric charge density; typically zero except for space–charge layer
J	electric current density: charge per area per time
c_{1i}	concentration of ith ions at inner surface of membrane
$c_{\text{in}i}$	concentration of ith ions in intracellular fluid
ξ	partition coefficient, defined by Eq. (3.7)
c_{2i}	concentration of ith ions at outer surface of membrane
$c_{\text{out}i}$	concentration of ions in extracellular fluid
w	width of neuron membrane
a_i, b_i, D	mathematically convenient groupings of terms; defined in Eqs. (3.5e) and (3.5f)
Δ	used to represent difference; e.g., $\Delta V = V2 - V1$
A	magnitude of control area of neural membrane
SC	total transmembrane current across a control area of neural membrane
C	membrane capacitance per unit area
G_i	resting membrane conductance to ith ionic species
g_i	active membrane conductance to ith ionic species
V_i	transmembrane equilibrium potential of ith ionic species
E	transmembrane electric potentials measured relative to resting level
τ	membrane time constant; equal to C/G
a_i	weighing factors for individual ion flows in a given active process, defined in Eq. (4.2), used in Chapter 4
E_{syn}	equilibrium potential of a given synaptic channel, defined in Eq. (4.2)
R	longitudinal resistance along neuron dendrite, soma, or axon
R_{ab}	effective longitudinal resistance between two contiguous compartments, a and b
l_i	length of ith compartment
	resistivity of intra- and extracellular fluid

Symbol	Definition
r_i	radius of ith compartment
r_o	effective radius of extracellular longitudinal current pathway
E_i	transmembrane potential of ith compartment
E_j^*	equilibrium potential of jth synaptic channel
i	membrane time constant of ith compartment
g_{ij}	active membrane conductance to jth synaptic channel in ith compartment
i_l	local current loops defined in Figure 5.2
a_i	A_i/A_s; ratio of area of ith compartment to area of soma, used around Eq. (5.3) and in Part II
I_i, I_e	total internal and external longitudinal current flows
R_i, R_e	total internal and external resistances per unit length
i	transmembrane current per unit area
r	local radius along neuron
x	longitudinal location along neuron
λ	neuron length constant, defined by Eq. (5.9c)
δ	Dirac delta function; infinitely large but infinitely narrow, so that the integral over its large value is 1
P	amplitude of synaptic conductance modulation, used in Eq. (5.11)
Y	electrical admittance
σ	conductivity of the intra- and extracellular fluid; equals the reciprocal of the resistivity,
$\bar{X}$	LaPlace transform of X
C_i, D_i	"constants" of integration over x; functions of s; used around Eq. (5.15)
A_l	cross-sectional area of the lth compartment, used around Eq. (5.15)
Δ, L, B	lengths of compartments define in Figure 5.7
k	number of dendrites in example problem defined in Figure 5.7
H, DEN, NUM	convenient groupings of terms defined in Eq. (5.17)
	dendritic leakage factor, defined in Eq. (5.18)
m	slope of applied current ramp
a_i	coefficients in infinite series solution for potential, defined in Eq. (5.19c)
η	time constant of accomodation, defined in Eq. (5.20a)
$\theta(t)$	neuron firing threshold
θ_o	resting firing threshold

b_i	coefficients in infinite series solution for threshold, defined in Eq. (5.20b)
r	convenient term, defined in Eq. (5.20c)
τ	dummy integration variable in convolution integral, Eq. (5.21a)
P	scale factor on amplitude of postsynaptic conductance modulation, Eq. (6.1)
α	decay rate factor on post-synaptic conductance modulation; Eq. (6.1)
n, m, h	intermediate channel functions in Hodgkin–Huxley model of action potential, Eq. (6.2)
α_s, β_s	rate parameters on n, m, h defined in Eq. (6.2c)
g_K	refractory conductance to potassium
GE, GI	excitatory and inhibitory synaptic conductance modulations
c	accomodative parameter, in [0,1]
b	magnitude of refractory conductance change to potassium
τ_k	time constant of decay of refractory potassium conductance change
S	spiking variable, 0 or 1
w	output firing rate
t^*	time of firing (latency)
I^*	current strength at time of firing
ES, ED	soma and dendrite potentials
SCS, SCD	stimulating current applied to soma, to dendrites
GDS	longitudinal conductance from soma to dendrites
GSD	longitudinal conductance to dendrites from soma
TS, TD	effective membrane time constant in soma, dendrites
GKS, GKD	refractory potassium conductance in soma, in dendrites
τ_{GK}, τ_{GKD}	time constant of decay of refractory potassium conductance in soma, in dendrites
THS	soma threshold
$TH0$	resting soma threshold
C	accomodative constant
TTH	time constant of accomodation
PS	soma potential with action potential, defined in Eq. (6.11)
GCA	conductance of dendritic membrane to calcium
ECA	equilibrium potential of calcium ions

Symbol	Definition
D	scaling factor for magnitude of calcium conductance modulation
THD	potential threshold for activation of calcium channels
τ_{GCA}	time constant of decay of calcium conductance
CA	intracellular calcium ion concentration
A	scaling factor for magnitude of calcium ion flux
τ_{CA}	time constant of decay of calcium ion concentration
$CA0$	threshold of calcium concentration to trigger potassium influx

Part II—Given in Approximate Order of Occurrence

Symbol	Definition
n_l	number of lth-level dendritic branches
$n_{l-1,l}$	number of lth-level dendritic branches emanating from a given single $(l-1)$th level branch
n_b	total number of dendritic branches
n	total number of compartments
a_j	A_j/A_s = area of jth compartment/area of soma compartment
E_j	transmembrane potential of jth compartment
t	time
R_{ji}	longitudinal resistance between j and i compartments
g_j	active conductance modulations in the jth compartment
g_k	active refractory conductance to potassium
SC	stimulating current applied to the soma
G	resting conductance per unit area
θ	resting threshold
	resistivity of the intracellular and extracellular fluid
r_{lj}	radius of the ljth branch
E^*	equilibrium potential of active conductance modulation
E_k	equilibrium potential of potassium ions
ξ_j	$\mathrm{d}E/\mathrm{d}t$ at threshold
$g_l^* E_l^*$	iterative grouping of terms defined in Eq. (9.3b)
G_l^*	iterative grouping of terms defined in Eq. (9.3b)
α_l	percentage of current from a lth branch that gets to the entrance to that $(l-1)$th branch to which it connects, defined in Eq. (9.4c)
γ_l	percentage of current from a lth branch that gets to the entrance of the soma compartment, defined in Eq. (9.5b)
D	effective leakage of current from soma into dendritic tree

b	magnitude of refractory potassium conductance elevation per output spike
ρ^*	universal characteristic function, defined by Eq. (9.6)
w_o	output firing rate
τ_k	time constant of recovery of refractory potassium conductance
s_j	amplitude of postsynaptic conductance modulation
s	subscript, refers to soma
δ_j	duration of postsynaptic conductance modulation
w_j	input firing rate in jth compartment
Δ	characteristic integration interval of a neuron; can typically be taken as the neuron's membrane time constant
w_o^*	characteristic output firing rate; equal to one-half the reciprocal of the neuron integration interval (approximately the membrane time constant); a measure of high "on" firing in a neuron.
a_i	effective synaptic weighting factor of input in the jth compartment, defined in Eq. (9.8d)
b_j	effective synaptic weighting factor for inhibitory input in the jth compartment, defined by Eq. (9.8d)
β	characteristic nondimensional number defined by Eq. (9.8d); measures ratio of current leakage from soma due to refractoriness to that due to flow into dendrites
w_i^*	characteristic input firing rate; equals the threshold firing rate of an input channel with average conductance input and synaptic weighting factor of value a; defined in Eq. (9.8f); shown later to estimate that level of irregular unstructured input which produces output firing at about the characteristic output firing rate, w_o^*
a	representative input synaptic weighting factor
X	X terms that may effect the membrane equation, but that are usually absent or ignored, such as ξ or SC
w	average input firing rate
w_{io}^*	value of w_i^* when $X = 0$
α_l^o, α_l^o	universal morphology parameters to characterize spatially selective activation for a canonical neuron, defined in Eq. (10.5)
a_l^o, a_s, b_l^o, b_s	universal synaptic weighting factors to characterize selective activation for a canonical neuron, defined in Eq. (10.5)

EN_n — envelope of linearly summed response to regularly spaced PSPs, defined in Eq. (11.3)

P — amplitude of indivudal PSP

SC_{eff} — imaginary step current that would produce a potential coinciding with the envelope, EN_n

w_{iu}^* — threshold input firing rate for temporally unstructured input arbitrary parameter taken as one or greater to buffer the suprathreshold surplus of highly structured input, defined in Eq. (11.4b)

P_n — amplitude of response to n simultaneous synaptic inputs

k — ratio of P_n to n times P

w_{is}^* — threshold input firing rate for temporally structured input; defined in Eq. (11.4g)

E_{ss} — steady state transmembrane potential

p — probability a single PSP will arrive in a given interval of width z

$p(x)$ — probability that x PSPs will arrive in a given integration interval, Δ

μ_x — mean number of PSPs that arrive in a given integration interval

x — standard deviation of number of PSPs that arrive in a given integration interval

k_x — number of standard deviations bewteen x and μ_x

w_{il}^* — value of input firing rate for unstructure Poisson input that produces output firing at a specified rate of $A(k^*)$; a lower limit of effective random input; defined by Eq. (12.8a)

w_{ih}^* — value of input firing that produces output firing at the characteristic output firing rate; an upper limit of effective input firing rate, defined by Eq. (12.8b)

Part III—Given in Approximate Order of Occurrence

b — increment of postfiring potassium conductance

B_t^i — the specific subset of cells that fire at time t in a bed of a memory trace

c — proportionality factor in model for accommodation

d_c — diminution factor for synaptic strengths within a given bed

d_l — synaptic strength of synapses in a bed

$d_{l\text{ex}}$	synaptic strength of excitatory synapses in a bed
$d_{l\text{in}}$	synaptic strength of inhibitory synapses in a bed
Δ_l	increment of synaptic strength
E	cell generator potential
E_{ex}	equilibrium potential of excitatory synapses
E_{in}	equilibrium potential of inhibitory synapses
E_K	equilibrium potential of potassium conductance
F_t^i	cells of B_t^i that fire at t in a given realization of a trace
GE	conductance change at excitatory synapses
GI	conductance change at inhibitory synapses
GK	conductance change for potassium
k	identify of population in which a given trace is embedded
L	number of time intervals in which distinct sets of bed cells fire; length of bed
m	number of populations in which a given trace is embedded
n	number of cells that fire at one time in a bed
n_f	number of trace cells in a given population that fire at one time in a realization
$n_{f\text{ex}}$	number of trace cells in an excitatory population that fire at one time in a realization
$n_{f\text{in}}$	number of trace cells in an inhibitory population that fire at one time in a realization
n_L	number of links to which a given link projects synaptic connections within a bed
n_r	number of nontrace (random) cells that fire at one time in a realization
q	particular time at which a given trace subset is active
R_t^j	cells that fire at t and are not in a trace; noise or random cells
s_{ee}	strength factor for excitatory to excitatory synapses
s_{ei}	strength factor for excitatory to inhibitory synapses
s_{ie}	strength factor for inhibitory to excitatory synapses
s_{ii}	strength factor for inhibitory to inhibitory synapses
S	spiking variable for neuron cell model
SC	stimulating current applied to cell model
t	time
t_a	initial time interval of a realization (≥ 1)
t_b	possible termination time of a noncyclic realization ($\leq L$)
T_{gk}	time constant of postfiring potassium conductance

T_{mem}	membrane time constant
TH	cell threshold
TH_o	resting cell threshold
A	ratio of network cells that have $\xi\# \geq \xi_{\text{crit}}$ at any one time
b	maximum signal level/cell threshold
B	factor defined in Eqs. (16.8) and (16.11)
c	occlusion factor for nonlinear conductance leakages in PSP summation, approximately 0.6
comp	"competing"
CT	"cross-talk"
CTi	"cross-talk of ith competing trace"
d	$\Sigma\, d_l$
D	$\Sigma\, (d_l^2)$
Δ	length of time interval in milliseconds, taken here as 1
$k(A)$	number of standard deviations between μ and ξ_{crit}
k_n	n/n_f
L	number of time intervals in which distinct sets of bed cells fire; length of bed
L_{mx}, L_{mn}	maximum L; minimum L
μ_U	mean value of the random variable U
m	number of standard deviations between $\mu + k^*\sigma$ and ξ_{crit}; usually applied with $k = 0$
n_{fi}	number of trace cells which fire at one time in a realization of the ith competing trace
N	number of cells in a single population
N_T, N_{Tmx}, N_{Tmn}	number of traces; maximum number of traces; minimum number of traces
N_p	peak value of N
O_x	the set of those links that share x cells with a given sender link
$p(x)$	probability that two independently chosen links share x cells
q	probablity of firing in a given time interval by random activation, $= (w_r^*\Delta)/1000$
r	n/N
r_p	n_f/N_p
R	random variable for random spontaneous firings of network cells
$R1$, $R2$	receiver links
s	convenient summation ration, defined as $S/(b^* S_{mx}^* s\#)$

$s\#$	average synaptic strength
sig	"signal"
S	amplification factor for summation of signal PSPs, modified for nonlinear conductance leakages
$S1, S2$	sender links
S_{mx}	amplification factor for summation of infinite train of unitary PSPs
S_{xl}	sets of those links activated by lth-order synapses by single links in O_x
S_s	amplification factor for summation of signal PSPs
σ_U	standard deviation of the random variable U
θ	cell threshold
U	mean value of the random variable U
w_{mx}	maximum normal firing rate of individual neurons
w_r	rate of spontaneous random activity in network cells
x	number of cells shared by two given links
y	number of links stored in a multiply embedded net
y_c	maximum allowable number of links such that $\xi \leq \gamma_{\text{crit}}$
y_p	maximum allowable number of links at N_p for net with random firings
z_{xl}	random variables for the number of S_{xl} links that have a given independently chosen cell as a member
Z_x	random variables for the number of y links which share x cells with a given sender link
ξ	random variable for the summed magnitude of cross-talk synaptic activation falling on a given cell at a given time
ξ_{crit}	critical value of ξ.

INDEX

NEURAL NETWORKS: FOUNDATIONS TO APPLICATIONS

Steven F. Zornetzer, Joel Davis, Clifford Lav, and Thomas McKenna, editors

Thomas McKenna, Joel Davis, and Steven F. Zornetzer, *Single Neuron Computation*

Randall D. Beer, Roy E. Ritzman, and Thomas McKenna, *Biological Neural Networks in Neuroethology and Robotics*

Ronald J. MacGregor, *Theoretical Mechanics of Biological Neural Networks*